DESIGNING OPTIMAL STRATEGIES FOR MINERAL EXPLORATION

DESIGNING OPTIMAL STRATEGIES FOR MINERAL EXPLORATION

J. G. De Geoffroy
Consultant, Geostatistics and Mining Exploration
Eden, New South Wales, Australia

and
T. K. Wignall
California Polytechnic State University
San Luis Obispo, California

PLENUM PRESS • NEW YORK AND LONDON

Library of Congress Cataloging in Publication Data

De Geoffroy, J. G.
　Designing optimal strategies for mineral exploration.

　Includes bibliographies and index.
　1. Prospecting—Data processing. 2. Prospecting—Mathematical models. 3. Mathematical optimization. I. Wignall, T. K. II. Title.
TN270.D4　1985　　　　　　　622′.1　　　　　　　85-19383
ISBN 0-306-41977-7

© 1985 Plenum Press, New York
A Division of Plenum Publishing Corporation
233 Spring Street, New York, N.Y. 10013

All rights reserved

No part of this book may be reproduced, stored in a retrieval system, or transmitted in any form or by any means, electronic, mechanical, photocopying, microfilming, recording, or otherwise, without written permission from the Publisher

Printed in the United States of America

PREFACE

Few knowledgeable people would deny that the field of mineral exploration is facing some difficult times in the foreseeable future. Among the woes, we can cite a worldwide economic uneasiness reflected by sluggish and at times widely fluctuating metal prices, global financial uncertainties, and relentless pressures on costs despite a substantial slowing down of the rate of inflation. Furthermore, management is forced to turn to more sophisticated and expensive technologies and to look farther afield to more remote regions, as the better-quality and more easily accessible ore deposits have now been revealed. This rather gloomy outlook should persuade explorationists to cast about for a new philosophy with which to guide mineral exploration through the challenging decades ahead.

Once already, in the early 1960s, a call for change had been heard (Ref. 30 in Chapter 1), when it became obvious that the prospecting methods of yesteryear, so successful in the past, could not keep up with the rapidly growing demand for minerals of the postwar period. The answer, a massive introduction of sophisticated geophysical and geochemical technologies backed by new geological models, proved spectacularly successful throughout the 1960s and the 1970s. But for both economic and technological reasons, the brisk pace of the last two decades has considerably slowed down in the early 1980s, as if a new threshold has been reached. We believe that the answer lies in a concerted effort to make an efficient use of the technology at hand, which can be translated into the earthy language of the man in the field as "getting more bang for the exploration buck," or into the more esoteric language of management circles as "maximizing the efficiency of operations under budget constraints"—in other words: *optimization,* which is what the new philosophy of mineral exploration should be about.

Optimization is nothing new in itself. It has been a household word in engineering and industry circles for quite some time. Optimization has been instrumental in the spectacular reduction in energy demand witnessed in the past few years, by providing a much more rational and efficient use of the resources at hand. Unfortunately, however, the new philosophy of optimization does not as yet seem to have made much inroad into the thinking of the mineral explorationists of the 1980s. A search of the English language literature on the mineral industry published during the past two decades revealed a great wealth of material on the optimization of the production section of the business, but only precious few journal articles on the optimization of the exploration section, although the necessary methodology and the required computational facilities are available.

Mineral exploration, being essentially a sequential procedure, is eminently suited for optimization, despite its inherent context of uncertainty and risk which require the use of probabilistic methods, as opposed to the deterministic approach that prevails in engineering. The optimization of mineral exploration as a whole and that of the various individual stages of the exploration sequence constitutes a rather forbiddingly large subject matter which would require several volumes to do justice to it. Because of format restrictions we are limiting ourselves to two practical problems of exploration planning and field investigation, including (1) the optimization of survey designs for the detection of specified types of mineral deposits, and (2) the optimal selection for drill testing purposes of exploration targets outlined by field surveys.

The first part of the book, comprising Chapters 1–5, describes the methodology of the optimization of the detection of ore deposits. The subject matter is covered in a descriptive manner with a very minimal mathematical content, so that explorationists of the older school with a modest mathematical and statistical background should not feel discouraged by arcane language and symbols. No mathematical proofs are offered, but interested readers are directed to carefully selected references to satisfy their curiosity. Chapter 5 briefly depicts how the theoretical methodology can be applied to the optimization of the airborne geophysical, ground geophysical, and drilling detection of specific types of ore deposits known to occur in North America.

The second part of the book covers the optimization of the detection of six types of ore deposits commonly sought in North America—(1) porphyry–Cu–Mo, (2) contact metasomatic, (3) Ni–Cu ultramafic, (4) volcanogenic massive sulfides, (5) Mississippi Valley-type Pb–Zn, and (6) vein gold deposits—in three geological regions of the North American continent, namely, the Cordillera Belt, the Precambrian Shield, and the Arctic Paleozoic Platform (Chapters 6–11). Finally, Chapter 12 covers the crucial problem of the optimal selection of exploration targets worthy of testing.

Because we have been critical of the rather offhanded manner with which references and selected readings have been dealt with in many recent books, we have devoted much time and attention to this important matter. In order to assist the more selective readers with their research into specific topics, we are submitting lists of pertinent titles at the end of each chapter. When warranted (Chapters 1, 2, 4, and 12), the references are further segregated into clearly labeled categories covering narrower topics.

PREFACE

One last word about the logo displayed on the book cover: being dissatisfied with the perennial geologists' picks and prospectors' shovels and gold pans which have been adorning the covers of books on mineral exploration for the past 50 years, we decided to produce a design more attuned to the times and the topic of the book. The parabola curve aptly and concisely portrays the mathematical optimization approach, a keystone of the book, as a proper link between modern exploration methods (aircraft towing a "geophysical bird") of worldwide scope (map of the world) and commercial success, which is symbolized by the outline of a mine headframe.

A forthcoming volume titled *Geomathematical Models for Optimizing Mineral Exploration* deals with the optimized search for the same six types of ore deposits as in the present volume, but covers four other regions of the world. These regions include Northern Europe, the Mediterranean, Australasia and East Asia, as well as the Appalachian Belt and the U.S. portion of the North American Paleozoic Platform, which could not be covered in this volume due to the lack of adequate data.

J. De Geoffroy
T. K. Wignall

Sydney, Australia

ACKNOWLEDGMENTS

Three kinds of elements are required in order to carry out a pioneering project of the type undertaken here: (1) inspiration, encouragement, and guidance; (2) good-quality data; and (3) computational facilities for the processing of the data and extraction of conclusions and inferences. It was our good fortune to be able to secure the assistance of several persons with respect to each of the three requisites mentioned above. We are very much indebted to Dr. F. P. Agterberg, Head, Geomathematics Section, Geological Survey of Canada, Ottawa, for his early interest and encouragement, and for helpful suggestions which steered this project along its present line, away from a less fruitful earlier one. However, we accept full and sole responsibility for any flaws, oversights, or conceptual errors which may appear in the text. Much gratitude is also owed to E. H. Rosenberg, Senior Editor, Plenum Publishing Corporation of New York, for his early interest in the project, and for providing much needed encouragement during the two arduous years which were required to complete the manuscript.

This project could not have come to fruition without the most generous assistance of D. A. Barr, Vice President, Du Pont of Canada Exploration, Vancouver and his staff, who provided almost one half of the whole database required for the coverage of six types of ore deposits of the North American continent. Much credit should be given to J. Korenic, Geologist, Du Pont of Canada, for so ably culling the required geometric data into a readily processable form. Among others who assisted in the arduous gathering of the necessary data, we would be most pleased to mention Dr. E. G. Pye, formerly Director, Ontario Geological Survey, Toronto, who most kindly provided excellent excerpts from departmental files regarding volcanogenic massive sulfide and nickel–copper deposits of Ontario. Finally, much thanks are owed to W. W. Moffat, P. Eng., of Willowdale, Ontario, and Gordon Moffat, Geologist, Calgary, Alberta, who generously helped us fill a number of very troublesome gaps in our database.

Because of the voracious appetite for computer time shown by pioneering projects of this kind, we could not have succeeded without the unflagging interest shown by Dr. B. S. Thornton, Dean, Faculty of Mathematical and Computing Sciences, the New South Wales Institute of Technology, Sydney, Australia, who most generously provided full use of the excellent computing facilities of his Department. We are also indebted to Jeff Pickering, Visual Aids Officer, and to Chas. Evans, Senior Technical Officer, both of the New South Wales Institute of Technology, for their excellent photographic and drafting work, which was required for many of the illustrations included in the book.

Finally, we are much indebted to Julie Symonds of Eden, New South Wales, Australia, for so ably handling the daunting task of typing the 188 statistical tables included in this book.

It is hoped that this book will stand as an excellent example of what can be accomplished through the fruitful collaboration of business, academe, and government agencies in several nations separated by many miles of ocean.

CONTENTS

Chapter One Optimizing Mineral Exploration

1.1. General Statement	1
1.2. The Mitigation of Uncertainty and Risk in Mineral Exploration	5
1.3. Conclusion: Scope of the Book	11
References and Selected Readings	12

Chapter Two Evaluation of the Probability of Detection of Mineral Deposits

2.1. The Detection Process	17
2.2. Geometric Considerations Involved in the Detection of Mineral Deposits	22
2.3. The Theory of Geometric Probabilities as a Foundation for the Evaluation of the Probability of Detection of Mineral Deposits	31
2.4. Evaluation of the Probability of Target Detection by Airborne Geophysical Surveys	33
2.5. Evaluation of the Probability of Target Detection by Ground Geophysical Surveys and Drilling Programs	42
2.6. Probability of Sequential Detection of Ore Deposits	50
References and Selected Readings	51

Chapter Three Cost of Detection

3.1. General Statement	55
3.2. Costing and Cost Estimates	56
3.3. Cost Functions	60
3.4. Tabulation of Cost Functions	68
References and Selected Readings	74

Chapter Four Optimizing Ore Detection

4.1. General Statement	75
4.2. Basic Theory of Optimization	77
4.3. Methodology of Optimization	81

| 4.4. Application of Optimization Theory to Ore Detection | 88 |
| References and Selected Readings | 92 |

Chapter Five Application of the Optimization Methodology to the Search for Six Types of Ore Deposits in North America

5.1. General Statement	95
5.2. Construction of a Database	95
5.3. Data Processing	97
5.4. Computation of Detection Probabilities	100
5.5. Design of Three Strategies for the Detection of Six Types of Ore Deposits in North America	104
References and Selected Readings	107

Chapter Six Designing Optimized Field Programs for the Detection of Porphyry–Cu–Mo Deposits of the North American Cordillera Belt

6.1. General Geological Background	109
6.2. Field Detection Methodology	111
6.3. Statistical Modeling of the Geometric Parameters of Porphyry–Cu–Mo Deposits and their Associated Pyritic Halos	114
6.4. Construction and Organization of Detection Probability Tables	117
6.5. Designing Three Strategies for the Search for Porphyry–Cu–Mo Deposits in the North American Cordillera Belt	118
References and Selected Readings	139

Chapter Seven Optimized Search for Four Types of Contact Metasomatic Deposits of the North American Cordillera Belt

7.1. General Geological Background	141
7.2. Field Detection Methodology	146
7.3. Probability of Detection of Cu–Fe Contact Metasomatic Deposits	148
7.4. Probability of Detection of Pb–Zn–Cu–Ag Contact Metasomatic Deposits	161
7.5. Probability of Detection of Cu–Mo–Au Contact Metasomatic Deposits	175
7.6. Probability of Detection of W–Mo Contact Metasomatic Deposits	188
7.7. Designing Three Strategies for the Detection of Four	

CONTENTS   xiii

    Types of Contact Metasomatic Deposits of the North
        American Cordillera 198
    References and Selected Readings 209

Chapter Eight Detection of Ni–Cu Ultramafic Deposits of the North American Shield by Optimized Geophysical Surveys and Drilling Programs

8.1. General Geological Background 211
8.2. Field Detection Methodology 213
8.3. Statistical Modeling of Geometric Parameters of Ni–Cu Deposits .. 215
8.4. Construction and Organization of Detection Probability Tables ... 218
8.5. Designing Three Strategies for the Detection of Ni–Cu Deposits of
     the North American Shield 219
8.6. Detection Probability and Optimization Tables 223
     References and Selected Readings 233

Chapter Nine Optimized Airborne and Ground Search for Volcanogenic Massive Sulfide Deposits of the North American Shield and Cordillera Belt

9.1. General Statement 235
9.2. General Geological Background: Volcanogenic Sulfide Deposits of
     the North American Shield 236
9.3. Field Detection Methodology 238
9.4. Statistical Modeling of Geometric Parameters of Shield
     Volcanogenic Sulfide Deposits 241
9.5. Construction and Organization of Detection Probability Tables for
     Shield Volcanogenic Sulfide Deposits 245
9.6. Design of Three Strategies for the Detection of Shield
     Volcanogenic Sulfide Deposits 246
9.7. Detection Probability and Optimization Tables for Shield
     Volcanogenic Sulfide Deposits 249
9.8. Geological Synopsis for Volcanogenic Sulfide Deposits of the
     North American Cordillera Belt 261
9.9. Statistical Modeling of Geometric Parameters of Cordillera
     Volcanogenic Sulfide Deposits 262
9.10. Construction and Organization of Detection Probability Tables
     for Cordillera Volcanogenic Sulfide Deposits 264
9.11. Designing Three Strategies for the Detection of Cordillera
     Volcanogenic Sulfide Deposits 264

9.12. Detection Probability and Optimization Tables for Cordillera
Volcanogenic Sulfide Deposits 267
References and Selected Readings 279

Chapter Ten   Designing Optimized Field Programs for the
Detection of Mississippi Valley-Type Pb–Zn
Deposits of the North American Arctic Paleozoic
Platform

10.1. General Geological Background 283
10.2. Field Detection Methodology 285
10.3. Statistical Modeling of the Geometric Parameters of North
American Arctic Deposits 288
10.4. Construction and Organization of Detection Probability Tables .. 290
10.5. Designing Three Strategies for the Detection of Arctic Mississippi
Valley-Type Pb–Zn Deposits 291
10.6. Detection Probability and Optimization Tables 294
References and Selected Readings 303

Chapter Eleven   Detection of Vein-Gold Deposits of the North
American Shield by Optimized Ground
Programs

11.1. General Geological Background 305
11.2. Field Detection Methodology 308
11.3. Statistical Modeling of Geometric Parameters of Vein-Gold
Deposits 310
11.4. Construction and Organization of Detection Probability Tables .. 312
11.5. Design of Three Strategies for the Detection of Archean Vein-
Gold Deposits of the North American Shield 314
11.6. Detection Probability and Optimization Tables 316
References and Selected Readings 323

Chapter Twelve   Optimal Selection of Exploration Targets for
Drill Testing

12.1. General Statement 325
12.2. Optimal Selection of Exploration Targets Based on
Control Locations 328

CONTENTS

12.3. Optimal Selection of Exploration Targets without
       Control Locations 340
       References and Selected Readings 346

Conclusion 351

Appendix: Measurement Conversion Table 353

Subject Index 355

Author Index 361

Listed Journals 363

CHAPTER ONE

OPTIMIZING MINERAL EXPLORATION

1.1. GENERAL STATEMENT

1.1.1. Purposes and Nature of Mineral Exploration

Mineral exploration can be defined as a scientific endeavor to discover and gain a natural resource which is difficult to find and of a nonrenewable nature. Mineral exploration encompasses only the search for, discovery, identification, and evaluation of the prize, but not its commercial development. The main purpose of exploration is success to be obtained at a reasonable cost in terms of expenditure of time, money, and skill.

Success in mineral exploration can be defined in two principal aspects; technical success and commercial success. Technical success is highlighted by the discovery of a mineral deposit sufficiently attractive to justify the expenditure required to establish the commercial potential of the deposit (tonnage, grade). The investigation leads to three possible outcomes: (1) inadequacy of the mineral deposit to meet the criteria required for a profitable operation in the foreseeable future; (2) qualified commercial success: the mineral deposit appears to be sub-economic at the present time and should be put on the "shelf" for future needs in more favorable circumstances; and finally, (3) commercial success: the mineral deposit does meet all criteria required for a profitable operation and is now described as an "ore body," based on the commonly accepted definition of "ore" as mineralized material that can be mined at a profit.

Technical success depends on two critical events; occurrence and detection. Occurrence of a specific type of mineral deposit in a restricted area is a state of nature on which the explorationist has control only to the extent of the selection of the area of investigation. Detection depends largely on the quality of the search, which implies (1) the selection of the technology and methodology best suited for the type of ore target and its environment, (2) proper planning and organization, including scheduling and logistics, and (3) availability of risk capital on reasonably attractive terms.

Commercial success relies on a third critical element, that of "economic worth," whose adjunction transforms a technical success into a commercial one. As in the case of occurrence, economic worth partly reflects a state of nature, i.e., the intensity and extent of geological processes, on which the explorationist has no control. But, again, many other factors of a geographic, economic, financial, and political nature are involved as well.

1.1.2. The Context of Mineral Exploration: Uncertainty and Risk

1.1.2.1. Uncertainty. It is a common experience that most mineral exploration decisions such as the selection of areas, technology of investigation, and exploration targets for testing, etc. are based on information that is limited both in quality and quantity. Most exploration situations lie between the two extremes of absolute certainty on the one hand, and total uncertainty on the other hand. In practice, we have to deal with partial availability of perfect information, or with imperfect information, or with a degree of ignorance about the quality of information. The resulting uncertainty leads to the use of much more complex decision models which introduce a greater possibility of occurrence of unfavorable outcomes than would be the case if certainty prevailed. This brings us to the concept of risk, which reflects the degree of unfavorability with which different possible outcomes are viewed by the decision maker.

1.1.2.2. Risk. That mineral exploration is a highly risky business is an opinion widely shared by the mining fraternity and the public at large. While holding the possibility of financial returns larger than that expected in most other business ventures, mineral exploration is also beset by a high probability of failure, which may lead to heavy financial losses. Risk may result from geological, technological, economic, and political uncertainty, as well as intangibles such as the quality of geological, managerial, and financial skills available in the corporate structure of exploration companies.

Geological risk is associated with the concept of occurrence and with the selection of areas of investigation and targets for testing. Furthermore, once a mineral deposit has been discovered, there is a geological risk associated with its appraisal whose validity is strongly affected by the geological variability within the deposit.

Technological risk is associated with possible problems in mine development due to poor rock conditions, etc., and in ore treatment due to unfavorable characteristics of the ore material. These risks must be considered early and can be reduced by preliminary testing, operation of pilot plants, etc. There is a much higher level of financial risk involved in the final commitment to development and production than in target selection for testing, because of the magnitude of financial resources which could be wasted in the former case owing to faulty geological advice. Economic risk is present at all stages of economic forecasting on which the production decision hinges, including prices, costs, market conditions, etc.

OPTIMIZING MINERAL EXPLORATION

Finally, political risk is ever present at all stages of exploration and development of mineral deposits; it may result from unforeseen restrictions in land tenure or access to mineral land, or changes in taxation rules, environmental protection, as well as labor laws. Perceived high political risk favors the search for small, high-grade, "bonanza" types of mineral deposits which are characterized by low investment levels, a short life, and a high payout. On the contrary, expected low political risk will enable the search for and development of large, low-grade deposits featuring low profit margins and long pay-back periods.

1.1.3. Quantification of Exploration Success in a Context of Uncertainty and Risk

1.1.3.1. The Probabilistic Concept of Exploration Success. The concept of probability is the most fruitful way of expressing quantitatively the degree of uncertainty and risk which permeates every stage of the mineral exploration decision-making process. A probabilistic assessment of uncertainty is intuitively understood in the context of the likelihood of occurrence of an event. The probability of an event is expressed by a number chosen in a numerical scale ranging from 0, standing for certainty of nonoccurrence, to unity, representing certainty of occurrence, while the fractions of unity in between reflect the degree of uncertainty attached to the occurrence of the event. The application of the principle may be further stretched to express quantitatively the degree of belief of an observer in the occurrence or nonoccurrence of an event, which leads us to the foundation of the theory of subjective probability. Finally, the same zero to unity scale may be used to measure the degree of knowledge about events in the context of imperfect or partial information which affect so many of the mineral exploration decisions.

In contrast to the previous "subjectivist" concept of probability, which is useful to measure uncertainty as viewed by an observer, another concept of probability, which is referred to as "objectivist" because it does not involve an observer, proves quite useful in the quantitative assessment of risk. It is known as the "frequentist" probability concept. Supposing that a desirable event called "success" occurs h times out of a possible total of n times in the course of a chance process, we can define the success ratio of the process as $R_s = h/n$. If the chance process is repeated many more times, so that n increases toward infinity, the success ratio tends toward a limit which is the probability of success P_s of the event. The complement of the probability of success, which is by definition written as $1 - P_s$, is called the probability of failure P_f. The concept of probability of failure provides a way of quantifying the occurrence of unfavorable events, i.e., the quantification of risk.

1.1.3.2. Historical Success Ratio as an Estimate of the Probability of Success. The concept of success ratio introduced above is of great assistance to the planning explorationist, because it provides an estimate of the probability of success of a planned type of program based on the record of past performances of similar programs in similar conditions. Both technical and commercial success ratios are useful to consider.

Generally, one finds that success ratios in oil exploration are much better documented than those of mineral exploration.[13] One of the first comprehensive studies of the success ratio in mineral exploration was that of Koulomzine and Dagenais,[8] published in 1959, which covered the postwar record of the Canadian mineral exploration, with particular reference to northwestern Quebec. They found that the overall Canadian technical success ratio is 0.116, while its commercial counterpart is only 0.063. At a regional scale, the beneficial role of geological screening based on conceptual models is well demonstrated by the increase of the technical success ratio from 0.116 to 0.233. At a local scale, for example in the Val d'Or region of northwestern Quebec, Koulomzine reports a technical success ratio as high as 0.410 due to a thorough screening of the most favorable geological areas.

1.1.3.3. Factoring the Probability of Success in Mineral Exploration. It can be intuitively understood that the occurrence of some type of event of interest could be affected by the condition that other types of events have occurred previously. In the probabilistic language, this is tantamount to say that the probability assigned to an event is conditional on the probability of occurrence of other events. The concept of conditionality is the foundation of the theory of Bayesian probabilities which we will touch on in Chapter 12 regarding the optimal selection of exploration targets for testing.

A basic tenet of the concept of conditional probability is the following law, which states that, if the events E and F are statistically independent, the probability of E and F occurring simultaneously equals the product of the probability of occurrence of E by that of F.

Applying now these concepts to mineral exploration situations, we find that the probability of technical success is conditional on the simultaneous occurrences of two events which are statistically independent, namely, geological occurrence and detection, so that we are justified in writing

$$\text{Probability of technical success} = (\text{Probability of geological occurrence}) \times (\text{Probability of detection})$$

Likewise, since technical success and economic worth are statistically independent, we can write

Probability of commercial success =
(Probability of technical success) × (Probability of economic worth)

Although the expressions written above are quite justified and very useful, it may be argued that mineral exploration is not really a one-stage procedure but has a sequential structure, an important point which will further develop below. Any well-planned exploration program consists of three stages, including (1) detection and delineation of exploration targets with the probability of success P_1, (2) investigation and testing of exploration targets with a probability of success P_2, and finally, (3) outlining ore targets within the tested exploration targets, with a probability of success P_3. Since the three events are statistically independent, we may write

Probability of commercial success =
(Probability P_1) × (Probability P_2) × (Probability P_3)

1.2. THE MITIGATION OF UNCERTAINTY AND RISK IN MINERAL EXPLORATION

1.2.1. Introduction

Uncertainty and risk are unfavorable features which are, unfortunately, an integral part of the mineral exploration business. It is therefore the hallmark of good management that every effort be made to minimize the magnitude of uncertainty and risk and mitigate their harmful effects at each stage of the ore search, whether it be planning, field execution of surveys, or final assessment of results. There are three types of approach to uncertainty and risk reduction: (1) repetition of trials,(2) sequential approach, and (3) optimization approach.

The first scheme, quite straightforward though potentially very costly, calls for the repetition of trials to improve the odds of success in the long run. The second one is more sophisticated and more resource-effective than the first one; it is based on a structured sequence of stages, each one depending on the results of the previous one and leading to the next one in a prescribed manner. The third approach takes advantage of the second one; it consists of maximizing a desired goal under cost or risk constraints at each stage of the sequential structure, resulting in a fully optimized overall procedure within the context of a prespecified corporate strategy.

1.2.2. Repetition of Trials

The concept of improving the odds of success by increasing the number of trials is quite intuitive and well recognized in everyday life. The "spreading of risk" over the largest possible number of independent ventures, another related approach, is the gist of any sound investment management which is well practiced in mineral exploration.

The theoretical justification of the repetition approach is found in the frequentist concept of the probability of success, which was mentioned above in Section 1.1.3.1. The concept is aptly illustrated by a heuristic model in the form

$$\text{(Probability of success)} = 1 - \exp(-kN)$$

where k is an empirically determined constant and N is the number of trials. It is clear that, if the number of trials N becomes quite large, the negative exponential term which represents the risk factor becomes quite small, and the probability of success nears its maximum value of unity. Obviously, if the number of trials increases, the overall cost increases in a parallel manner, quickly reaching budget ceiling before the probability of success reaches its maximum of unity; attempts to do so could lead to the "gambler's ruin," which plainly underscores the serious limitations of the first approach.

1.2.3. The Sequential Approach

1.2.3.1. Summary. The sequential approach to uncertainty and risk reduction is an alternative avenue which is more sustainable costwise than the previous one. The sequentiality of the structure of mineral exploration should be considered in two main aspects, including the time domain (chronologic sequentiality) and the space domain (spatial sequentiality). Both aspects are integral parts of all well-planned exploration programs. Figure 1.1 summarizes in a diagrammatic style the complex interrelationships between the two aspects of ore search sequentiality. The ore search is divided up in a somewhat arbitrary manner into seven stages, each affected by three types of decision making.

1.2.3.2. Sequentiality in the Time Domain. It is well recognized, albeit in an intuitive manner only, that the mineral exploration procedure has to be undertaken in a logical time sequence, otherwise the risk of financial loss will markedly increase. The generally accepted sequence starts with large-scale regional coverages first from the air, to be followed by ground geological and geochemical reconnaissance surveys. The next stage includes more detailed geological, geochemical, and ground geophysical surveys to delineate exploration

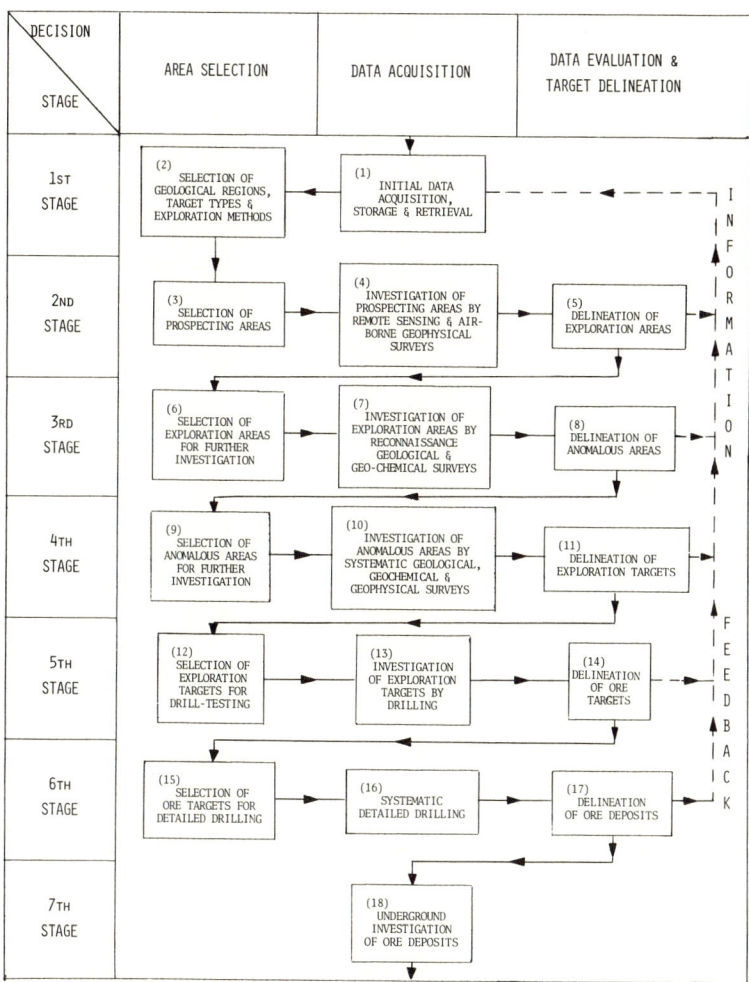

FIGURE 1.1
Flow Diagram of Mineral Exploration Sequence.

targets. The final stage consists in the detailed three-dimensional investigation of selected exploration targets by drilling, before underground exploration is decided on. It should be noted that each stage is, in turn, structured along a logical time sequence in such a way as to enmesh closely with the previous stage and the next one.

The uncertainty and ensuing risk factor may be alleviated by means of the sequential organization of the flow of various types of information. Three main types of information are to be considered: (1) prior information based on case histories and acquired from data storage at the initial planning stage, (2) newly acquired information which is gained as a result of the execution of the field programs, and (3) feedback information arising from both success and failure in field exploration, which is recycled to storage for future use at subsequent stages of the same program or in other programs. As the amount and quality of information increase progressively at each stage as compared to the previous one, both uncertainty and risk are being gradually mitigated.

Many factors of diverse nature, including geographic, particularly climatic, as well as economic and political ones, have to be reckoned with in the overall planning, because of their disruptive influence on the proper time sequence of activities. As modern ore search spreads out farther afield to investigate remote regions affected by harsh climatic conditions, it is obvious that a special allowance has to be made for these conditions when considering the logistics and scheduling of activities. For example, airborne surveys may be carried out on a year-round basis in most regions but the tropical ones, where the monsoon season is quite unsuitable for systematic flying. Ground surveys should be preferably carried out during the winter season in both sub-tropical deserts and subarctic regions, or during the dry season in regions affected by monsoonal climates, for reasons of greater logistics convenience and crew efficiency.

1.2.3.3. Sequentiality in the Space Domain. Sequentiality in the space domain is a second feature of the organization of mineral exploration which greatly assists risk reduction by means of two approaches, "scanning" and "skimming." Scanning is a sequential reduction of acreage under investigation from larger acreages of lesser geological merit to smaller acreages of upgraded potential which are to be searched by methodologies of increasing resolution at an increasing cost per unit of area. As shown by Brant,[17] the rate of acreage reduction is higher than that of increase of unit cost, thus slowing the rate of escalation of the overall cost.

The result of scanning is illustrated by the rightmost column of Figure 1.1, as the acreage of the investigated area decreases from a few thousands of square miles at stage one to only a few acres or even fractions of acre at stage six. Scanning is also accompanied by skimming at each stage of the sequence, for the purpose of reducing the number of targets outlined at one stage to be subjected to more detailed investigations at the following stage.

1.2.4. The Optimizing Approach

1.2.4.1. Rationale of the Optimizing Approach. It is appropriate at this stage to pause and briefly look back at the ground covered so far as to the nature and structure of the mineral exploration business. Its aim is the search for and discovery of nonrenewable resources which can be extracted at a profit. The search requires the making of rational decisions which involve the expenditure of large amounts of money, time, and skills in anticipation of expected profits. The search takes place in an environment of uncertainty and risk whose costly effects may be alleviated by structuring the procedure of exploration in a sequential manner.

At each stage of the sequence, decisions have to be made which require the explorationist to weigh up and compare different courses of action. Because of the context of uncertainty in which mineral exploration operates, the actual consequences of the various options are not fully known in advance. The explorationist must be satisfied to make a choice, which, if not the best choice in absolute terms, is the *best achievable* course of action under the prevailing constraints of time, budget, and skills available and within the context of prespecified corporate policies. This is the definition of the optimization approach in a nutshell.

Optimization is a an objective and pragmatic procedure which codifies the selection of options in conditions of uncertainty by the use of quantified criteria and formal rules of action in order to lead to a rational choice compatible with corporate goals. Through the use of the Dynamic Programming approach to be described in Chapter 4, the overall efficiency of the mineral exploration sequence may be maximized by optimizing each stage, while taking full advantage of the optimization of the previous stages. This is the essence of the new philosophy of optimization which should be guiding the mineral exploration of the 1980s and 1990s towards new records of success emulating those of the past two decades.

1.2.4.2. Optimization of the Preliminary Planning Phase. In the preliminary planning phase, the initial decisions of critical importance belong to the field of subjective optimization which relies on the qualitative or semi-quantified judgment of individuals or panels of experts. These decisions cover the selection of regions to be searched and types of ore deposits to be targeted, as well as methodologies to be used for the search.

Subsequent decisions are amenable to the quantitative optimization approach which is described in Chapter 4. The first decision deals with the optimal selection of prospecting areas within the previously selected region; this may be formally handled by a multivariate statistical procedure based on a geostatistical model.[41,42] The second set of decisions involves the quantitative optimization of field survey designs with which to cover the optimally selected prospecting areas for the detection of expected ore deposits. This problem of detection optimization is the major topic of the present book. Finally, there arises a problem of optimal allocation of resources including time, money, and skills between the various stages of the exploration sequence, which is within the scope of the new science of Operations Research. (See Chapter 4.)

1.2.4.3. Optimization of the Field Implementation Phase. The types of decisions which have to be taken in the field implementation phase are rather different in nature from those made by the planners. The decision maker has to be quite flexible; at the end of each stage, decisions have to be made on the spot and have to be tailored to field results which were unforeseen in the preliminary planning phase.

One example is the decision whether to terminate a survey or extend it further in order to acquire additional data. Zero information is zero cost, but entails a large risk factor. Complete information is unattainable because of its prohibitive cost and the special nature and structure of geo-data. Therefore, an optimal amount and quality of information has to be sought for as a compromise between the two extreme cases mentioned above. The formal solution of this optimal choice problem has been attempted by several Operations Research workers. They use the Dynamic Programming approach combined with Bayesian decision theory in a graphical manner (decision tree diagrams).

Another problem of optimal choice under uncertainty arises after the completion of field surveys. The sequential combination of scanning and skimming mentioned above leads to the delineation of a number of exploration targets which is still far too large for full investigation by drilling within the bounds of the allocated budget. Two approaches to the optimization of the selection of exploration targets are considered in Chapter 12 of the book. The first one, relying on control locations which have been previously investigated, is suitable for established mining districts. The second one, recently devised by the second writer, does not require the availability of control locations, which is an obvious advantage when dealing with poorly prospected regions.

Finally, once the exploration targets of maximum merit have been selected within geological and budgetary parameters, the optimal allocation of funds between selected targets for drill testing has to be dealt with. Celasun (Ref. 21 in Chapter 4) and other workers in the field of Operations Research have investigated the matter in a formal analytical manner.

1.2.4.4. Optimization of the Evaluation Phase. The final phase of the surface exploration sequence covers the evaluation of results of the drill testing of optimally selected exploration targets. Once a target has been successfully tested and a mineral deposit has been detected and approximately outlined, an evaluation of the commercial potential (tonnage, grade) of the mineralization is called for, before making a decision regarding the underground development of the deposit.

The evaluation has to be based on closely spaced sampling by surface drilling, which is time-consuming and very expensive. Once again, we are faced with a problem of optimal survey design as in the preliminary planning phase. However, in this case the gist of the matter is different: instead of seeking to optimize detection efficiency, we are now interested in optimizing sampling precision. The degree of geological variability prevailing throughout the mineral deposit dictates the degree of precision required for the estimates of tonnage and grade. The desired precision level, in turn, prescribes the number of samples required, and, therefore, the sample spacing and corresponding coverage cost. A few workers have investigated this important optimization problem (Refs. 19 and 28 in Chapter 4).

1.3. CONCLUSION: SCOPE OF THE BOOK

Because of the prevalence of uncertainty and risk which affect every aspect of the mineral exploration business, it is proper to view the whole ore search procedure in a probabilistic context. The goal of mineral exploration is therefore expressed in terms of probability of success, which, in turn, is factored into three main components, namely, probability of geological occurrence, probability of detection, and probability of economic worth. A full coverage of the evaluation of each of the three components could not be considered within the one-volume format required here: for example, an adequate treatment of the probability of occurrence alone would require a full-size book. We have therefore decided to restrict the topic of the present volume to only one component of the probability of success, that of the detection probability, which is fully investigated in its theoretical aspects in Chapter 2 and in its practical implications in Chapters 6–11. Carefully selected references are provided at the end of this chapter for the readers interested in delving further into the very important topics of occurrence and economic worth.

The brief coverage offered in Section 1.2.4 amply demonstrates the breadth of scope of the optimization approach in the mineral exploration business. Despite the rather scant attention devoted to date to the optimization of mineral exploration, a full treatment of the subject, however desirable, is not possible within the constraints of a single-volume format. Only two of seven basic types of optimal choices arising in the course of ore search are fully covered in the present book. These comprise (a) the optimization of field survey designs in the preliminary planning phase dealt with in Chapters 2, 3, and 4, and (b) the optimization of the selection of exploration targets for testing at the end of implementation phase, which is treated in Chapter 12. The methodology of the optimization of detection, which is described in Chapters 2–4, is then applied to the search for six common types of ore deposits in North America (Chapters 5–11).

REFERENCES AND SELECTED READINGS

Success in Mineral Exploration

1. BORCH, K. H., 1972, *The Economics of Uncertainty,* Princeton University Press, Princeton, New Jersey.
2. BROWN, G. A., 1970, The evaluation of risk in mining ventures, *Can. Inst. Min. Metall. Bull.* **63**(701), 1165–1171.
3. COOPER, D. O., DAVIDSON, L. B., and REIM, K. M., 1973, Simplified financial and risk analysis for mineral exploration, Annual Symposium on Computer Applications No. 11, Vol. 1, pp. B1–B14, Tucson, Arizona.
4. DE GEOFFROY, J., and WIGNALL, T. K., 1974, Evaluating that exploration project, *Can. Min. J.* **95**(5), 42–44.
5. GRAYSON, C. J., JR., 1960, *Decisions under Uncertainty: Drilling Decisions by Oil and Gas Operators;* Harvard University, Graduate School of Business Administration, Cambridge, Massachusetts.
6. GRIFFITHS, J. C., 1967, Mathematical exploration strategy and decision making, *Seventh World Petroleum Congress Proceedings, Mexico City,* pp. 599–604.
7. KNIGHT, F. H., 1921, *Uncertainty and Profit,* Chapter 2, Houghton-Mifflin, New York.
8. KOULOMZINE, T., and DAGENAIS, R. W., 1959, Statistical determination of the chances of success in mineral exploration in Canada, *Can. Min. J.* **74**(4), 107–110.
9. LEBIS, A. P., 1978, The economics of mining and other risky ventures, *Can. Inst. Min. Metall. Bull.* **71**(791), 169–171.
10. MACKENZIE, B. W., 1973, Corporate exploration strategies, Application of Computer Methods in the Mineral Industry Symposium, Johannesburg, pp. 1–8.
11. MACKENZIE, B. W., 1981, Looking for the improbable needle in a haystack: The economics of base metal exploration in Canada, *Can. Inst. Min. Metall. Bull.* **74**(830), 115–125.
12. MOORE, P. G., 1972, *Risk in Business Decision,* Chaps. 1, 5, 6, 7, and 11, Longman, London.
13. PETERS, C. W., 1978, *Exploration and Mining Geology,* Wiley, New York.
14. PRATT, J. W., RAIFFA, H., and SCHLAIFER, R., 1964, The foundations of decision under uncertainty: An elementary exposition, *J. Am. Stat. Assoc.* **59**, 353–375.
15. ROSCOE, W. E., 1971, Probability of an exploration discovery in Canada, *Can. Inst. Min. Metall. Bull.* **64**,(707), 134–137.

Evaluation of the Probability of Ore Occurrence: Univariate Modeling

16. BOZDAR, L. B., and KITCHENHAM, B. A., 1972, Statistical appraisal of the occurrence of lead mines in the Northern Pennines, *Trans. Inst. Min. Metall.* **81,** B183–B187.
17. BRANT, A. A., 1968, The pre-evaluation of the possible profitability of exploration prospects, *Miner. Deposita* **3,** 1–17.
18. BROWN, D., and ROTHERY, P., 1978, Randomness and local regularity of points in a plane, *Biometrika* **65,** 115–122.
19. CLARK, P. J., 1956, Grouping in spatial distributions, *Science* **123,** 373–375.
20. DREW, M. W., 1974, A deposit distribution model for uranium, Programs Analysis Unit Report 10-74, U.K. Atomic Energy Authority, London.
21. EBDON, D., 1977, *Statistics in Geography: A Practical Approach,* Chap. 7, Blackwell, Oxford.
22. GRIFFITHS, J. C., 1962, Frequency distributions of some natural resources materials, Bulletin No. 63, Mineral Industry Experimental Station, University of Pennsylvania, pp. 174–198.
23. JOHNSON, N. I., and KOTZ, S., 1970, *Discrete Univariate Distributions,* Vol. 1, Houghton-Mifflin, Boston.
24. MacDOUGAL, E. B., 1976, *Computer Programming for Spatial Problems,* E. Arnold, London.
25. McCONNELL, M., 1966, Quadrat methods in map analysis, Discussion Paper No. 3, Department of Geography, University of Iowa.
26. MILLER, R. L., and KAHN, J. S., 1962, *Statistical Analysis in the Geological Sciences,* Chap. 16, Wiley, New York.
27. NEFT, D. S., 1966, Statistical analysis for areal distribution, Monograph 2, Regional Science Research Institute, Philadelphia.
28. ROGERS, A., 1974, *Statistical Analysis of Spatial Dispersion: the Quadrat Method,* Chap. 1, Pion, London.
29. SARMA, D. D., 1979, An exploration strategy for prospecting with a case study on copper prospects at Ingladahl, India, *Miner. Deposita* **14,** 263–279.
30. SLICHTER, L. B., 1960, The need for a new philosophy of prospecting, *Min. Eng. (New York)* June, 570–576.
31. WILLIAMSON, E., and BRETHERTON, M. H., 1963, *Tables of the Negative Binomial Probability Distribution,* Wiley, New York.

Evaluation of the Probability of Ore Occurrence: Multivariate Modeling

32. AGTERBERG, F. P., 1970, Multivariate prediction equations in geology, *J. Math. Geol.* **2,** 319–324.
33. AGTERBERG, F. P., 1971, A probability index for detecting favourable geological environments, *Can. Inst. Min. Metall. Spec. Vol.* **12,** 82–91.
34. AGTERBERG, F. P., 1974, Automatic contouring of geological maps to detect targets areas for mineral exploration, *J. Math. Geol.* **6,** 373–395.
35. AGTERBERG, F. P., 1977, Frequency distributions and spatial variability of geological variables, *Proceedings of the A.P.C.O.M. Symposium No. 14,* New York, pp. 287–298.
36. AGTERBERG, F. P., CHUNG, C. F., DIVI, S. R., and EADE, K. E., 1981, Preliminary geomathematical analysis of geological, mineral occurrence and geophysical data, Southern

District of Keewatin, Northwest Territories, Canada, *Geol. Surv. Can.*, Open File Report No. 778, pp. 1–29.
37. CHUNG, C. F., 1978, Computer program for logistic model to estimate the probability of occurrence of discrete events, *Geol. Surv. Can. Pap.* No. 78-11.
38. CHUNG, C. F., and AGTERBERG, F. P., 1980, Regression models for estimating mineral resources from geological map data, *J. Math. Geol.* **12**, 458–473.
39. COX, D. R., 1970, *Analysis of Binary Data*, Methuen, London.
40. DE GEOFFROY, J., and WIGNALL, T. K. W., 1970, Application of statistical decision techniques to the selection of prospecting areas and drilling targets in regional exploration, *Can. Inst. Min. Metall. Bull.* **63**(699), 893–899.
41. DE GEOFFROY, J., and WIGNALL, T. K., 1971, A probabilistic appraisal of mineral resources in a portion of the Grenville Province of the Canadian Shield, *Econ. Geol.* **66**, 466–479.
42. DE GEOFFROY, J., and WIGNALL, T. K., 1973, Design of a statistical data processing system to assist regional exploration planning; *Can. Min. J.* **94**(11), 30–35; **94**(12), 35–36.
43. HARRIS, D. P., 1965, Multivariate statistical analysis: A decision tool for mineral exploration; Computer Applications in Mining & Exploration Symposium, Vol. 1, pp. C1–C35, University of Arizona, Tucson, Arizona.
44. HARRIS, D. P., 1969, Alaska's base and precious metals resources: A probabilistic regional appraisal, *Q. Col. Sch. Mines* **64**(3), 295–327.
45. HARRIS, D. P., and AGTERBERG, F. P., 1981, The appraisal of mineral resources, *Econ. Geol.* **75**, 897–938.

Evaluation of the Probability of Ore Occurrence: Subjective Approach

46. BARRY, G. S., and FREYMAN, A. J., 1970, Mineral endowment of the Canadian Northwest: A subjective probability assessment, *Can. Inst. Min. Metall. Bull.* **63**(700), 1031–1038.
47. ELLIS, J., HARRIS, D. P., and VAN WIE, N., 1975, A subjective probability appraisal of uranium resources in the State of New Mexico, U. S. Energy Research and Development Administration, Grand Junction, Colorado.
48. GOLABI, K., and LAMONT, A., 1981, A probabilistic approach to the assessment of uranium resources, *J. Math. Geol.* **13**(6).
49. HARRIS, D. P., 1973, A subjective probability appraisal of metal endowment of Northern Sonora, Mexico, *Econ. Geol.* **68**(2), 222–242.

Evaluation of the Probability of Economic Worth: Univariate Modeling

50. AGTERBERG, F. P., 1980, Lognormal models for several metals in selected areas of Canada, Reprint, *Mem. BRGM*, No. 106, pp. 83–90.
51. AGTERBERG, F. P., and DIVI, S. R., 1978, A statistical model for the distribution of copper, lead and zinc in the Canadian Appalachian Region, *Econ. Geol.* **73**, 230–245.
52. ALLAIS, M., 1957, Methods of appraising economic prospects of mining exploration over large territories, *Manag. Sci.* **3**, 285–347.
53. DIVI, S. R., 1980, Deposit modeling and resources estimation of stratiform massive sulfide deposits in Canada, *Comput. Geosci.* **6**(2), 163–174.
54. DE GEOFFROY, J., and WU, S. M., 1970, A statistical study of ore occurrences in the Greenstone Belts of the Canadian Shield, *Econ. Geol.* **65**, 496–504.

55. SANGSTER, D. F., 1980, Quantitative characteristics of volcanogenic massive sulfide deposits, *Can. Inst. Min. Metall. Bull.* **73**(814), 74–81.
56. SINCLAIR, A. J., 1974, Probability graphs of ore tonnage in mining camps: A guide to exploration, *Can. Inst. Min. Metall. Bull.* **67**(749), 71–75.
57. SINGER, D. A., COX, D. P., and DREW, L. J., 1975, Grade and tonnage relationships among copper deposits, *U.S. Geol. Surv. Prof. Pap.* **907-A,** A1–A11.

CHAPTER TWO

EVALUATION OF THE PROBABILITY OF DETECTION OF MINERAL DEPOSITS

2.1. THE DETECTION PROCESS

2.1.1. Introduction

The principal purpose of field programs is the acquisition of information which will lead to the detection of mineral deposits. There are two types of approach to the problem of detection, namely, the direct and indirect methods, which are commonly used simultaneously or sequentially to best advantage. Table 2.1 summarizes the main aspects of the two paths with respect to the types of target sought and detection environment, as well as the methodology and procedures involved.

The direct approach seeks to detect the mineral deposit itself, by direct observation. The observation is based on visual recognition, either unaided as in ground prospecting or mapping, or aided by photogeology or remote sensing techniques from airborne platforms, or by mechanical probes, such as drilling, in three-dimensional investigations. The second method requires a more elaborate treatment to be given below, because of its indirect nature.

2.1.2. The Indirect Approach to the Detection of Mineral Deposits

The indirect approach seeks to detect mineral deposits, or closely associated features such as halos, in an indirect manner, through local changes called "anomalies," by sampling the continuum of the regional geophysical or geochemical environment. There are two main types of sampling procedures; continuous and discrete sampling.

Continuous sampling surveys are feasible only from airborne platforms. The data are recorded as graphs, or on photographic films, or on magnetic tapes, while sweeping the ground along regularly spaced profile lines. The data are then subjected to the process of "digitization," which calls for the sampling of continuous recordings at regular intervals in order to extract representative discrete data for convenient processing. Discrete sampling is conducted on the ground at various points provided by the topography (outcrops, stream beds,

TABLE 2.1
Methodology of the Detection of Mineral Deposits.

Detection approach	Detection method	Type of target sought	Detection environment	Procedures involved
direct detection	prospecting	ore deposit	surface outcrops and "floats"	visual recognition + physical sampling + laboratory analyses
	trenching		subsurface and bedrock	physical sampling + laboratory analyses + visual recognition
	surface drilling		underground bedrock	physical sampling + laboratory analyses + visual recognition
	tunneling and underground drilling		underground bedrock	physical sampling + laboratory analyses + visual recognition
indirect detection	geophysical surveys		underground bedrock	measurements by sensors + mathematical data processing and evaluation
		primary mineralogic halos (magnetite, sulfides, etc)	underground bedrock	measurements by sensors + mathematical data processing and evaluation
	geochemical surveys	primary geochemical halos	underground bedrock	physical sampling + laboratory analyses + statistical data processing and evaluation
		secondary geochemical halos	surface (soil, water, plants)	physical sampling + laboratory analyses + statistical data processing and evaluation

etc.) in reconnaissance surveys, or at control grid intersections, as is the case for all systematic surveys and drilling programs.

The indirect detection approach may be used on two different types of targets: mineral deposits themselves through their geophysical signatures and geophysical or geochemical halos associated with the deposits, which are larger and thus easier to detect than the deposits. Following the detection and delineation of the halos, a second stage of investigation is carried out to locate the expected mineral deposits, either indirectly by further geophysical surveys, or directly by systematic drilling, within the outline of the halos.

Halos of two types, namely, primary and secondary, are known to occur in association with many types of mineral deposits. The primary halos were formed at the same time as the mineral deposits and show a close spatial relationship with them. There are primary halos whose mineralogical make-up includes pyrite, pyrrhotite, magnetite, or radioactive elements which are all detectable by geophysical methods. Bedrock sampling followed by multielement laboratory analysis is used to detect primary geochemical halos formed by the syn-ore dispersion of minor elements.

The secondary halos arise from the reworking of primary halos and mineral deposits by secondary chemical, physical, and biological processes. The resulting halos are much larger and easier to detect than the primary ones; unfortunately, they are not closely related to the causative body spatially, which makes the second stage of the ore search much riskier than in the case of primary halos.

Table 2.2 presents a summary coverage of the principal techniques of indirect detection used in most modern exploration programs. We do not believe it is necessary to go into further details, because the subject matter has been well covered in many general books,[9,15,17] as well as in more specialized ones.[4,11,12]

2.1.3. The Nature and Structure of the Indirect Detection Process

The indirect detection of mineral deposits is a sequentially structured process consisting of three main stages: recording, resolution and processing, and recognition and confirmation. At the initial stage, the intersection of the mineral deposit by a sensor generates what is referred to as "messages" in information theory parlance.[16] The structure of messages is always complex and is made up of two main portions: (1) a portion of interest to the explorationist called "signal," which is generated by the mineral deposit itself, and (2) a portion called "noise," generated by the geo-environment surrounding the ore deposit, and by extraneous bodies lying between the mineral deposit and the recording platform.

TABLE 2.2
Methodology of the Indirect Detection of Mineral Deposits.

	Indirect Detection of Ore Deposits by Geophysical Sensors					Indirect Detection of Ore Deposits by Sampling of Geochemical Halos						
Approach	Method	Techniques		Parameter	Application	Type of environment	Geological processes involved	Litho- sampling	Techniques and sampling medium			
									Pedo- sampling	Hydro- sampling	Bio- sampling	Atmo- sampling
Natural	"potential" fields	magnetometric		Earth magnetic field	-airborne -ground	Soil- atmosphere interface						aerosols and vapors
		gravimetric		Acceleration of gravity	-airborne -ground							
		radiometric		natural gamma radiation	-airborne -ground	Biosphere						plant tissues
	nonpotential	spontaneous polarization		natural electro- chemical action	-ground	Surficial deposits	secondary dispersion		transported soil	lake water		
		AFMAG	frequency domain	natural electro- magnetic pulses	-airborne -ground					stream water		
Artificial	applied	electric	resistivity mise a la masse	resistivity or conductivity	-ground -ground				residual soil	stream silt		
		electro magnetic	frequency domain	induced electro- magnetic field	-airborne -ground					ground water		
			time domain (INPUT)		-airborne	Lithosphere	primary dispersion	bedrock				
		induced polariza- tion	frequency domain	apparent resistivity								
			time domain	transient voltage	ground	Ore deposit	hydrother- mal or mag- matic con- centration					

Noise is a detrimental feature which may distort, mask, or suppress signals making up the signature of mineral deposits. The nature of noise is quite varied and includes (a) white noise of a random nature (b) thermal noise generated by the recording instrumentation, and (c) geological noise generated by the geo-environment.

Messages are either translated as readings visually observable and recorded manually by an operator or are recorded automatically on graphs or more likely on magnetic tapes. On most occasions, several messages are recorded and have to be separated. The ability of sensors and recorders to distinguish between messages that are closely associated either in the space or time domain is referred to as "power of resolution."[18] The separated messages are then subjected to a special form of statistical processing, known as "filtering" in information theory parlance, which separates the signal portion from the undesirable noise portion and enhances the signal strength.[10,16,21]

Some of the filtered signals are generated by causative bodies of no interest to the explorationist, although they resemble and may be confused with the signatures of mineral deposits, thus requiring discrimination. The discrimination is achieved in an objective manner in a multivariate context by means of a statistical method known as "pattern recognition" (Ref. 11 in Chapter 12). A classifying model is "trained" on signals generated by known ore deposits of the required type. The signals generated by the expected target are statistically compared with the control signals and are classified either as target or spurious at a prespecified level of confidence.

Since there is always some degree of uncertainty involved in the detection of a mineral deposit, explorationists are justified in requiring that the targets be intersected several times, at least twice, by the sensor in order to confirm the detection.

2.1.4. Factors Involved in the Detection Process

Many complexly interrelated factors are involved in the detection process, whether direct or indirect. Some factors are of a geological or economic nature and should be viewed in a probabilistic context; others are technological and methodological in nature and should be considered in a deterministic perspective. Most of these factors interact in a sequential manner.

For the most part, geological factors involved in the detection process are not under the explorationist's control. Among these are the characteristics of the expected prizes. Their mineralogical, geophysical, and geochemical properties are derived by analogy from those of known deposits of the selected type. Others, such as geometric and economic parameters, and spatial distribution parameters are derived by statistical modeling of known deposits.

The geological characteristics of the expected prizes will largely dictate the choice of technology best suited to assist the detection of mineral deposits of the selected type. The explorationist will choose the type and design of sensor or probe with regard to performance parameters such as penetration, sensitivity to the three types of noise mentioned above, resolution, discrimination, and lateral coverage.[13]

The combination of geological and technological factors will, in turn, prescribe the selection of the most suitable methodology. The latter is defined as the most appropriate sequence and mix of techniques to ensure the maximization of the probability of detection of expected ore deposits.

2.2. GEOMETRIC CONSIDERATIONS INVOLVED IN THE DETECTION OF MINERAL DEPOSITS

2.2.1. Introduction

As indicated above, intersection of the expected target by a sensor or a probe is the prerequisite of detection. The intersection requirement is essentially a matter of spatial relationships which are based on geometric considerations. These considerations involve not only the intrinsic geometric characteristics of expected deposit, search grids, and detectors, but also their mutual configurations.

The geometric characteristics of the expected deposit include size, shape, and attitude considered in a three-dimensional context, and are probabilistic in nature. Grid geometry is described by shape and spacing parameters which are deterministic in nature, as are the parameters describing the detector geometry (depth of penetration and inclination). The configuration of the expected target and the control grid is a very important factor which can be viewed in two aspects: a dimensional one, expressed by the ratio of the longest dimension of the target over the grid spacing, and a directional one, expressed by the angle between the grid and the direction of the longest dimension of the target (strike). Another important directional consideration is the attitude of the detector with respect to the target.

2.2.2. Geometric Characteristics of Ore Deposits

2.2.2.1. Nature of Geometric Parameters Describing Ore Deposits. The descriptors of the geometry of ore deposits may be grouped into two categories: dimensional parameters and attitudinal parameters. The dimensional

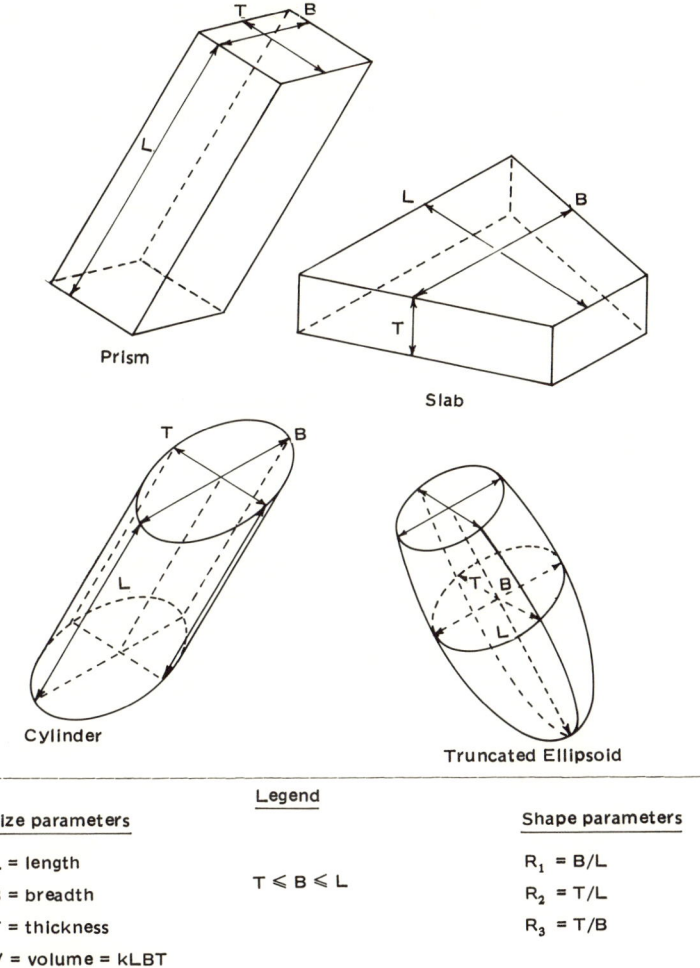

FIGURE 2.1
Size and Shape Parameters of Ore Deposits.

parameters are illustrated in Figure 2.1; they include length, breadth, and thickness, which can be grouped for greater convenience of processing into ratios such as shape ratios, or products such as areas of principal sections or volumes.

The attitudinal parameters are angular in nature and include the strike orientation, dip angle, and angle of plunge. The strike orientation is measured clockwise from the true north within 0–180 degrees. The true dip angle (a_0) is measured within 0–90 degrees below the horizontal, within the vertical plane which is perpendicular to the strike direction. The apparent dip angle (a), measured in any other vertical plane, is related to the true dip angle by the following relationship which is tabulated in Table 2.3.:

$$\tan a = \tan a_0 \times \sin d$$

where d is the angle measured between the vertical plane considered and that containing the strike line of the target. Finally, the angle of plunge measures the inclination of the ore deposit within the vertical plane containing the strike line, in the same manner as the dip.

TABLE 2.3
Conversion of True Dip to Apparent Dip Angles in Degrees.

Angle of deviation from strike line in degrees	True dip angle in degrees							
	5	10	20	30	40	50	60	70	80
5	0	1	2	3	4	6	9	13	26
10	1	2	4	6	8	12	17	26	45
15	1	3	5	8	12	17	24	35	56
20	2	3	7	11	16	22	31	43	63
25	2	4	9	14	20	27	36	49	67
30	3	5	10	16	23	31	41	54	71
35	3	6	12	18	26	34	45	58	73
40	3	6	13	20	28	37	48	60	75
45	4	7	14	22	31	40	51	63	76
50	4	8	16	24	33	42	53	65	77
55	4	8	17	25	35	44	55	66	78
60	4	9	17	27	36	46	56	67	78
65	5	9	18	28	37	47	58	68	79
70	5	9	19	28	38	48	58	69	79
75	5	10	19	29	39	49	59	69	80
80	5	10	20	30	40	50	60	70	80
85	5	10	20	30	40	50	60	70	80

Apparent dip angles in degrees

2.2.2.2. Factors Involved in the Definition of Geometric Parameters of Ore Deposits. The determination of dimensional parameters of ore deposits is affected by a complex interaction of geological, economic, and technological factors, whereas the attitudinal parameters are essentially governed by geological factors and are simpler to measure. The natural geometry of a *mineral deposit* as expressed by dimensions and attitude is governed by geological limits which may be stratigraphic, structural or petrographic, and geochemical in nature. The actual dimensions of *ore deposits,* however, do not coincide with that of the mineral deposits because economic considerations have to be introduced. The "cutoff" grade, below which the mineralized material cannot be extracted at a profit and thus is not ore, is the key economic factor affecting the determination of the dimensional parameters of ore deposits. An unfortunate result is that the dimensions of ore deposits will expand or shrink within the maximum geological dimensions of the mineralization, reflecting the fluctuations of the economic environment.

Finally, technological considerations have to be taken into account in addition to the geological and economic ones. For example, the minimum stoping width or height requirements for easy access of the extractive machinery are an important factor modifying the economic dimensions of ore deposits. Ore dilution caused by weak hanging walls is a very detrimental factor affecting the dimensions of ore deposits.

2.2.2.3. Statistical Modeling of Geometric Parameters. Because the geometry of the expected prizes is not known beforehand, we have to rely on probabilistic estimates of geometric parameters which are derived by the statistical modeling of known ore deposits of the same genetic type as that of the expected target. The purpose of model fitting is to establish on the basis of statistical tests (χ-squared goodness of fit tests) which theoretical model best describes the observed frequency distribution of parameter measurements obtained from known ore deposits. The procedure and pitfalls of statistical modeling, as well as the statistical properties of the most commonly used models, are well described in Refs. 22, 24, 26, 27, and 28.

The normal model, a cornerstone of modern Statistics, is useful when dealing with dip angles. A related model, known as circular normal is used to advantage to model orientated angular parameters such as strike directions. The distributions of dimensional parameters of ore deposits are satisfactorily fitted by lognormal models. The lognormal model is related to the normal model through a logarithmic transformation: in other words, data are said to be lognormally distributed if the distribution of their logarithms is normal. Figure 2.2 illustrates the fitting of lognormal models to the observed distributions of geo-

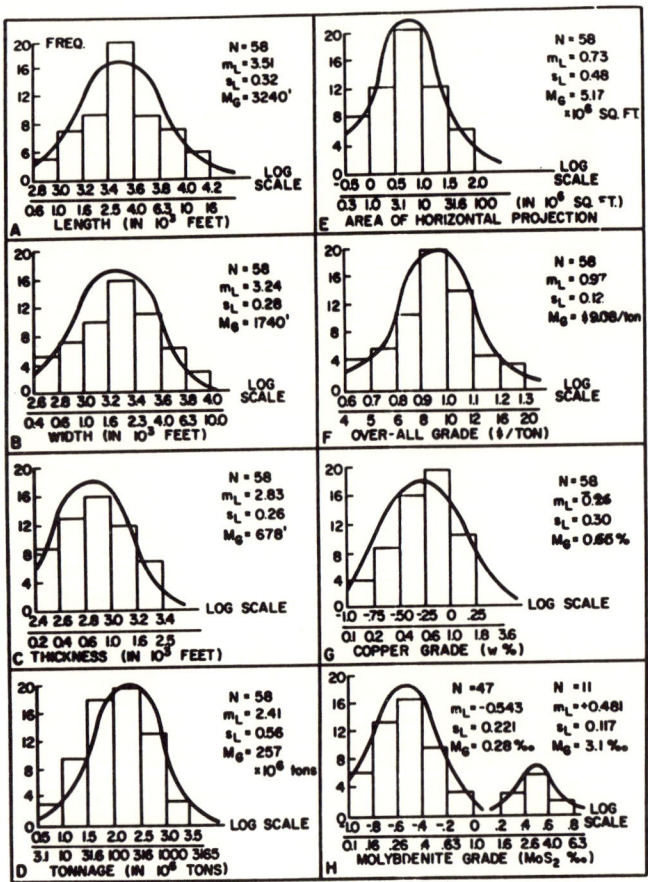

FIGURE 2.2
Histograms of Distributions of Geometric Parameters of Porphyry–Cu–Mo Deposits of the North American Cordillera Belt and Fitted Lognormal Models.

metric parameters of porphyry Cu–Mo deposits of the Cordillera Belt of the Americas.[25]

2.2.3. Geometry of Ore Deposits in the Context of Detection

2.2.3.1. Introduction. Since the main topic of the book is the detection of ore deposits, it is appropriate to consider not so much the geometry of the deposits themselves but that of the target they offer to detection. Detection is obtained from a horizontal plane, either the ground, or the flight plane for airborne surveys. It is restricted to a specific vertical range below the observation plane, and involves the configuration of detector and ore deposit.

2.2.3.2. Vertical and Angled Direct Detection. We will first consider the case of vertical detection by drilling. The target geometry for detection purposes is defined by the dimensions of the vertical projection on the horizontal plane of the portion of the ore deposit which lies within the range of penetration of the probe. For practical purposes, 300 feet would be an appropriate figure for the depth of most preliminary drilling programs.

The four parameters required for the calculation of the probability of detection are the target length L_t, target shape ratio $R_t = B_t/L_t$, the true dip a_0, and the plunge c. Figure 2.3 illustrates how the target parameters L_t, B_t, and R_t are obtained in terms of the dimensional parameters L_h and B_h of the horizontal

FIGURE 2.3
Relationship between Target Breadth, Dip Angle, and Detection Range in the Case of Vertical Detection of Ore Deposits.

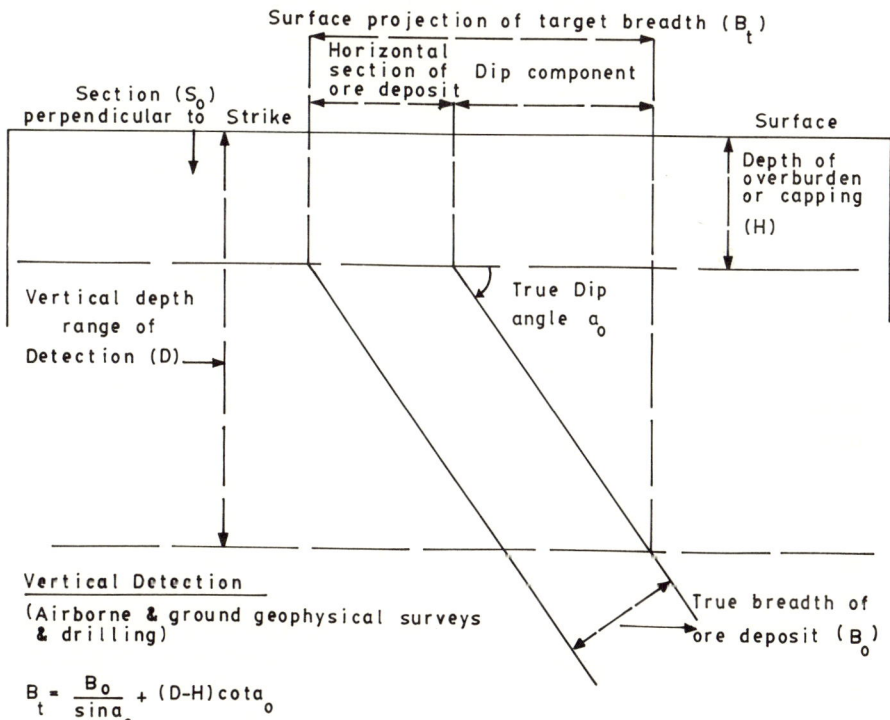

section of the ore deposit within the range D, and the attitude parameters (true dip and plunge). The relationships are as follows:

$$L_t = L_h + D \cot c$$
$$B_t = B_h + D \cot a_0 \quad \text{with } B_h = B_0/\sin(a_0)$$

B_0 being the "true width" in common drilling parlance, and

$$R_t = B_t/L_t$$

If the deposits are subvertical (dip and plunge in the 85–90 degree range), the three target parameters L_t, B_t, and R_t are approximately equal to the parameters L_h, B_h, and R_h of the horizontal section of the deposits.

Let us consider now the case of angled detection by drilling to the same vertical depth D as in the previous case. The diagram of Figure 2.4 shows that the dip component of the target breadth B_t is inflated by the addition of a term involving the drilling angle b in the following manner:

$$B_t = B_h + D \cot a_0 + D \cot b$$

This results in a much larger shape ratio R_t, which substantially boosts the magnitude of the probability of detection. If we consider subvertical deposits, the term cot a_0 becomes equal to zero and the dip component included in the calculation of the target breadth is reduced to that introduced by angled drilling.

2.2.3.3. Vertical Indirect Detection. The previous section dealt with direct detection by vertical or angled drilling. We now consider the indirect detection of mineral deposits by geophysical sensors, which is essentially a vertical detection situation.

The vertical distance r between the sensor platform and the mineral deposit to be detected strongly affects the strength s of the signal generated by the deposit, as expressed by the following relationship:

$$S = k/r^n$$

the exponent n being an integer depending on the type of detection technology and shape of the deposit.[13,15] For example, $n = 1$ for magnetic methods, $n = 2$ for gravity surveys, and $n = 3$ for electromagnetic methods.

The nature of the spatial relationship between geophysical signatures and the causative mineral deposits is of a critical importance in the indirect detection

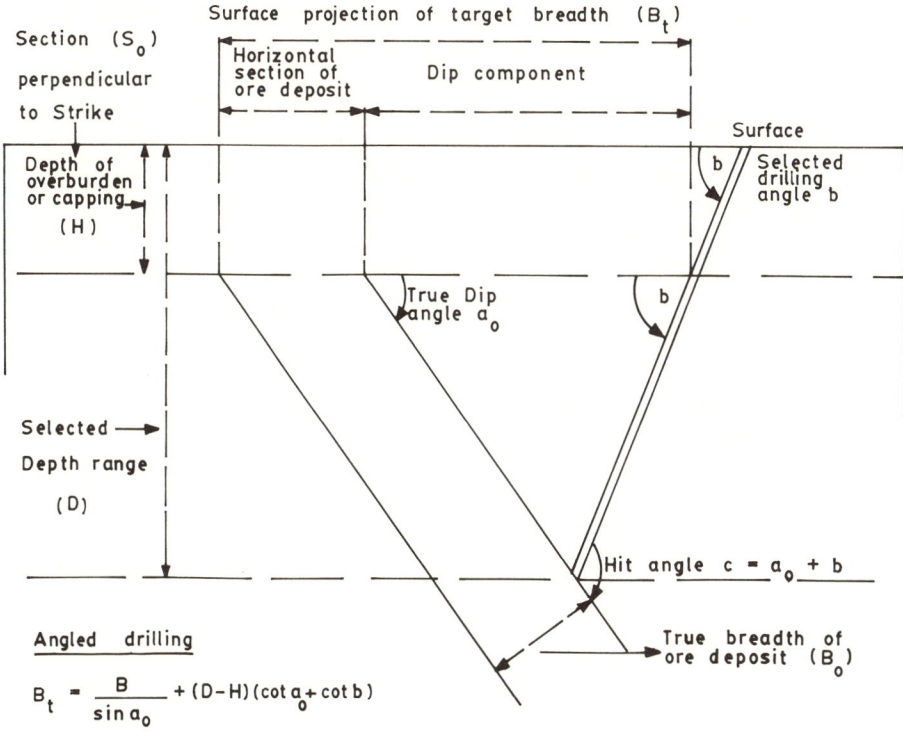

FIGURE 2.4
Relationship between Target Breadth, Dip Angle, and Detection Range in the Case of Angled Detection of Ore Deposits.

process. All types of signatures essentially reflect a contrast of geophysical properties between the ore deposit and the encasing wallrock. When "potential field" methods (gravity, magnetics, radiometrics) are used, a large portion of the signal intensity consists of an "edge effect" generated by the contrast, while the remainder is attributed to a "volume effect" reflecting the intrinsic geometry of the mineral deposit. Through mathematical processing involving the calculation of the second derivative of the potential functions, such as "continuation," which can be applied to all three methods, or "reduction to the Pole" for magnetic surveys only, it is possible to ensure a very close spatial coincidence between signature and causative body.[2,8,20] If we deal with artificially generated fields (electromagnetics and induced polarization, in both frequency and time domains), however, the edge effect makes up most of the signal intensity and the contribution of the volume effect is quite small.

In practice, most airborne or ground geophysical surveys rely on a combination of techniques of the two types, measuring natural potential or artificially induced fields. The criterion of success becomes a multiple spatial coincidence of various types of signatures. For example, out of 33 well-documented discoveries of volcanogenic massive sulfide deposits in northwestern Quebec, as many as 29, or 83% of the total, showed good multiple coincidence and spatial correlation with anomalies; similar results based on ten discoveries were recorded in the new volcanogenic sulfide district of Western Ontario.

In conclusion, if we consider indirect vertical detection, we can legitimately assume that the geometry of the target defined by combined geophysical signatures closely compares with that of the target generated by the portion of the causative mineral deposit which lies within the detection range.

2.2.4. Grid Geometry in the Context of Detection

2.2.4.1. Intrinsic Grid Geometry. An accurate recording of the location of samples or sampling profiles is an essential requirement for all systematic airborne or ground surveys or drilling programs in order to pinpoint the exact location of any discovery. Airborne surveys are flown by aircraft along imaginary lines located later from photographic recordings, while ground surveys and drilling programs are carried out at the intersections of actual lines materialized by equidistant markers.

Three parameters, (1) type of design, (2) shape, and (3) spacing, are required to describe the geometry of control grids. The designs are of two kinds: (1) parallel lines, mainly used for continuous sampling by airborne surveys, and (2) orthogonally intersecting lines, chiefly used for ground surveys and drilling programs. In several theoretical studies, however, other geometric designs such as rhombic, triangular, and hexagonal point-nets are considered for ground search programs.[34,39,45,47]

The third geometric parameter of control grids is the spacing between lines or between points. Parallel grids are defined by the spacing s between lines. The size of orthogonal grids is defined as $s \times s$ for square-shaped grids, and as $s \times t$, or s and $w = s/t$, for rectangular grids. When considering ground sampling, one should bear in mind that the number of sampling points and, therefore, the sampling cost are inversely proportional to the square of the grid spacing; i.e., if the grid spacing is cut by half, the sampling cost will quadruple.

2.2.4.2. Geometric Relationship Between Target and Grid. Two main considerations are involved when selecting the spacing of a control grid in the search for a specific type of mineral deposit. The first one is related to the geometry of the target and can be viewed in two aspects, namely, dimensional and directional. The second one is related to the size of the area to be covered and the cost implications.

The dimensional aspect of the geometric relationship between target and control grid can be simply quantified by the ratio U_t of the longest dimension of the target, L_t, over the grid spacing s. The ratio U_t is of critical importance in the calculation of probabilities of detection by both airborne and ground surveys. The shape ratio of the target, $R_t = B_t/L_t$, plays an important role in the calculation of probabilities of detection by discrete ground sampling, but is of lesser importance in continuous sampling airborne surveys.

Next to the dimensional aspect of the target–grid relationship, the directional aspect has a substantial effect on detection probabilities. The directional factor may be simply expressed by the angle between the directions of the longest dimension of the target L_t and the lines of parallel grids, or either side of square grids, or the longest side of rectangular grids.

2.3. THE THEORY OF GEOMETRIC PROBABILITIES AS A FOUNDATION FOR THE EVALUATION OF THE PROBABILITY OF DETECTION OF MINERAL DEPOSITS

2.3.1. Rationale of the Application of Geometric Probabilities to the Detection of Mineral Deposits

On the one hand, the standard theory of probabilities is concerned with random variables, which are quantities or sets of quantities taking numerical values in some sets of possibilities, subject to certain required conditions. On the other hand, the theory of geometric probabilities deals with random elements which are not quantities but geometrical objects, including points, lines, and convex planar figures. A principal theme of the theory of geometric probabilities is the relationship between the various geometrical objects, as expressed by intersection, confusion, coverage, etc.

As indicated above, the detection of mineral deposits results from the intersection of the target by a detector moved along a control grid; therefore, we are faced with a geometric problem. On the one hand, the parameters defining grid geometry and detector attitude are deterministic in nature. On the other hand, the geometric characteristics of expected targets (size, shape, attitude, and location) are essentially probabilistic in nature, and are expressed by expected values and their fiducial intervals or probability distributions. Accordingly, it seems quite appropriate to introduce the theory of geometric probabilities as a foundation for the calculation of the probabilities of detection of mineral deposits.

2.3.2. History of the Development of the Geometric Probability Theory

One of the earliest examples of the consideration of geometric probabilities is the famous Buffon problem of the 18th century. A needle of length L is thrown at random on a horizontal plane on which parallel lines are ruled a unit distance apart. Buffon formulated the expression of the probability that the needle intersects the lines.

More recently, during the last two decades, workers such as Kendall,[36] Savinskii,[40] Singer,[45] and Solomon[50] have studied the geometric probability aspects of some much more complex problems of intersection and coverage. The problem of coverage involves the calculation of probabilities that certain planar geometric figures of known shape and position are covered by other geometric figures whose position is random. One much-studied example involves the coverage of a planar lattice of points of known geometry by some convex figure such as a circle, or an ellipse, whose position and orientation are random. The probability of interest involves the number of points of the lattice lying within the ellipse.

2.3.3. Application of the Theory of Geometric Probabilities to the Detection of Mineral Deposits: Survey of the Literature

There was a noticeable lack of interest in both the theory of geometric probabilities and its applications prior to World War II. Following the impetus of research done on practical war-time problems, interest in the application of geometric probabilities spread to mineral exploration in the early postwar period.

In the mid 1950s, Agocs[30] was among the first workers to apply the solution of Buffon's problem to the calculation of probabilities of detection of mineral deposits by airborne surveys. Slichter[19] attacked the problem of detection of targets of varied shapes (annular, rectangular, and circular) by vertical drilling on square grids. Since that time, most of the applied geometric probability research concerned systematic drilling programs. Some eight years later, Celasun[31] expanded Slichter's work by introducing grids of varied shapes (triangular, rhombic, and rectangular) to guide the search for rectangular or elliptical targets of random or preferred orientation. At about the same time, Ellis (Ref. 25 in Chapter 4) tackled the more realistic problem of detection by vertical drilling on square grids of parallelipipedic targets in the three-dimensional space rather than in the planar context of the previous workers.

In 1965, the first publication of Savinskii's probability tables for the detection of elliptical targets by discrete sampling grids[40] was an important step in the application of the theory of geometric probabilities to field exploration. A few years later, Drew[33] was among the first workers to consider the effect of grid orientation on the detection of targets by ground surveys. Two years later, Singer[45] published a very important set of probability tables covering the detection of elliptical targets of random or known orientation by point-nets of square, rectangular, and hexagonal shapes, which was much more useful for field exploration than the earlier Savinskii tables.

More recently, in the mid and late 1970s, additional mathematical treatments of the problem of target detection by drilling were presented by Singer,[47,48] Shurygin,[42,43] and Drew.[34] Sinclair[44] is one of the few workers to have touched on the detection of geochemical halos by point-net sampling in a nonmathematical way. In 1983, the second writer developed the computer program OPTGRID to improve on the Singer ELLIPGRID program of 1972[46] for the ground detection of elliptical targets by square point-nets. Detection probability tables based on the OPTGRID program are included in Section 2.5.5 of this chapter, and in all six chapters dealing with the detection of six types of ore deposits in North America.

In the mean time, after a gap of some 20 years which witnessed a massive use of airborne geophysical surveys, interest in the application of the theory of geometric probabilities to airborne ore detection was renewed. McCammon[38] derives the expressions of the probability of detection of targets of varied shapes by continuous sampling on parallel, orthogonal, and rhombic grids from concepts involving conditional probabilities. Later on, Chung[32] expanded McCammon's work by considering the grid–target orientation factor, as well as lateral coverage along strips rather than flight lines.

2.4. EVALUATION OF THE PROBABILITY OF TARGET DETECTION BY AIRBORNE GEOPHYSICAL SURVEYS

2.4.1. Introduction

Since the early 1950s, as most of the more easily detectable mineral deposits had been discovered by ground prospecting, a much favored strategy for the early stages of regional exploration programs consists of the pattern searching for hidden mineral deposits by aircraft carrying continuously recording sensors of various kinds. A mineral deposit is considered as detected if the sensor intersects the target at least once, and the detection is considered as confirmed if the target is intersected by the sensor at least twice. The matter involved in the calculation of detection probabilities is a problem of continuous geometric probabilities. At this stage we are interested only in selecting the grid spacing and orientation which will maximize the probability of detection. Cost considerations will be entertained later in Chapter 3.

2.4.2. Parameters of the Study

The geometric considerations involved in the calculation of the probabilities of target detection by continuous sampling airborne surveys include (a) target geometry, (b) target–detector configuration, (c) grid geometry, and (d) target–grid configuration. The target geometry parameters required for the calculation of the probability of target detection include the length L_t, shape ratio $R_t = B_t/L_t$, strike orientation, and true dip.

The vertical distance between sensor and target is the most important aspect of the target–sensor configuration; it includes two components: flight elevation above the ground and depth of the center of the target below ground. The flight elevation component varies with the type of survey and nature of the topography. Generally, the flight elevation is kept within a range of 700–1000 feet for magnetic surveys, but is reduced to a 300–400-foot range for radiometric surveys and between 100 and 400 feet for electromagnetic surveys. At the Whistle Mine, Ontario, Slichter[19] showed that the signal strength generated by a sulfide orebody is reduced by 400% when the flight elevation is increased by only 50%, from 500 to 750 feet. The average depth penetration below ground is 300 feet for most methods, with the exception of radiometric surveys.[13,15]

The most commonly used type of grid design is the parallel one, with spacing s and orientation at right angles with the expected strike direction of the target. However, as pointed out by Chung,[32] the lack of parallelism and equidistance of the flight lines causes difficulties in detection and interpretation. The matter of three-dimensional navigational control is of the utmost importance. When the likely orientation of the expected target is not known, square grid designs may be used to advantage despite the greater coverage cost, because it reduces the probability of missing the target. If it proves too costly to fly a regular square pattern, a compromise rectangular design with spacing s and shape ratio $w = s/t$ smaller than unity may prove advantageous. A rectangular grid design can be usefully considered as intermediate between the parallel design ($w = 0$) and the square grid design ($w = 1$).

The target-grid configuration is expressed by dimensional and directional parameters. The dimensional parameter is the ratio U_t of the longest dimension of the expected target L_t over the grid size s. Agocs[30] illustrates the marked effect of the variation of spacing, from 1/4 to 1 mile, on the geometry and structure of a large aeromagnetic anomaly outlining the Marmora, Ontario, magnetite deposit. Henderson[8] shows similar results for the aeromagnetic coverage of a portion of the State of Pennsylvania. The directional parameter is the angle between the direction of the longest dimension of the target L_t and the flight lines of a parallel grid or sides of orthogonal grids.

2.4.3. Calculation of the Probabilities of Target Detection by Airborne Surveys

2.4.3.1. Scope of Study. The three following cases are covered by the present study:

1. calculation of probability of single and confirmed detection of randomly oriented targets of varied shapes by airborne surveys on grids of diverse designs;
2. calculation of probability of single detection of oriented targets of varied shapes by airborne surveys on parallel grids;
3. calculation of the probability of single detection of randomly oriented linear targets by airborne surveys covering parallel strips.

The expressions of detection probabilities pertaining to the three cases listed above are assembled in Table 2.4 for the convenience of the reader. Originally, Agocs[30] listed many of them without mathematical proof; this was later provided by McCammon[38] based on the concepts of conditional probabilities. Later, Chung[32] formulated the expressions related to the detection of linearly oriented targets by airborne surveys on parallel grids and along parallel strips.

2.4.3.2. Probabilities of Detection and Confirmed Detection of Randomly Oriented Targets of Varied Shapes by Airborne Surveys on Parallel Grids. The right half of Table 2.5 lists the probabilities of detection of randomly oriented targets with three shape ratios by airborne geophysical surveys on parallel grids for values of the ratio U_t = Longest target dimension (L_t) over grid spacing s that vary between 0.1 and 2.0, based on the expressions listed in the upper tier of Table 2.4. As noted from Tables 2.4 and 2.5, there is a

TABLE 2.4
Expressions of Probabilities of Detection of Targets of Varied Shapes by Airborne Geophysical Surveys on Varied Types of Grids.

Probability of at Least One Intersection of Randomly Oriented Targets of Various Shapes by Parallel or Orthogonal Grids of Lines

Grid characteristics	Target/grid geometry	Linear shape "linear" probability component	Circular shape "circular" probability component	Elliptical shape Composite probability
parallel spacing (S)	$0 < U \leq 1$	$P_l = 2U/\pi$	$P_c = U$	$P_e = (1 - R)P_l + RP_c$
	$U > 1$	$P_l = \left[\frac{2}{\pi} U\left(1 - \sqrt{1 - \frac{1}{U^2}}\right) + A\cos\frac{1}{U}\right]$	$P_c = 1$	R = target breadth/target length U = target length/grid size P_l = "linear" probability component P_c = "circular" probability component
square size (SxS)	$0 < U < 1$	$P_l = \frac{U}{\pi}(4 - U)$	$P_c = U(2 - U)$	
rectangular size (SxT)	$W = \frac{S}{T} \leq 1$	$P_l = \frac{U}{\pi} 2(W + 1) - UW$	$P_c = U[W + 1) - UW]$	

Probability of at Least One Intersection of a Linear or Elliptical Target with Preferred Orientation (α) by a Parallel Grid

Grid characteristics	Target/grid geometry	Linear shape		Elliptical shape
parallel spacing (S)	$0 < U \leq$	$P_l = \frac{2U}{\pi} \cos 2\alpha$		$P_e = (1 - R)P_l + RU$

Probability of at Least Two Intersections of Randomly Oriented Targets of Various Shapes by a Parallel Grid

Grid characteristics	Target/grid geometry	Linear shape	Circular shape	Elliptical shape
parallel spacing (S)	$U \geq 1$	$P_l = \frac{2}{\pi}\left[\cot an(a) + \left(a - \frac{\pi}{2}\right)\right]$ with $a = A\sin\frac{1}{U}$	$P_c = 1$	$P_e = (1 - R)P_l + R$

Probability of at Least One Intersection of Randomly Oriented Targets of Various Shapes by Parallel Strips

Grid characteristics	Target/grid geometry	Linear shape "linear" probability component	Circular shape "circular" probability component	Elliptical shape composite probability
strip central line spacing (S)				
strip width (W)	$0 < U \leq 1 - \frac{W}{S}$	$P_l = \frac{2U}{\pi} + \frac{W}{S}$	$P_c = U + \frac{W}{S}$	$P_e = (1 - R)P_l + RP_c$

Probability of at Least One Intersection of a Linear or Elliptical Target with a Preferred Orientation (α) by Parallel Strips

Grid characteristics	Target/grid geometry	Linear shape "linear" probability component		Elliptical shape composite probability
strip central line spacing (S)				
strip width (W)	$0 < U \leq 1 - \frac{W}{S}$	$P_l = \frac{2U}{\pi} \times \cos 2\alpha + \frac{W}{S}$		$P_e = (1 - R)P_l + R\left(U + \frac{W}{S}\right)$

TABLE 2.5
Probabilities of Detection of Randomly Oriented Targets of Varied Shapes by Airborne Surveys.

Target length = L feet, and breadth = B feet
R is defined as the ratio B/L in the interval 0⟨R⟨1
U is defined as the ratio Length/Grid spacing (S)

U=L/S	Detection Square grid (S)x(S) feet			Detection Parallel grid with spacing(S)ft		
	R=0.10	R=0.50	R=1.0	R=0.10	R=0.50	R=1.0
0.100	0.131	0.157	0.190	0.067	0.082	0.100
0.200	0.254	0.301	0.360	0.135	0.164	0.200
0.300	0.369	0.432	0.510	0.202	0.245	0.300
0.400	0.477	0.549	0.640	0.269	0.327	0.400
0.500	0.576	0.654	0.750	0.336	0.409	0.500
0.600	0.668	0.745	0.840	0.404	0.491	0.600
0.700	0.753	0.823	0.910	0.471	0.573	0.700
0.800	0.829	0.887	0.960	0.538	0.655	0.800
0.900	0.898	0.939	0.990	0.606	0.736	0.900
1.000	0.965	0.978	1.000	0.673	0.818	1.000
	Confirmed detection Parallel grid with spacing(S)ft.					
1.100	0.116	0.509	1.000	0.714	0.841	1.000
1.200	0.144	0.525	(*)	0.743	0.857	(*)
1.300	0.179	0.544	(*)	0.766	0.870	(*)
1.400	0.217	0.565	(*)	0.785	0.881	(*)
1.500	0.259	0.588	(*)	0.801	0.889	(*)
1.600	0.302	0.612	(*)	0.814	0.897	(*)
1.700	0.348	0.638	(*)	0.826	0.903	(*)
1.800	0.395	0.664	(*)	0.836	0.909	(*)
1.900	0.443	0.691	(*)	0.845	0.914	(*)
2.000	0.492	0.718	(*)	0.854	0.919	(*)

N.B. (*) indicates that all probabilities = 1.000

discontinuity for $L_t = s$, or $U_t = 1$, beyond which another expression of probability of detection has to be used. It can be readily noted that, for a given value of U_t, the probabilities of detection increase substantially when the shape changes from linear to circular.

The lower left corner of Table 2.5 lists the probabilities of confirmed detection (at least two target intersections by the sensor) of randomly oriented targets of three kinds of shape by airborne surveys on parallel grids. The variation of the U_t ratio is restricted to the 1.0–2.0 range, because it is intuitively understood that the probability of confirmed detection is always equal to zero when the ratio U_t is smaller than unity.

A comparison of the lower left and right halves of Table 2.5 illustrates the drastic reduction of magnitude of the probability of detection of linear targets for specified values of U_t, when confirmation is required. For example, the reduction is 85% when $U_t = 1.1$ and 43% when $U_t = 2.0$. If we consider elliptical targets, the reduction is not so great: it is only 40% for $U_t = 1.0$ and 22% when $U_t = 2.0$.

Since most airborne geophysical sensors, and particularly the radiometric ones, respond to the effect of a narrow strip of land on either side of the actual flight lines instead of that of the flight lines alone,[13,32] it is appropriate to consider the effect of lateral coverage on the detection probabilities. Chung shows that if w is the width of strips and s the spacing between the edges of adjoining strips, the probabilities of detection are inflated by the ratio w/s. (See the fourth tier of Table 2.4.)

2.4.3.3. Probabilities of Detection of Randomly Oriented Targets of Varied Shapes by Airborne Geophysical Surveys on Orthogonal Grids. When the expected orientation of the target is not known, it may be preferable to use an orthogonal grid design rather than a parallel one. The probabilities of detection of targets of three kinds of shape for values of U_t ranging between 0.1 and 1.0 are listed in the upper left corner of Table 2.5 for square grid designs.

A comparison between the probabilities of detection for parallel grids and those for square grids for similar target shapes and values of the ratio U_t shows a substantial improvement of probabilities when using square control grids. For example, the improvement is 95% for $U_t = 0.1$ and down to 43% for $U_t = 1.0$ for linear targets. In the case of circular targets the figures are respectively, 90% and 0%.

When U_t lies in the interval of 1 to square root of 2, the expression used for the calculation of probability of detection of a linear target by airborne surveys on square grids is the same as that used for parallel grids when U_t is greater

PROBABILITY OF DETECTION OF MINERAL DEPOSITS

than unity. (See the left corner of the first tier of Table 2.4.) When $(U)_t$ is greater than the square root of 2, the probability of detection is always unity. Similarly, the probability of detection of a circular target by a survey on a square grid is always unity when U_t is greater than 1.0, while the probabilities of detection of elliptical targets are bracketed by those of the other two types of shape. The expressions of the probabilities of detection of linear and circular targets by rectangular grids defined by the spacing s and ratio $w = s/t$ are shown in the bottom portion of the first tier of Table 2.4.

2.4.3.4. Generalized Aerial Search on Various Types of Grids. As mentioned above, the probabilities of detection of an elliptical target can be interpolated by considering any ellipse as intermediate between a line and a circle. It was also indicated at the end of Section 2.4.2 that an airborne search flown on a rectangular grid may be considered as intermediate between a search on a parallel grid and one on a square grid. We now have four limiting cases including two for target shape and two for grid shape. The most general case, that of a search for an elliptical target (R in the interval 0–1) on a rectangular grid with shape factor w in the interval 0–1, will be bracketed by the four extreme cases, as illustrated by Figure 2.5.

The probabilities of detection can be calculated by combining the two interpolation expressions listed below for each value of the ratio U_t. If P_1, P_c,

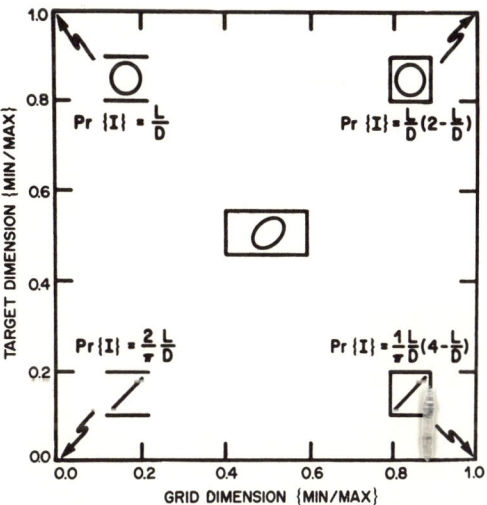

FIGURE 2.5

Generalized Search: Detection of Elliptical Randomly Oriented Targets by Airborne Surveys on Orthogonal Grids. [From McCammon, R. B.[38] Figure 4, by permission of Plenum Publishing Corporation, New York.]

and P_e are the probabilities of detection of linear, circular, and elliptical targets for a specified type of grid, the relationship between the three is as follows:

$$P_e = (1 - R)P_1 + RP_c$$

R being the target shape ratio in the interval 0–1. If P_p, P_s, P_r are the probabilities of detection of a target of specified shape R by surveys on parallel, square, and rectangular grids, the expression relating the three is written as

$$P_r = (1 - w)P_p + wP_s$$

where w is the grid shape ratio in the interval 0–1.

2.4.4. Probabilities of Detection of Oriented Targets of Varied Shapes by Airborne Surveys on Parallel Grids

Agocs' and McCammon's studies on which the above is based considered only the case of randomly oriented targets. Chung[32] was the first worker to consider the effect of target orientation on the probabilities of detection of linear targets by continuous sampling surveys on parallel grids. The second tier of Table 2.4 shows the expression of the probability of detection of an oriented linear target, as derived by Chung from Marriot's work,[37] when U_t lies in the 0–1 interval. The angle a is that between the flight lines and the expected strike direction of the target. In practice, however, the direction of the strike is known only within a confidence interval of, say, + or − 10 degrees. The correct way of calculating the probability of detection corresponding to the confidence interval for the strike direction is to integrate the expression

$$\int_{-\pi/18}^{+\pi/18} 2U/3\pi \cos(2a)\, da$$

If the angle a is taken as 90 degrees (flight lines perpendicular to the expected strike within its confidence interval), the correct expression of the probability of detection of a linear target is

$$P_1 = 2U/\pi + 6/\pi^2[\sin(\pi)/9]$$

The diagram shown in Figure 2.6 illustrates the effect of varying target–grid orientations on the magnitude of the probability of detection of a linear target.

PROBABILITY OF DETECTION OF MINERAL DEPOSITS

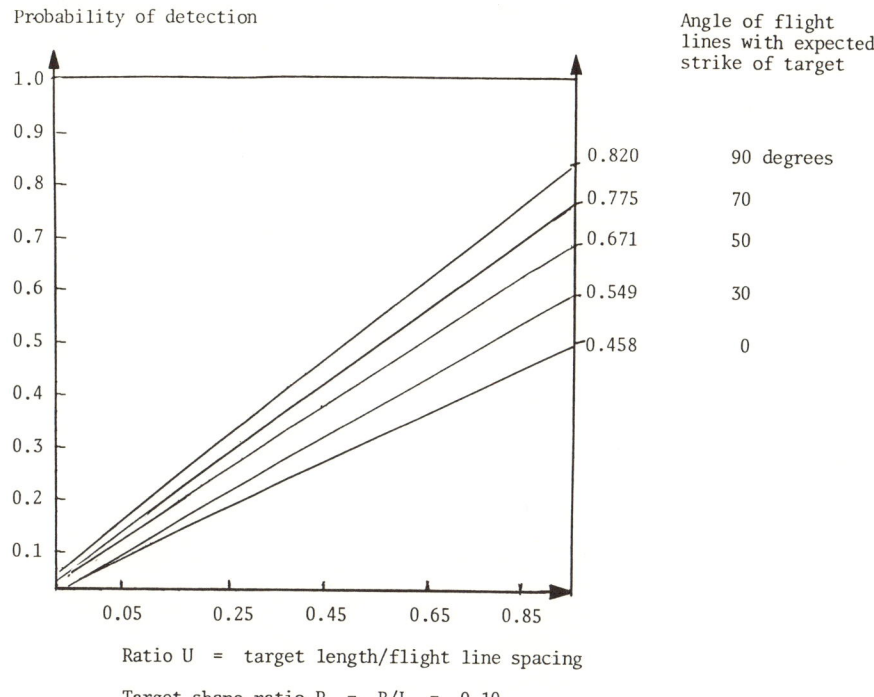

FIGURE 2.6
Probabilities of Detection of Linear Targets by Airborne Surveys on Parallel Grids Oriented at Different Angles with Respect to Targets.

For example for $(U_t) = 1$ the probability of detection increases by 80% when the grid is rotated from 0 to 90 degrees with respect to the expected strike.

Circular targets do not exhibit any preferred orientation because of the symmetry, so that the probability of detection is always equal to the ratio U_t. The probability of detection of an elliptical target whose long axis is making an angle of a degrees with the grid is obtained by combining the expressions of the probability of detection of linear and circular targets as indicated in the previous Section 2.4.3.4. For example, the increase in the probability of detection of an elliptical target with shape ratio $R_b = 0.50$ is 30% when the grid is rotated from 0 to 90 degrees with respect to the long axis of the target if $(U_t) = 0.50$.

2.5. EVALUATION OF THE PROBABILITY OF TARGET DETECTION BY GROUND GEOPHYSICAL SURVEYS AND DRILLING PROGRAMS

2.5.1. Introduction

The widespread use of discrete sampling programs on ground grids preceded that of continuous sampling by airborne surveys along parallel profiles. Most of the work done on the application of the theory of geometric probabilities to ground exploration during the past three decades pertained to drilling programs rather than geophysical or geochemical surveys, because of the incentive to reduce the very high cost of systematic drilling.

2.5.2. Parameters of the Study

2.5.2.1. Target Geometry and Target–Detector Configuration. Targets are assumed to be elliptical in shape with a major axis L_t and a minor axis B_t. The parameters required for the calculation of detection probabilities are L_t and the shape ratio $R_t = B_t/L_t$, as well as the true dip and the direction of the major axis with respect to true north. A vertical detection range of 300 feet is assumed for geophysical surveys and for vertical or angled drilling programs alike. Note that, for practical reasons, surface angled drilling is restricted to the 35–90 degree range of inclination.

2.5.2.2. Grid Geometry and Grid–Target Configuration. Ground grids for discrete sampling are considered as point-nets made up of contiguous cells which have a regular geometric shape. The three parameters defining the grid are the shape and size of the cells and the orientation of the net with respect to the expected strike direction of the target. The sampling is carried out at the nodes of the net. Much research work has been done on drilling grid shapes in order to determine the most cost-effective designs. The shapes considered vary from rhombic or lozenge[31,39,42,43] to hexagonal and derived equilateral triangular.[34,45,46] However, in practice, only orthogonal, and particularly square grids, are considered for field work.

The dimensional relationship between target and grid, which is an important consideration in the calculation of the probabilities of detection, is simply expressed by the ratio U_t = longest dimension of the target L_t over the grid spacing s. Drill hole spacing requirements are a function of the geology and the nature of the program. The spacing should be subjected to a continuing statistical analysis in order to reduce progressively the size of the net and adjust it to the succession of objectives of the search.

In the case of ground geophysical surveys, the spacing between readings should be adjusted to the anticipated depth and size of the target sought for. This is particularly important in the case of the high-cost induced polarization surveys, where the spacing of the electrodes as well as their configuration are of critical importance.

The directional aspect of the target–grid relationship has to be considered. It is advantageous to lay a geophysical grid at right angles to the geological grain of the survey area, because there is a greater geological contrast across the strike than along it, thus enhancing the magnitude of the geophysical response.[12] However, we will find that the probability of detection of an elliptical target by a ground sampling survey on a square grid is maximized if the grid is orientated at angles varying from 18 to 45 degrees with respect to the long axis L_t of the target, depending on the shape ratio R_t of the target. If we consider rhombic or rectangular grids, the detection probabilities are maximized when the longest dimensions of both grid-cell and target coincide.

2.5.3. Methodology of the Calculation of the Probabilities of Detection of Elliptical Targets by Discrete Sampling on Square Grids

2.5.3.1. Introduction. Because of the discrete nature of the sampling, the mathematical aspects of the methodology of detection probability calculation are vastly more complicated than in the case of detection by continuous sampling airborne surveys. The 1969 Singer tables[45] and ELLIPGRID computer program[46] considerably improved on Savinskii's contribution to the problem of detection of elliptical targets by point-net sampling.[40] However, in 1982, the second writer of this book perfected the OPTGRID computer program which, while based on same general geometric foundation as Singer's, is more efficient and accurate and better attuned to field exploration requirements than the former programs.

2.5.3.2. Detection of Oriented Elliptical Targets. The OPTGRID approach is more efficient than previously published programs, because it makes use of information implicitly contained in the previous studies but not taken advantage of, which results in the maximization of detection probabilities. In the first stage of the overall optimization of target detection based on the dynamic programming approach (see Chapter 4), it was found that detection probabilities can be maximized depending on the target shape ratio $R_t = B_t/L_t$ by orienting the grid with respect to the longest dimension of the target L_t in the following manner:

- for R_t less than 0.15, the "maximal" angle is 18–20 degrees;
- for R_t in the 0.15–0.30 interval, the "maximal" angle is 30 degrees;
- for R_t greater than 0.30, the "maximal" angle is 45 degrees.

If we consider the confirmed detection requirement (at least two target intersections by the detector), the "maximal" angle is zero degrees since such a configuration makes the best use of the full length of the target.

The OPTGRID approach is also more realistic than the previous ones, which assume that the strike direction of the target is accurately known. However, this is not the case in practical field exploration situations. Generally, the strike direction is approximately known within a confidence interval of, say, ±10 degrees about a mean orientation. In the OPTGRID program, the grid is orientated at the optimal angle with the target mean strike direction according to the target shape, and probabilities of detection are calculated within the full 20 degree confidence interval by integration.

Unfortunately, the integration by standard analytical methods is intractable because of the very complex nature of the expression of the probability of detection in terms of grid and target parameters. A graphical method based on Simpson's rule provides an approximation to the integration. In the Simpson method, any small arc (P_1, P_2), however complex in shape, which is limited by two ordinates corresponding to an elementary increment h of the variate x, is replaced by an elemental arc of parabola passing through the points P_1 and P_2. Thus the area comprised between the two ordinates, the parabola arc and the x axis, is a numerical approximation of the value of the integral of the very complicated function $F(x)$ in the interval $(x, x + h)$. This can be easily extended to any finite arc of curve of any complicated shape corresponding to the (a, b) interval for the x variate, which can be divided into n intervals of width h equal to $(b - a)/n$. This leads to the Simpson expression

$$\int_a^b F(x)\, dx \# h/3 \Big[F(a) + 4F(a + h) + 2F(a + 2h) + 4F(a + 3h) + \cdots$$
$$+ 2F(a + (n - 2)h) + 4F(a + (n - 1)h) + F(b) \Big]$$

which can be very easily programmed on the computer.

The introduction of the Simpson rule to take into account the confidence interval for the target–grid orientation improves greatly the accuracy and practicality for field use of the detection probability calculations. This is illustrated by the following example considering the detection of an elliptical target of shape $R_t = 0.25$. According to previous results, we find that the optimal orientation of the grid is 30 degrees plus or minus 10 degrees with respect to the long axis (L_t) of the target. Thus the interval of variation of the orientation angle is from 20 to 40 degrees, or $\pi/9$ to $2\pi/9$ if expressed in radians. The $\pi/9$ interval is divided into 100 elemental intervals, each $\pi/900$ wide. The computer is programmed to calculate the area under the curve between the x axis and the $\pi/9$ and $2\pi/9$ ordinates by means of the Simpson's approximation formula. The probability of detection, which represents the mean height of the probability function within the interval $\pi/9 = 0.348$, is then easily obtained by dividing the Simpson result by the coefficient 0.348.

2.5.3.3. Detection of Randomly Oriented Elliptical Targets. Since the long axis L_t of the target is randomly oriented within the horizontal plane, there is no maximal orientation of the grid which would enhance the probability of detection, as was the case in Section 2.5.3.2. Because of the symmetry introduced by the use of a square grid, it is sufficient to restrict the scope of the study to the 0–$\pi/4$ range of grid–target orientations.

Instead of applying the Simpson rule to the interval from $+10$ to -10 degrees about the optimal grid orientation, we apply it to the full interval extending from 0 to $\pi/4$. The elemental increment h is equal to $\pi/400$. The value of the probability of detection is obtained by dividing the area calculated by means of the Simpson rule by $\pi/4 = 0.785$.

2.5.4. Application of the Methodology to the Construction of Detection Probability Tables

2.5.4.1. Detection Probabilities for Oriented and Randomly Oriented Targets of Varied Shapes. The probabilities of detection of linear ($R_t = 0.10$), elliptical ($R_t = 0.50$), and subcircular ($R_t = 0.90$) targets with longest dimension L_t by surveys on square grids of size s are calculated by the method described above for varying values of the grid–target dimensional ratio $A_t = L_t/2s$ in the 0–2 range. The probabilities of detection of oriented and randomly oriented targets are displayed in the left portion of Tables 2.6 and 2.7, respectively.

When A_t is less than 0.5, i.e., L_t is less than s, the probabilities of detection of oriented and randomly oriented targets are equal for similar values of R_t and A_t. However, when A_t is greater than 0.5, the probabilities of detection of oriented targets are substantially larger than that of randomly oriented targets of similar

TABLE 2.6
Probabilities of Detection of Oriented Targets of Varied Shapes by Ground Surveys or Vertical Drilling Programs.

Survey design: square grid with spacings S by S feet
Elliptical targets with expected major axis = L feet, and expected minor axis = B feet
R is defined as the ratio B/L in the interval 0(R(=1
A is defined as the ratio semi-major axis/grid size = L/2S in the interval 0.5(A(2

	Detection			Confirmed detection		
A=L/2S	R=0.10	R=0.50	R=0.90	R=0.10	R=0.50	R=0.90
0.50	0.079	0.393	0.707	0.000	0.012	0.016
0.55	0.095	0.475	0.839	0.000	0.040	0.053
0.60	0.113	0.565	0.919	0.002	0.079	0.109
0.65	0.133	0.662	0.969	0.005	0.128	0.235
0.70	0.154	0.756	0.997	0.009	0.252	0.398
0.80	0.201	0.911	(*)	0.021	0.411	0.908
0.90	0.254	0.993	(*)	0.040	0.622	(*)
1.00	0.314	(*)	(*)	0.067	0.974	(*)
1.10	0.380	(*)	(*)	0.100	(*)	(*)
1.20	0.452	(*)	(*)	0.136	(*)	(*)
1.30	0.531	(*)	(*)	0.174	(*)	(*)
1.40	0.613	(*)	(*)	0.214	(*)	(*)
1.50	0.695	(*)	(*)	0.254	(*)	(*)
1.60	0.777	(*)	(*)	0.295	(*)	(*)
1.70	0.850	(*)	(*)	0.335	(*)	(*)
1.80	0.910	(*)	(*)	0.375	(*)	(*)
1.90	0.955	(*)	(*)	0.413	(*)	(*)
2.00	0.984	(*)	(*)	0.449	(*)	(*)

(*) Probabilities = 1.000

N.B. (1): If A (0.5 all probabilites are the same as randomly orientated targets, and may be read from that table

N.B. (2): The detection probability is maximized when the grid is laid out at the following angles with the expected target strike:
18 degrees + or - 10, when R (0.15
30 degrees + or - 10, when 0.15 (R (0.50
45 degrees + or - 10, when 0.50 (R (1.00

N.B. (3): The probability of confirmed detection is maximized when the grid is laid out at an angle of 0 degrees + or - 10 with the expected target strike.

TABLE 2.7
Probabilities of Detection of Randomly Oriented Targets of Varied Shapes by Ground Surveys or Vertical Drilling Programs.

Survey design: square grid with spacings S by S feet
Elliptical targets with expected major axis = L feet, and expected minor axis = B feet
R is defined as the ratio B/L in the interval 0(R(=1
A is defined as the ratio semi-major axis/grid size = L/2S in the interval 0(A(=2

A = 1/2S	Detection			Confirmed detection		
	R=0.10	R=0.50	R=0.90	R=0.10	R=0.50	R=0.90
0.10	0.003	0.016	0.028	0.000	0.000	0.000
0.20	0.013	0.063	0.113	(**)	(**)	(**)
0.30	0.028	0.141	0.254	(**)	(**)	(**)
0.40	0.050	0.251	0.452	(**)	(**)	(**)
0.50	0.079	0.393	0.707	0.000	0.003	0.022
0.60	0.113	0.553	0.916	0.000	0.013	0.102
0.70	0.152	0.713	0.994	0.002	0.057	0.391
0.80	0.196	0.854	1.000	0.005	0.151	0.817
0.90	0.244	0.947	(*)	0.011	0.473	(*)
1.00	0.295	0.986	(*)	0.019	0.875	(*)
1.10	0.349	1.000	(*)	0.030	0.994	(*)
1.20	0.406	(*)	(*)	0.045	(*)	(*)
1.30	0.463	(*)	(*)	0.064	(*)	(*)
1.40	0.520	(*)	(*)	0.088	(*)	(*)
1.50	0.576	(*)	(*)	0.118	(*)	(*)
1.60	0.631	(*)	(*)	0.153	(*)	(*)
1.70	0.683	(*)	(*)	0.195	(*)	(*)
1.80	0.731	(*)	(*)	0.245	(*)	(*)
1.90	0.773	(*)	(*)	0.303	(*)	(*)
2.00	0.807	(*)	(*)	0.417	(*)	(*)

N.B. (*) Probabilities = 1.000 (**) Probabilities = 0.000

shape for similar A_t values. For example, the improvement is 6% for $A_t = 1$, rising to 20% for $A_t = 1.5$, and to as much as 25% for $A_t = 2$ in the case of linear targets $R_t = 0.10$. For elliptical targets $R_t = 0.50$, however, the maximum increase is only 7%, obtained when $A_t = 0.8$, and there is hardly any improvement when we deal with subcircular targets, as could be intuitively appreciated.

2.5.4.2. Confirmed Detection Probabilities for Oriented and Randomly Oriented Targets of Varied Shapes. The probabilities of confirmed detection (at least two intersections of the target by the detector) are tabulated in the right-hand portions of Table 2.6 (oriented targets) and Table 2.7 (randomly oriented targets). Obviously, when A_t is less than 0.5, or L_t less than s, the probabilities of confirmed detection are always nil in both cases.

A comparison of both sides of each table shows that the probabilities of confirmed detection are always much smaller than their single detection counterparts, as could be intuitively expected. For example, the probability of confirmed detection is only 2% of that of single detection for $A_t = 0.60$, increasing to a maximum of 45% for $A_t = 2$, when we consider oriented linear targets $R_t = 0.10$. In the case of randomly oriented targets, the probability of confirmed detection is only 2.5% of that of single detection when $A_t = 0.8$, reaching a maximum of 52% for $A_t = 2$.

We can conclude that the grid size required to obtain a specified level of probability of confirmed detection should be much smaller than in the case of single detection for specified values of R_t and A_t, therefore considerably boosting coverage costs. This could be easily expected: extra assurance costs extra money.

2.5.5. Influence of Orientation Errors on Detection Probabilities

In practical field exploration situations, it is usually not possible to specify exactly the strike orientation of an expected target. The best we can do in statistical terms is to specify a confidence interval about an expected orientation angle, as stated in Section 2.5.3.2. A topic of considerable interest is to find out how the probabilities of detection are affected by errors of varying magnitudes on target–grid orientation, if we assume that the other target parameters R_t and L_t are known without error.

The results of Mickey and Jespersen' investigation of the topic[39] are summarized by a set of curves graphed in Figure 2.7. The curves were obtained by varying the angular error from 0 to 90 degrees against the shape ratio R_t, for specified values of the probability of detection. These curves may be considered as contours representing a "sensitivity surface," i.e., the sensitivity of detection probability to orientation errors for various target shapes.

Mickey's conclusion was that the detection probability is affected only to a minor degree by even large errors of target orientation specification for a wide range of conditions. For example, the orientation of the long axis of an ellipse of shape ratio $R_t = 0.50$ has to be in error by as much as 48 degrees for the probability of detection to fall from 0.75 to 0.70. But the detection probability becomes increasingly, though moderately, sensitive to orientation errors, as the

PROBABILITY OF DETECTION OF MINERAL DEPOSITS

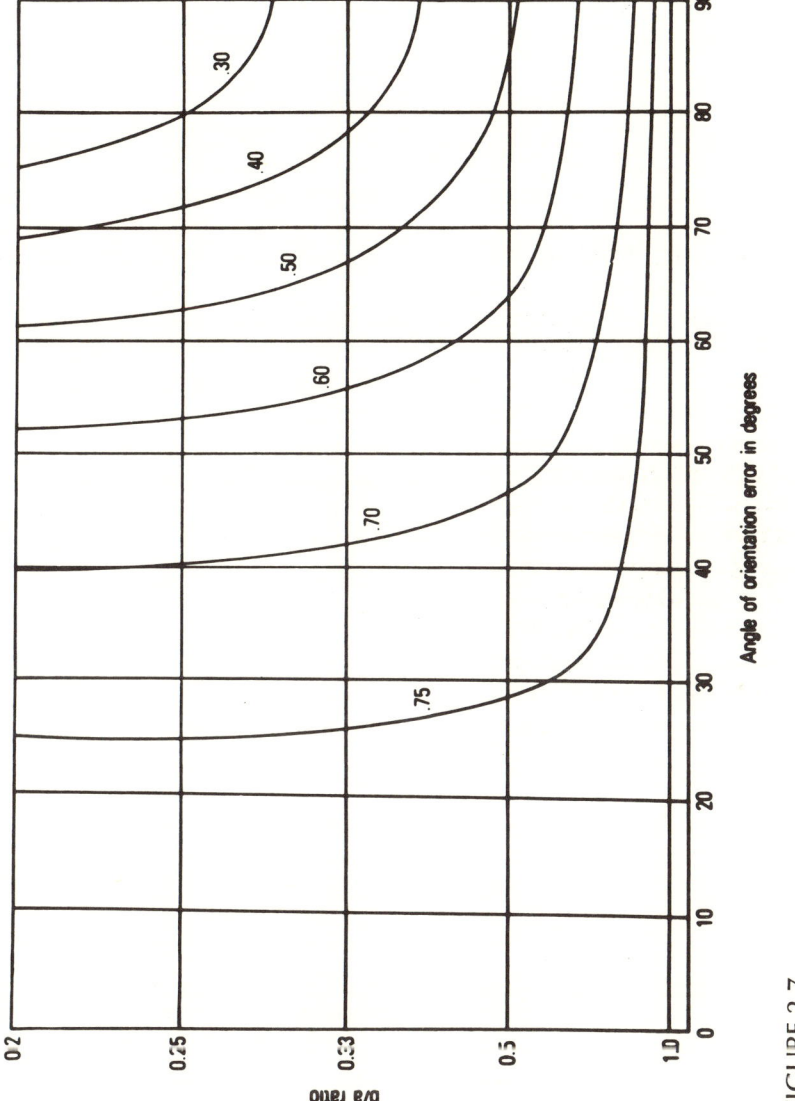

FIGURE 2.7
Contours of Sensitivity of Detection Probability to Errors in Target Orientation. [From Drew, L. J.[34] Figure 5, by permission of Plenum Publishing Corporation, New York.]

target becomes more elongated. For example, if the shape ratio $R_t = 0.20$, the orientation error has to be 30 degrees in order to reduce the probability of detection from 0.75 to 0.70, instead of 48 degrees for the more rounded target shape.

2.6. PROBABILITY OF SEQUENTIAL DETECTION OF ORE DEPOSITS

2.6.1. Introduction

In the previous sections of this chapter, we considered only single-stage detection procedures, such as airborne geophysical surveys, ground geophysical surveys, or drilling programs. It would be fruitful, however, to consider a commonly used two-stage sequential approach to the field detection of ore deposits. The procedure requires the detection of halos of the kinds known to be associated with specified types of ore deposits, as a first stage, to be followed by the detection within the halo of the causative ore deposit.

The methodology of the calculation of the probability of sequential detection is divided into three steps: (1) calculation of the probability of detection of the halo, (2) calculation of the probability of detection of the ore deposit within the halo, and (3) calculation of the overall probability of success based on (1) and (2).

2.6.2. Probability of Detection of Halos

There are several types of primary halos which surround mineral deposits as three-dimensional envelopes. Some of them are detectable by airborne or ground geophysical surveys, such as pyritic and magnetite halos associated with porphyry–Cu–Mo deposits, magnetic halos surrounding contact-metasomatic deposits, magnetic and radioactive halos associated with syenitic intrusions or diatremes. Primary geochemical halos of minor elements can be easily detected by systematic bedrock sampling. The procedure of calculation of the probability of detection of these two types of halos is similar to that used in relation to the detection of mineral deposits themselves, as described in Sections 2.5.3 and 2.5.4.

The detection of secondary geochemical halos is not considered here, because halos and causative bodies are not closely related spatially, so that an intermediate stage, generally a geophysical survey, has to be introduced before the final drilling stage.

2.6.3. Probability of Detection of Mineral Deposits within Halos

If an ore deposit of area A_t lies entirely within a larger area A_h which is limited by any type of convex perimeter, the probability that a sampling location randomly chosen within the area A_h will intersect the ore deposit is calculated according to the theory of geometric probabilities as

$$P_s = A_t/A_h$$

Assuming that both areas are elliptically shaped, their extent is easily calculated as the product of the lengths of the major and minor axes by $\pi/4$.

It should be noted that the probability of detection calculated in the manner described above is a *minimum value* which can be easily exceeded at a substantially higher cost by systematic grid sampling within the perimeter of the halo. In the latter case the probability of detection is calculated as shown above in Sections 2.5.3 and 2.5.4.

2.6.4. Overall Probability of Sequential Detection

One of the basic laws of the standard theory of probabilities states that if events E and F are statistically independent, the probability of E and F occurring simultaneouly is the product of the probabilities of occurrence of E and F. (See Section 1.1.3.3 of Chapter 1.) By applying this law to our sequential situation, we can express the overall probability of sequential detection as the product of the probability of detection of the halo by that of detection of the ore deposit within the halo, both being calculated as shown in Sections 2.5.3 and 2.5.4.

REFERENCES AND SELECTED READINGS

Detection: General Topics

1. BIRDSALL, T. G., 1965, The theory of signal detectability in *Information Theory in Psychology*, pp. 391–402, Free Press, Glencoe, Illinois.
2. CRIBB, J. L., and VICKERS, I. R., 1973, Experience with transformation of aeromagnetic data to the Pole, *Bull. Austr. Soc. Expl. Geophys.* **4** (September), 1–8.
3. CUMING, J. D., and WICKLUND, A. P., 1978, *Diamond Drill Handbook*, J. K. Smit, Toronto, Canada.
4. ELLIOT, I. L., and FLETCHER, W. K. (Ed.), 1975, *Geochemical Exploration 1974*, Elsevier, Amsterdam.

5. GREGORY, A. F., 1967, Remote sensing in the search for metallic ores: A review of current practices and future potential, *Geol. Surv. Can. Econ. Geol. Rep.* **26**, pp. 511–526.
6. GOETZ, A. F. H., ROCK, B. N., and GOWAN, L. C., 1983, Remote sensing for exploration: An overview, *Econ. Geol.* **78**, 573–590.
7. GREENWOOD, J. E. W., 1965, Air photographs in economic mineral exploration, *Geol. Surv. Can. Pap.* **65–6**.
8. HENDERSON, R. G., and ZIETZ, I., 1949, The upward continuation of anomalies in total magnetic intensity fields, *Geophysics* **14**, 517–533.
9. KUZWART, M., and BOHMER, M., 1978, *Prospecting and Exploration of Mineral Deposits*, Elsevier, Amsterdam.
10. LEE, Y. W., 1960, Statistical Theory of Communication, Wiley, New York.
11. LEVINSON, A. A., 1974, *Introduction to Exploration Geochemistry*, Appl. Publ. Maywood, Illinois.
12. PARASNIS, D. S., 1974, *Mining Geophysics*, Elsevier, Amsterdam.
13. PATERSON, N. R., 1971, Airborne electromagnetic methods as applied to the search for sulfide deposits, *Can. Inst. Min. Metall. Bull.* **64**(705), 29–38.
14. PEMBERTON, R. H., 1962, Airborne electromagnetics in review, *Geophysics* **27**, 691–713.
15. PETERS, W. C., 1978, *Exploration and Mining Geology*, Chap. 8, Wiley, New York.
16. RAISBECK, G., 1963, *Information Theory: An Introduction for Scientists and Engineers;* MIT Press, Cambridge, Massachusetts.
17. REEDMAN, J. H., 1979, *Techniques in Mineral Exploration*, Applied Science Publishers, London.
18. ROSENBERG, P., 1971, Resolution, detectability, and recognizability, *Photogramm. Eng* **37**, 1255–1258.
19. SLICHTER, L. B., 1955, Geophysics applied to prospecting for ores, *Econ. Geol. Jubilee Volume* **50**, 885–969.
20. SPECTOR, A., 1971, Aeromagnetic map interpretation with the aid of the digital computer, *Can. Inst. Min. Metall. Bull.* **64**(711), 27–34.
21. ZURFLUEH, E. G., 1967, Applications of two-dimensional linear wavelength filtering, *Geophysics* **32**, 1015–1033.

Statistical Modeling of Geometric Parameters of Ore Deposits

22. AGTERBERG, F. P., 1974, *Geomathematics*, Chaps. 7, 11, and 15, Elsevier, Amsterdam.
23. AITCHISON, J., and BROWN, J. A. C., 1966, The lognormal distribution with special references to its use in economics; Cambridge University Press, Cambridge.
24. DAVIS, J. C., 1973, *Statistics and Data Analysis in Geology*, Wiley, New York.
25. DE GEOFFROY, J., and WIGNALL, T. K., 1973, Statistical models for porphyry Cu–Mo deposits of the Cordillera Belt of North and South America, *Can. Inst. Min. Metall. Bull.* **66**(735), 84–90.
26. JOHNSON, N. I., and KOTZ, S., 1970, *Continuous Univariate Distributions*, Vol. 2, Houghton-Mifflin, Boston.
27. KOCH, G. S., and LINK, R. F., 1970, *Statistical Analysis of Geological Data*, Wiley, New York.
28. KRUMBEIN, W. C., and GRAYBILL, F. A., 1965, *Introduction to Statistical Models in Geology*, McGraw-Hill, New York.
29. WALPOLE, R. E., 1974, *Introduction to Statistics*, Macmillan, New York.

Evaluation of the Probability of Detection of Ore Deposits

30. AGOCS, W. B., 1955, Line spacing effect and determination of optimum spacing illustrated by Marmora, Ontario, magnetic anomaly, *Geophysics* **20**(4), 871–885.
31. CELASUN, M., 1964, The allocation of funds to reconnaissance drilling programs, *Q. Col. Sch. Mines* **59**(4), 169–186.
32. CHUNG, C. F., 1981, Application of the Buffon needle problem and its extensions to parallel line search sampling schemes, *J. Math. Geol.* **13**(5), 371–390.
33. DREW, L. J., 1967, Grid-drilling exploration and its application to the search for petroleum, *Econ. Geol.* **62**, 698–710.
34. DREW, L. J., 1979, Pattern drilling exploration: Optimum pattern types and hole spacing when searching for elliptical targets, *J. Math. Geol.* **11**(2), 223–254.
35. GRIFFITHS, J. C., 1966, Grid spacing and success ratios in exploration for natural resources, Mineral Industry Experimental Station, Pennsylvania State University, Special Publication No 1.
36. KENDALL, M. G., and MORAN, P. A. P., 1963, *Geometrical Probability,* Chas. Griffin, London.
37. MARRIOT, F. H. C., 1969, Associated directions, *Biometrics* **25**(4), 775–776.
38. McCAMMON, R. B., 1977, Target intersection probabilities for parallel lines and continuous grid types of search, *J. Math. Geol.* **9**(4), 369–382.
39. MICKEY, M. R., and JESPERSEN, H. W., 1954, Some statistical problems of uranium exploration, U. S. Atomic Energy Commission Report RME-3105.
40. SAVINSKII, I. D., 1965, *Probability Tables for Locating Elliptical Underground Masses with a Rectangular Grid,* Consultants Bureau, New York.
41. SCHUENMEYER, J. H., and DREW, L. J., 1977, An exploratory drilling exhaustion sequence plot programme, *Comput. Geosci.* **3**(4), 617–632.
42. SHURYGIN, A. M., 1976, Discovery of deposits of given size in boreholes with preselected probability, *J. Math. Geol.* **8**(1), 85–88.
43. SHURYGIN, A. M., 1976, The probability of finding deposits and some optimal search grids, *J. Math. Geol.* **8**(3), 323–330.
44. SINCLAIR, A. J., 1975, Some considerations regarding grid orientation and sample spacing, in *Geochemical Exploration 1974,* pp. 133–140, Elsevier, Amsterdam.
45. SINGER, D. A., and WICKMAN, F. E., 1969, Probability tables for locating elliptical targets with square, rectangular and hexagonal point nets, Pennsylvania State University Mineral Industry Experimental Station, Spec. Publ. No 1–69.
46. SINGER, D. A., 1972, ELLIPGRID, a FORTRAN IV program for calculating the probability of success in locating elliptical targets with square, rectangular and hexagonal grids, *Geocom. Programs* No. 4, 1–16.
47. SINGER, D. A., 1975, Relative efficiencies for square and triangular grids in the search for elliptically shaped resources targets, *J. Res. U. S. Geol. Surv.* **3**, 163–167.
48. SINGER, D. A., 1976, RESIN, a FORTRAN IV program for determining the area of influence of samples or drill holes in resources target search, *Comput. Geosci.* **2**, 249–260.
49. SINGER, D. A., and DREW, L. J., 1976, The area of influence of an exploratory drill hole, *Econ. Geol.* **71**, 642–647.
50. SOLOMON, H., 1978, *Geometric Probability,* Soc. Ind. Appl. Math. (S.I.A.M.), Pennsylvania.

CHAPTER THREE

COST OF DETECTION

3.1. GENERAL STATEMENT

3.1.1. Introducing Cost Considerations in Ore Detection

In the previous chapter our main concern was the calculation of the probability of detection of expected targets in terms of the geometry and relative configuration of targets, detectors, and control grids. The emphasis was on the maximization of detection, but no consideration was given to the cost required to achieve the set goal.

Since mineral exploration is a business enterprise and has to be considered in its economic context, it is obvious that cost considerations will have to be introduced before any course of action is taken. In most cases, cost will be a major factor restricting the range of available options, if we are to stay within prespecified budget limits.

Cost may be directly quantified and expressed in monetary units, as is the most common practice, or in time units. Cost may also represent the quantification of concepts of error, risk, or any other events which are detrimental to the acquisition of a reward.

3.1.2. Definitions and Terminology

In the common usage, terms such as price, cost, and expenditure are often rather loosely used in an interchangeable manner, at the risk of some confusion which should be clarified.

Price is the number of monetary units which has to be provided in exchange for specified amounts of goods or services or reward. The rate is theoretically fixed by the law of supply and demand, but is more often than not administratively set to suit specific policies. Price is neutral: it does not imply any connotation of activity or effort to acquire a reward.

Cost carries the implication of planned activity. Cost is the product of the monetary price per unit by the number of units of goods or services which the planner expects to be required to carry out a specified activity. Any complex activity such as mineral exploration can be broken into a number of component activities, each carrying a component cost. Program costing or cost estimation is an exercise of major importance in planning.

Expenditure belongs to the execution stage. Expenditure is the number of monetary units actually expended in carrying out a planned activity for which a preliminary cost estimate had been made. Expenditure is strictly limited by the allocation of resources decided on at the planning stage.

3.2. COSTING AND COST ESTIMATES

3.2.1. Introduction

Most of the planning stage of any mineral exploration program is taken up by the organization of activities and the cost estimations of actual expenditures. Costs cannot be determined exactly beforehand: they have to be estimated, because mineral exploration functions in a context of uncertainty affecting the execution and outcome of planned activities.

Cost estimation is generally based on the "analogy" principle. Initial estimates are made by comparison with well-documented case histories of programs which were carried out in economic and geographic environments similar to that of the planned project. More finely tuned estimates follow. They take into account local conditions and are indexed for time and geographic variations. There are two well-recognized methods of costing projects: unit costing, and ratio costing.

3.2.2. Cost Estimates

Cost estimation follows a well-established sequence starting with "order of magnitude" estimates, followed by "preliminary estimates," and ending with more refined "definitive estimates." Each stage has a different degree of precision requirement attached to the cost estimates. At the initial stage, one could be satisfied with a 30%–40% error tolerance, to be cut down to 20% for the intermediate stage and further down to the 5%–10% range at the final stage.

Order of magnitude and preliminary estimates are essentially based on the analogy principle with a relatively minor share of local input. They are documented mainly from "outside" sources of information such as the publications of State or national Bureaus of Mines, Departments of Trade and Commerce,[11] buyer's guides, trade journals, etc., which report on projects all over the world. Initial estimates are mainly based on "unit cost" information obtained from as wide a range of sources as possible, in order to provide an average figure bracketed by low and high estimates.

The definitive estimates should have a much larger local content than the initial estimates. They depend mainly on "inside" sources within the same company or administration, or on information obtained from other companies operating in the same district or adjoining regions. Final estimates may be further refined by indexation of unit costing to take into account time and geographic variations, thus leading to the ratio costing approach.

3.2.3. Unit Costing

At the initial stage, as we are dealing with "order of magnitude" estimates, overall unit costing is generally favored, covering a project as a whole or individual targets within the project. More accurate preliminary estimates are based on the breaking down of a project into a sequence of basic component activities. Each component is characterized by a specific unit cost attached to it, and the contribution of the component to the overall cost is obtained by multiplying the unit cost by the magnitude of the activity.

Unit costs may be considered in a spatial context: as cost per unit of distance or area covered or traveled, or as cost per unit of distance or volume worked. Unit costs may be considered in a time context, based on hours (flying), days, or months (salaries, charter transportation, etc.). Finally they may be considered in a task context, such as cost per station, cost per sample, etc. A schedule of unit costs expressed in 1982 U.S. dollars covering airborne and ground surveys and physical exploration as it applies to remote regions of North America was compiled from various public and private sources and is displayed in Table 3.1.

3.2.4. Indexation of Unit Costs

Uncertainty associated with cost estimation based on the analogy principle is well known, but it may be mitigated by the practice of indexing unit costs for time and geographic variations. Time variations generally result from the superimposition of short-term business cycles on longer-term underlying trends such as inflation, etc. Unit costs are usually reported in current monetary units (dollar) for a specified year or period of years. If we want to apply the unit cost to today's estimation problem, an adjustment must be made so that the cost is reported in real or constant dollars for a base year. An index most commonly used for that purpose is the Gross National Product Implicit Deflator, or Inflator, depending of the base year chosen. The index is based on the total value of goods and services constituting the Gross National Product of a nation during a specified period. An alternative is the Wholesale Price Index, which is based on the changes in wholesale prices of a large range of products and commodities within a specified period. It should be noted that the two indices do not, as a rule, behave in the same manner.[4]

Geographic cost variations reflect many factors including local economic and geographic conditions. For instance, the cost multiplier for a specified type

TABLE 3.1
Schedule of Exploration Unit Costs for Remote North American Locations (U.S. Dollars).

Airborne surveys

Reconnaissance work:

Remote Sensing: Colour photo (including Infra-red)	$150/sq. mile
Infra-red Scanner	$520/sq. miles
Magnetic	$ 50/line mile
Magnetic + Radiometric	$ 75/line mile

Detailed work: fixed wing aircraft:

E.M.	$ 80/line mile
E.M. + magnetic	$100/line mile

Helicopter:

E.M.	$100/line mile
E.M. + Magnetic	$130/line mile

Ground surveys

Reconnaissance:

Geological mapping:	$150/sq. mile
Geochemical drainage survey	$100/sq. mile
Gravity	$300/sq. mile

Detailed Work:

Gridding: open country	$ 50/line mile
dense bush	$120/line mile
Geological survey	$500/sq. mile
Geochemical (Rock or soil sampling)	$1000/sq. mile
Laboratory analysis:	$4 per sample
Geophysical surveys:	
Magnetic	$200/line mile
E.M.	$250/line mile
Radiometric	$120/line mile
Self Potential	$120/line mile
Gravimetric	$500/line mile
I.P.	$600/line mile

Bedrock exploration

Trenching:	$3/cubic foot
Drilling: percussion: reconnaissance	$10/linear foot
systematic	$6/linear foot
Diamond Drilling: reconnaissance	$35/linear foot
systematic	$20/linear foot
Geophysical logging of drill hole:	$5/linear foot

of survey coverage of a unit of area in Alaska or on the Arctic Coast may be, respectively, 3 times or 5 times the cost of a similar coverage in the western U.S.A. The practice of indexation makes it possible to compare meaningfully unit costs for similar activities carried out in different areas and in different periods of time.

3.2.5. Ratio Costing

The more finely tuned final cost estimates are greatly assisted by the ratio costing approach, which gets around the difficulties caused by time and geographic variations without resorting to indexation. In the previous approach of unit costing, the analogy principle was applied to the unit costs themselves for each component of the whole program. In the ratio costing approach, the analogy principle is applied to the percentage of the total cost which is attached to each component of the program. The rationale of ratio costing is that, while individual unit costs may considerably vary over time and geographically, the mix of types of costs and their relative contributions to the total cost of the project can be expected to vary very little over long periods of time, unless some drastic technological change occurs.

The very detailed cost data provided by Metz[11] for 23 exploration projects carried out during the past decade in three regions of Alaska provided the base material for an example of ratio costing which follows. The 23 programs were grouped into three categories: 3 regional reconnaissance, 6 localized reconnaissance, and 14 localized systematic coverage programs. The cost of each type of program was broken down into four main elements: salaries, transport, camp and provisions, and exploration. Transport was subdivided into air and ground, and exploration into four types of activities comprising geological and geochemical surveys, geophysical surveys, drilling, and miscellany, including staking.

If we consider the reconnaissance programs, we find that the overall cost of regional coverages is about 10 times that of localized coverage, while the cost mix is about the same: infrastructure cost share of 90% against actual exploration share of 10%. The overall cost of localized systematic coverage is about 25 times that of localized reconnaissance, but the cost structures are quite different.

The infrastructure share of the overall cost falls from 90% to 65% when moving from localized reconnaissance to localized systematic coverages. On an itemized basis, we find that the salaries share falls from 36% of the total to 23%, the air transport share remains at 24% in both types of coverages, while that of camp and provisions falls from 25% to only 12%. In contrast, the share of total exploration rises from 10% to 35%, including a rise of the drilling share from 0% to 25% of the total expenditures.

Providing that similar data bases are available for other regions of North America, ratio costing of the type described above for Alaska should greatly assist exploration management in establishing accurate definitive estimates for their projects.

3.3. COST FUNCTIONS

3.3.1. Introduction: Concept of Cost Function

The concept of cost function is based on the recognition that each type of exploration activity has a cost structure based on two main components; one is distance-related, the other is sampling-related. The structure of each component may be rather complex: for instance, the distance-related component may cover such items as traveling to and within the survey area, as well as the construction of control grids. The principles of ratio-costing are combined with unit-costing in order to construct a simplified model of a specific type of exploration activity in terms of the two components mentioned above. The model is called a cost function. Three types of models are considered in this chapter: (1) overall regional exploration, (2) airborne geophysical surveys, and (3) discrete ground sampling programs including drilling. They were assembled on Table 3.2 for the readers' convenience.

3.3.2. Cost Function Models for Regional Exploration

The distance factor directly or indirectly affects all stages of regional exploration from the initial reconnaissance stage to the target drill-testing stage. As a result, the overall cost function for regional exploration may be expressed as a linear function of the distance from the center of the prospecting area to the nearest transport facilities (railhead, road, airbase, etc.). This type of model has been used by the writers for various regional programs in the eastern Canadian Shield, and is written as follows:

$$C = c(1 + kD)$$

where C is the cost function, c is the estimate of the cost of exploration of a unit of area, D is the distance from the center of the unit area to the nearest

COST OF DETECTION

transportation facility in miles, and k is a coefficient of proportionality determined empirically. For instance, in the central portion of the Superior Province, we had $c = 0.635$ $ million for a 100-square-mile cell, and $k = 0.0182$.

Another type of model, more flexible than the previous one because it incorporates two coefficients instead of one, was used in the Grenville region and is written as follows:

$$C = c(1 + k_1 d_1 + k_2 d_2)$$

where $c = 0.265$ $ million to cover a 100-square-mile cell, d_1 = distance from the center of the cell to the nearest road, d_2 = distance from the center of the whole prospecting area to the nearest railhead. The coefficients k_1 and k_2 were determined empirically as equal to 0.0056 and 0.0275, respectively.

Finally, another type of model of an exponential nature has been used also in the Superior Province, and is made up as follows:

$$C = c(1 + kd^a)$$

where $c = 0.165$ $ million for the coverage of a 36-square-mile cell, k is a coefficient in the range 0–1, in this case 0.086, and a is an exponent in the range of 1–2, determined empirically in our program as 1.12.

3.3.3. Cost Function for Airborne Geophysical Surveys

The unit cost of airborne geophysical surveys depends on the type of aircraft used, helicopters being much more expensive to run and maintain than fixed-wing aircraft, and on the survey technology. Combined surveys are generally favored; the most popular combination includes magnetic, electromagnetic, and radiometric instrumentation.

Since geophysical airborne surveys are run along flight lines on a continuous sampling basis, the sampling- and distance-related components are confounded. There are, however, additional distance-related cost components which reflect (a) the distance from aircraft base to local operation base (ferrying), (b) distance from local operational base to survey area, and (c) the cost of surveying-in the fiducial points required for the control of the flight lines.

There are two main types of grids commonly used as control for airborne geophysical surveys: parallel grids with spacing s, and orthogonal grids, either square with dimensions s by s, or rectangular with dimensions s by t, with t greater than s.

TABLE 3.2
Tabulation of Expressions of Cost Functions for Various Stages of the Mineral Exploration Sequence.

Stage of Exploration Sequence	Methods and Techniques of Coverage	Parameters of Cost Functions	Type and Geometry of Survey Control	Cost Function Model for Area (A)	Cost Function Model for One Mile Square ($L = 1 = 1$)
Regional Coverage	Airborne Geophysical Surveys (Magnetic, Electromagnetic) INPUT, (Radiometric)	area covered = A sq. mi. area length(L) and width(ℓ) perimeter $P = 2(L+\ell)$ control grid spacing $= S$, (parallel lines, square grid) or S, T, (rectangular grid) C_o = unit cost ($/line mile)	parallel lines or strips (S)	$C = C_o\left(\frac{A}{S} + \frac{P}{2}\right)$	$C_1 = C_o\left(\frac{1}{S} + 2\right)$
			square grid (S x S)	$C = C_o\left(\frac{2A}{S} + \frac{P}{2}\right)$	$C_1 = 2C_o\left(\frac{1}{S} + 1\right)$
			rectangular grid (S x T)	$C = C_o\left[A\left(\frac{1}{S} + \frac{1}{T}\right) + \frac{P}{2}\right]$	$C_1 = C_o\left(\frac{1}{S} + \frac{1}{T} + 2\right)$
	Ground Reconnaissance: Geological: Geochemical: bedrock Sampling, Stream Geophysical: (gravimetric)	C_o = overhead cost C_d = cost of travelling to station C_s = cost of observation at station n = number of stations	no systematic grid	$C = C_o + \sqrt{n}\, C_d + nkC_s$	(not considered)

Selective and Local Coverage	Systematic Ground Surveys (geophysical, geochemical)	area covered = A = $L \times \ell$ perimeter = $2(L + \ell)$ grid spacings: S,T C_d = "distance" unit cost C_s = sampling cost at station $C_d = 20\ C_s$	square grid (S × S)	$C = C_s\left(\frac{\ell}{S} + 1\right)\left[\frac{L}{S} + 20L + 1\right]$	$C_1 = C_s\left(\frac{1}{S} + 1\right)\left(\frac{1}{S} + 21\right)$
			rectangular grid (S × T)	$C = C_s\left(\frac{\ell}{T} + 1\right)\left[\frac{L}{S} + 20L + 1\right]$	$C_1 = C_s\left(\frac{1}{T} + 1\right)\left(\frac{1}{S} + 21\right)$
	Systematic Drilling	area covered = A = $L \times \ell$ perimeter = $2(L + \ell)$ grid spacings : S,T C_d = "distance" unit cost C_s = sampling cost at station $C_s = 50\ C_d$	square grid (S × S)	$C = C_s\left(\frac{\ell}{S} + 1\right)\left[\frac{L}{S} + 0.02L + 1\right]$	$C_1 = C_s\left(\frac{1}{S} + 1\right)\left(\frac{1}{S} + 1.02\right)$
			rectangular grid (S × T)	$C = C_s\left(\frac{\ell}{T} + 1\right)\left[\frac{L}{S} + 0.02L + 1\right]$	$C_1 = C_s\left(\frac{1}{T} + 1\right)\left(\frac{1}{S} + 1.02\right)$
Overall Exploration Sequence		D = distance from centre of exploration area to nearest transportation facilities. A = area in sq. miles C_o = overall cost of coverage of unit area in easily accessible regions	(not considered)	$C = C_o \times A \times (1 + K.D^\alpha)$ $0 < K \leqslant 1$ $1 \leqslant \alpha \leqslant 2$	(not considered)

Parallel Grids. Considering a rectangular area of A square miles with dimensions L and l, the total mileage of flight lines with spacing s miles required to cover the area is expressed as

$$M = A/s + P/2$$

in terms of the area A and perimeter P, with $A = Ll$, and $P = 2(L + l)$; the cost function is then written as:

$$C = c(A/s + P/2)$$

with c the unit cost for the combined geophysical surveys.

If we choose a survey area covering one mile square, we have $L = l = 1$ mile, and $A = 1$ mile square, so that the cost function becomes more simply

$$C_1 = c(1/s + 2)$$

Rectangular Grids. The total mileage of flight lines run in a rectangular pattern with spacings s and t miles which is required to cover an area of A square miles with a perimeter P miles is

$$M = A(1/s + 1/t) + P/2$$

and the cost function becomes

$$C = c[A(1/s + 1/t) + P/2]$$

The simplified expression for the coverage of a one mile square area is

$$C_1 = c(1/s + 1/t + 2)$$

Square Grids. Since both spacing s and t are equal, the expression of the cost function becomes

$$C = c(2A/s + P/s)$$

for the coverage of an area A square miles, and

$$C_1 = 2c(1/s + 1)$$

for the coverage of a one mile square area.

3.3.4. Cost Functions for Ground Surveys and Drilling Programs

3.3.4.1. Introduction. All systematic ground surveys require the laying out of a permanent control grid for the accurate location of each sampling point. The structure of the cost function is thus quite different from that pertaining to the continuous sampling airborne surveys of the previous section 3.3.3. We have to deal with a sampling-related component and a distance-related component. The latter, in turn, comprises grid-related and traveling-related subcomponents. All three terms of the cost structure are weighted by different unit costs.

Supposing we wish to cover an area of A square miles rectangular in shape with dimensions L and l miles by means of a discrete sampling rectangular grid with spacing s and t miles; the expression of the distance-related cost component is

$$L(l/t + 1)$$

and that of the sampling-related component is

$$(l/t + 1)(L/s + 1)$$

If c_d and c_s are the unit costs for the distance- and traveling-related components, respectively, the most general expression of the cost function for discrete ground sampling becomes

$$C = (l/t + 1)[Lc_d + (L/s + 1)c_s] \qquad (1)$$

3.3.4.2. Cost Function for Ground Surveys. Based on past experience, it is generally agreed that the distance-related component of the cost structure is a significant item, the largest share of which is taken by the cost of establishing the control grid for the survey. We estimate that the \$/mile unit cost c_d of the distance-related component is 20 times the \$/station unit cost of the sampling-related component c_s.

Therefore the general cost expression (1) becomes

$$C = c_s(l/t + 1)(L/s + 20L + 1) \qquad (2)$$

for the coverage of an A square-mile area of dimensions L and l mile. For the coverage of a 1-mile-square-area, the expression becomes

$$C_1 = c_s(1/t + 1)(1/s + 21) \qquad (3)$$

The expressions (2) and (3) are further simplified when we deal with square grids where $s = t$ miles. The general cost expression for the coverage of an area of A square miles becomes

$$C = c_s(l/s + 1)(L/s + 20L + 1) \qquad (4)$$

and for the coverage of a 1-mile-square area, we now have

$$C_1 = c_s(1/s + 1)(1/s + 21) \qquad (5)$$

3.3.4.3. Cost Function for Drilling Programs. The structure of the cost function for drilling programs is quite different from that of the ground surveys, because of the marked predominance of the sampling-related cost component over the distance-related one.

The sampling component is a function of the grid spacing, as illustrated by Figure 3.1, and of the depth of holes, as well as being affected by the unit cost (cost per linear foot) which varies with the type of drilling. If we take the unit cost for "churn" drilling, the cheapest of all drilling methods, as unity, then the unit cost for noncoring percussion drilling is about 2.5 and that for the coring diamond drilling method is about 7.

The distance-related cost component of the function includes two subcomponents. One reflects the traveling cost between setups (moving), and the second one that of establishing the control grid for the drilling program and the surveying of collar locations. Generally the unit cost for moving is substantially larger than that pertaining to the control grid.

Based on past experience, we can simplify the expression of the cost function for drilling programs by assuming that the \$/hole unit cost for the sampling component c_s is about 50 times the \$/mile unit cost for the whole distance component c_d, so that the most general expression for a rectangular grid (1) becomes

$$C = c_s(l/t + 1)(L/s + 0.02L + 1) \qquad (6)$$

for an area of A square miles with dimensions L and l miles. The expression (6) is further simplified if we deal with a 1-mile-square area, as $L = l = 1$ mile, and becomes

$$C_1 = c_s(1/t + 1)(1/s + 1.02) \qquad (7)$$

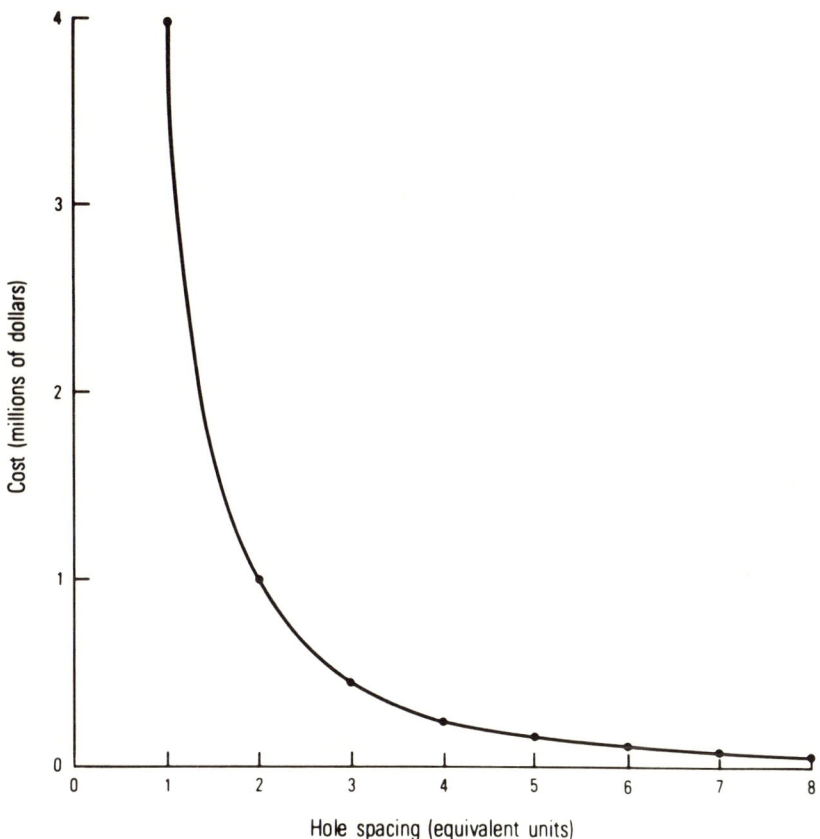

FIGURE 3.1
Graph of Drilling Cost versus Hole Spacing when Holes Cost $4000 Each. [From Drew, L. J.[34] (of Chapter 2), Figure 8, by permission of Plenum Publishing Corporation, New York.]

If we choose a square grid with spacings s by s miles instead of a rectangular one, then expression (7) is further simplified and becomes

$$C_1 = c_s(1/s + 1)(1/s + 1.02) \tag{8}$$

So far we have considered only vertical drilling programs. However, angled drilling should be favored when searching for dipping targets, providing that the

68 CHAPTER THREE

likely direction of the dip can be ascertained. The shallowest surface drilling angle which can be efficiently used in the present state of the technology is 35 degrees. If we drill a hole with an inclination of b degrees from the horizontal, the unit cost is affected by the angle factor: $1/\sin b$, which is greater than unity. A 35-degree angled-hole will cost 75% more to drill than a vertical hole to the same vertical depth. For 55-degree angled-drilling, the relative cost increase is 22%, and for 65-degree drilling, it is only 10%.

3.4. TABULATION OF COST FUNCTIONS

3.4.1. Foreword

For the readers' convenience, we tabulated five cost functions covering the three main types of field programs, including airborne surveys, ground surveys, and drilling programs. Tables 3.3 and 3.4 cover airborne geophysical surveys on parallel and square grids, based on combined magnetic, electromagnetic, and radiometric technologies, which are costed at 70 1982 U.S. dollars per line mile for fixed-wing craft, and $100 per line mile for helicopter surveys.

Table 3.5 covers combined geophysical surveys including the following commonly used combinations: magnetic + electromagnetic, magnetic + gravity, magnetic + induced polarization which are costed at 6 1982 U.S. dollars per station. Similarly, we are using the $6 station cost for combined multielement geochemical surveys including processing laboratory costs. Table 3.6 deals with vertical drilling on square grids to a depth of 300 feet. Percussion holes are costed at 2000 1982 U.S. dollars and diamond drill holes at $6000. The costs shown in Table 3.6 for vertical drilling are adjusted by means of the angle factor: $1/\sin 55$ when considering 55-degree angled drilling (Table 3.7).

3.4.2. Cost Function Tables

The cost functions pertaining to airborne geophysical surveys are to be found on pages 69 and 70. The cost function for ground geophysical and geochemical surveys is tabulated on page 71, while those dealing with vertical and angled drilling appear on pages 72 and 73.

COST OF DETECTION 69

TABLE 3.3
Cost Function for Airborne Surveys on Parallel Grids in U.S. Dollars per Mile Square.

Survey design: parallel flight lines with spacing S feet = s miles
Cost function: $C = (1/s + 2)c$, where c is unit cost in $/line mile

S in feet	Grid spacing s in miles	C in $ when c = $70/line ml.	C in $ when c = $100/line ml.
200	0.038	1988	2840
400	0.076	1064	1520
600	0.114	756	1080
800	0.152	602	860
1000	0.189	509	728
1200	0.227	448	640
1400	0.265	404	577
1600	0.303	371	530
1800	0.341	345	493
2000	0.379	324	464
2200	0.417	308	440
2400	0.455	294	420
2600	0.492	282	403
2800	0.530	272	389
3000	0.568	263	376
3200	0.606	255	365
3400	0.644	248	355
3600	0.682	242	347
3800	0.720	237	339
4000	0.758	232	332
4200	0.795	228	326
4400	0.833	224	320
4600	0.871	220	315
4800	0.909	217	310
5000	0.947	213	306
5200	0.985	211	302

TABLE 3.4
Cost Function for Airborne Surveys on Square Grids in U.S. Dollars per Mile Square.

Survey design: square grid with spacings S by S feet = s by s miles
Cost function: $C = 2(1/s + 1)c$, where c is unit cost in $/line mile

| Grid spacing | | C in $ when | C in $ when |
S feet	s miles	c = $70	c = $100
200	0.038	3836.00	5480.00
400	0.076	1988.00	2840.00
600	0.114	1372.00	1960.00
800	0.152	1064.00	1520.00
1000	0.189	879.20	1256.00
1200	0.227	756.00	1080.00
1400	0.265	668.00	954.29
1600	0.303	602.00	860.00
1800	0.341	550.67	786.67
2000	0.379	509.60	728.00
2200	0.417	476.00	680.00
2400	0.455	448.00	640.00
2600	0.492	424.31	606.15
2800	0.530	404.00	577.14
3000	0.568	386.40	552.00
3200	0.606	371.00	530.00
3400	0.644	357.41	510.59
3600	0.682	345.33	493.33
3800	0.720	334.53	477.89
4000	0.758	324.80	464.00
4200	0.795	316.00	451.43
4400	0.833	308.00	440.00
4600	0.871	300.70	429.57
4800	0.909	294.00	420.00
5000	0.947	287.84	411.20
5200	0.985	282.15	403.08

COST OF DETECTION 71

TABLE 3.5
Cost Function for Ground Geophysical or Geochemical Surveys on Square Grids in U.S. Dollars per Mile Square.

```
Survey design:  square grid with spacings S by S feet = s by s miles
Cost function C = (1/s + 1) (1/s + 21)c, where c is the cost in $ at
each station
```

Grid spacing		C when
S feet	s miles	c = $6
100	0.019	23822.64
150	0.028	12206.64
200	0.038	7792.56
250	0.047	5590.17
300	0.057	4307.76
350	0.066	3482.79
400	0.076	2913.84
450	0.085	2500.83
500	0.095	2189.00
550	0.104	1946.16
600	0.114	1752.24
650	0.123	1594.15
700	0.133	1463.03
750	0.142	1352.65
800	0.152	1258.56
850	0.161	1177.47
900	0.170	1106.91
950	0.180	1044.98
1000	0.189	990.23
1050	0.199	941.49
1100	0.208	897.84
1150	0.218	858.53
1200	0.227	822.96
1250	0.237	790.62
1300	0.246	761.10
1350	0.256	734.05
1400	0.265	709.17
1450	0.275	686.22
1500	0.284	664.98
1550	0.294	645.28
1600	0.303	626.94
1650	0.313	609.84
1700	0.322	593.86
1750	0.331	578.88
1800	0.341	564.83

TABLE 3.6
Cost Function for Vertical Drilling Programs on Square Grids in U.S. Dollars per Mile Square.

Survey design: square grid with spacings S by S feet = s by s miles
Cost function $C = (1/s + 1) \cdot (1/s + 1.02) \cdot c$, where c is the cost in thousands of \$ of a 300 foot vertical drill-hole

Grid spacing		C (percussion)	C (diamond)
S feet	s miles	c = 2 (\$thou.)	c = 6 (\$thou.)
100	0.019	5791.00	12373.00
200	0.038	1502.62	4507.85
400	0.076	403.85	1211.54
600	0.114	192.47	577.42
800	0.152	115.82	347.47
1000	0.189	79.13	237.38
1200	0.227	58.54	175.61
1400	0.265	45.72	137.17
1600	0.303	37.15	111.46
1800	0.341	31.10	93.30
2000	0.379	26.64	79.93
2200	0.417	23.26	69.77
2400	0.455	20.61	61.82
2600	0.492	18.49	55.48
2800	0.530	16.77	50.31
3000	0.568	15.35	46.04
3200	0.606	14.15	42.45
3400	0.644	13.14	39.41
3600	0.682	12.27	36.80
3800	0.720	11.51	34.54
4000	0.758	10.86	32.57
4200	0.795	10.28	30.84
4400	0.833	9.77	29.30
4600	0.871	9.31	27.94
4800	0.909	8.90	26.71
5000	0.947	8.54	25.61
5200	0.985	8.20	24.61

TABLE 3.7
Cost Function for 55-degree Angled Drilling Programs on Square Grids in U.S. Dollars per Mile Square.

Survey design: square grid with spacings S by S feet = s by s miles
Cost function $C = (1/s + 1) \cdot (1/s + 1.02) \cdot c/\sin(55)$, where c is the cost in thousands of $ of a 300 foot vertical drill-hole

Grid spacing		C (percussion)	C (diamond)
S feet	s miles	c = 2 ($thou.)	c = 6 ($thou.)
200	0.038	1834.36	5503.07
400	0.076	493.01	1479.02
600	0.114	234.96	704.89
800	0.152	141.39	424.18
1000	0.189	96.60	289.79
1200	0.227	71.46	214.38
1400	0.265	55.82	167.46
1600	0.303	45.35	136.06
1800	0.341	37.97	113.90
2000	0.379	32.53	97.58
2200	0.417	28.39	85.17
2400	0.455	25.16	75.47
2600	0.492	22.57	67.72
2800	0.530	20.47	61.42
3000	0.568	18.73	56.20
3200	0.606	17.28	51.83
3400	0.644	16.04	48.11
3600	0.682	14.98	44.93
3800	0.720	14.06	42.17
4000	0.758	13.25	39.76
4200	0.795	12.55	37.65
4400	0.833	11.92	35.77
4600	0.871	11.37	34.10
4800	0.909	10.87	32.61
5000	0.947	10.42	31.26

REFERENCES AND SELECTED READINGS

1. BAILLY, P. A., 1964, Methods, costs, land requirements and organization in regional exploration for base metals, Preprint, A.I.M.E. Fairbanks meeting, Alaska, March 18–21.
2. CALLAWAY, H. M., 1950, Expense of exploration, *Econ. Geol.* **45,** 328–330.
3. COX, J. L., 1961, Cutting costs through operations research, *Min. Congr. J.* **47,** 45–46.
4. CRANSTONE, D. A., and MARTIN, H. L., 1973, Are ore discovery costs increasing? *Can. Min. J.* **94**(4), 53–64.
5. DE GEOFFROY, J., and WIGNALL, T. K., 1973, Design of statistical data processing system to assist regional exploration planning, *Can. Min. J.* **94**(11), 30–35; **94**(12), 35–36.
6. DE GEOFFROY, J., and WIGNALL, T. K., 1974, Evaluating that exploration project, *Can. Min. J.* **95**(5), 42–44.
7. DERRY, D. R., 1970, Exploration expenditure, discovery rate and methods, *Can. Inst. Min. Metall. Bull.* **63**(694), 362–366.
8. DERRY, D. R., and BOOTH, J. K. B., 1978, Mineral discoveries and exploration expenditures (1966–1976), *Min. Mag.* **138**(5), 430–433.
9. GRANT, F. S., 1971, Some thoughts on the next decade in mineral exploration, *Geoexploration* **9,** 63–77.
10. KUZWART, M., and BOHMER, M., 1978, *Prospecting and Exploration of Mineral Deposits,* Elsevier, Amsterdam.
11. METZ, P. A., and CAMPBELL, B. W., 1982, Cost of exploration for metallic minerals in Alaska, 1982, M.I.R.L. Report No. 56, Department of Commerce and Economic Development, State of Alaska, Fairbanks.
12. PERRY, A. J., 1968, Organization and costs of mineral exploration in the Southwest U.S.A., Pacific S.W. Mining Industry Conference, A.I.M.E., May 1968.
13. PETERS, C. W., 1959, Cost of exploration for mineral raw material, *Cost Eng.* **4**(3), July.
14. PETERS, C. W., 1967, Cost and value of drill hole information, *Min. Congr. J.* **53**(1), 56–59.
15. SARMA, D. D., 1979, An exploration strategy for prospecting with a case study on copper prospects at Ingladahl, India, *Miner. Deposita,* **14,** 263–279.
16. SLICHTER, L. B., 1955, Geophysics applied to prospecting for ores, *Econ. Geol. 50th Anniversary Volume* **50,** 885–969.

CHAPTER FOUR

OPTIMIZING ORE DETECTION

4.1. GENERAL STATEMENT

4.1.1. Rationale of the Optimization of Ore Detection

In the previous chapters, we first introduced the concept of probability of detection (Chapter 2) and showed how the theory of geometric probabilities could be put to use to compute the detection probabilities for various types of field programs in terms of target, control grid, and detector geometry and mutual configurations. Obviously, in the economic context in which mineral exploration operates, we have to take into account the amount of effort required to obtain detection of the expected prizes. There has to be a balancing of the detection reward against its cost.

An inspection of the detection probability tables exhibited in Chapter 2 (Tables 2.3, 2.4, and 2.5) will readily confirm the state of affairs described above. We can see in the lower portion of the tables that, as the grid spacing increases, or, conversely, as the coverage cost per unit of area decreases, the probability of detection falls very gradually toward zero. But if we look at the upper portion of the tables, we find that, as the grid spacing diminishes, and conversely as coverage costs rise rapidly, the probability of detection improves only very slowly until it reaches its peak value of unity.

This is well described graphically by the heuristic model shown in Figure 4.1, which relates in a simple manner the probability of detection of any type of mineral resources to the search expenditure, as follows:

$$P_d = 1 - \exp(-Kx)$$

where P_d is the probability of detection, x is the expenditure in dollars for the saturation coverage of a unit of area, and K is a parameter to be determined empirically. It can be seen from the left portion of the graph that the probability of detection rises rapidly to 0.50 or more for relatively modest increments of exploration expenditures. In the right portion of the graph, however, when the probability of detection exceeds 0.80, we find ourselves in a situation of diminishing returns, where large increments of expenditures result in ever-decreasing

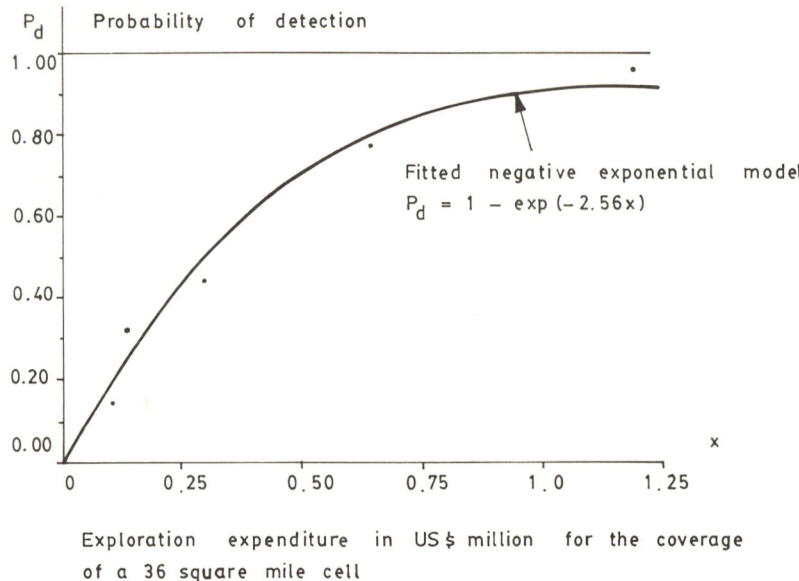

FIGURE 4.1
Graph of Probability of Detection versus Exploration Expenditure.

improvements in probability of detection, until its maximum value of unity is reached.

It is obvious that a tradeoff has to be sought between the legitimate desire to maximize the detection of the prizes on the one hand, and the harsh realities of budget stringency on the other hand. If the budget allocation is too low, the mesh of the net will be too coarse by necessity and the prize will fall through and go undetected. If the mesh is too fine, the cost of the search will be so great as to lead possibly to the "gambler's ruin," or to a deplorable but not unknown situation when the cost of the investigation exceeds the value of the prize. The practical recognition of these exigencies intuitively leads to the concept of the optimization of ore search.

4.1.2. Survey of Literature on Detection Optimization

The optimization of detection is a critical step in the overall optimal planning of mineral exploration. It follows optimal choices covering ore deposit types to be targeted, regions to be covered, and prospecting areas to be selected for

OPTIMIZING ORE DETECTION 77

investigation within the regions, and methodology of coverage. However, if one uses the volume of publications to gauge the level of awareness and interest in specific topics, it appears that relatively little consideration has been given to date to the topic of detection optimization in mineral exploration.

The results of a diligent search of commonly available English language literature proved rather disappointing. Among the enormous mass of papers related to mineral exploration published during the past three decades, only a handful, perhaps not more than 25, deal with the topic of field exploration optimization. The majority are about drilling, leaving only very few dealing with geophysical or geochemical survey optimization. Most of the 25 are included in our list of references and selected readings.

The topic of airborne survey optimization has been very briefly touched on by Agocs (Chapter 2, Ref. 22) and McCammon (Chapter 2, Ref. 30), in the context of the calculation of the probabilities of detection. Otherwise, the matter does not appear to have been investigated in the literature. The optimization of ground geophysical surveys has been mentioned by Brant (Chapter 1, Ref. 15) and by Cobb (Chapter 4, Ref. 13). The optimization of geochemical surveys has been touched on by Sinclair (Chapter 2, Ref. 36), and treated by Kelley (Chapter 4, Ref. 30) and by the first writer (Chapter 4, Ref. 22).

The optimization of systematic drilling programs has received the bulk of what little attention has been devoted to date to field exploration optimization, because of the incentive provided by the mitigation of the very high expenditure required by the search for "blind" ore targets. Slichter[32] was among the first workers to consider this problem, in 1955. At least six more workers published on the same topic during the following two decades. The most recent papers are those by Shurygin[31] in 1976 and by Drew[24] in 1979.

4.2. BASIC THEORY OF OPTIMIZATION

4.2.1. Fundamental Concepts and Definitions

Most business situations, including mineral exploration, require management to choose the best possible course of action from a rather large number of options without perfect knowledge of the possible outcomes. In order to deal with the complex problems of optimal choice under uncertainty, several important concepts have to be established and defined. The three main concepts of interest are goals, criteria, and strategies.

Goals are most often of a qualitative and subjective nature. They are defined within the framework of corporate policies and resources. Goals may call for the acquisition of some type of reward that can be directly translated into monetary units, for example, gross value, gross profit, etc. But in other cases, different kinds of rewards are sought such as effectiveness, efficiency, precision, etc., which are indirectly translated into monetary units at the outcome.

The quantification of the qualitative goals into criteria is a necessary step of the optimization of management's decision making. Criteria may be simple expressions, or composite ones such as products and ratios or differences which are better suited to deal with more complex situations.

The third basic tenet of the optimization of business decision making is the concept of strategy. After defining qualitative goals and quantifying them into criteria of choice, management has to establish a set of rules to follow, which will unambiguously tell the decision maker how to deal with the criteria in order to choose that course of action which is optimal in the prespecified context of corporate policies and resources. This set of rules is what is referred to as "strategies."

4.2.2. Criteria

4.2.2.1. Structure of Criteria. Criteria may be simply structured by representing a single quantity such as gross value, cost, probability of success, risk of error as expressed statistically by the variance or coefficient of variation, or risk of failure as expressed by the complement of the probability of success. Criteria may be structured as composite expressions such as the product of two of the quantities listed above. For example, the expected monetary value (EMV) criterion proposed by Grayson (Chapter 1, Ref. 5) is the product of gross value times the probability of success; the expected loss criterion is the product of cost times the probability of failure (complement of the probability of success). Ratios of two single quantities are also commonly used as criteria, for example, the effectiveness ratio, which is the ratio of the number of targets actually detected over the total number of expected targets in the area under investigation. Another ratio, used by Slichter[32] as an exploration criterion, is his "prospecting success ratio," which compares the gross value of the prizes to the cost of their detection.

Finally, more complex and flexible structures, such as differences between some of the composite expressions listed above, are generally more useful when dealing with actual business situations. The general design is that of a difference between a reward function and a cost function, both expressed in terms of common variates and units. Two useful difference criteria are referred to as the "efficiency criterion" and the "payoff criterion" in the present study.

4.2.2.2. Efficiency Criterion. The efficiency criterion is a convenient and flexible type of criterion which is quite suitable for most business, engineering, and mineral exploration situations. It is a difference criterion based on a heuristic model which implies a comparison between a reward and the cost required to acquire it.

The efficiency criterion may be simply expressed in an analytical manner as follows:

$$\text{Efficiency function } (x,y,z,) = \text{Reward function } (x,y,z) - K \cdot \text{Cost function}(x,y,z)$$

$(x,y,z,)$ being common variates expressed in common units such as dollars, spacing in feet, etc., and K being a scaling factor. The reward function may be expressing gross value, probability of success, precision, etc.

The precision reward function may be used to advantage in engineering design situations as well as in mineral exploration. We can cite the design of optimal geochemical sampling programs, or the design of drill sampling plans for newly discovered ore deposits as good examples of the application of the precision reward concept. The reward function is expressed in terms of the number of samples per unit area required to obtain a prespecified level of precision, which is translated into sample spacing. Likewise, the cost function is expressed in terms of sample spacing. Kelley[30] gives an example of the application of the criterion to the optimization of geochemical programs, while Hewlett[28] describes the optimization of the sampling of a porphyry copper deposit in the Southwest U.S.A., and Roubens[19] optimizes the sampling of a coal bed in the U.K.

The detection reward function is that used throughout this study to optimize the search for six types of ore deposits in North America, by means of airborne geophysical, ground geophysical, and drilling programs. The criterion pits the probability of detection expressed in terms of grid spacings, or grid orientation with respect to target, or detector orientation with respect to target, against a cost function also expressed in terms of grid spacings or angular intervals.

4.2.2.3. Payoff Criterion. The payoff criterion is the difference expression which is most commonly used in general business situations. It is also useful in certain aspects of mineral exploration planning. The reward function is the expected monetary value already mentioned. The EMV is expressed as the economic value of an event (gross value, gross profit) weighted by the probability of occurrence of that event (probability of success). The EMV reward function is matched against a "loss function" which is another product-type expression referred to above as "expected loss." It is expressed as the cost of failure weighted by the probability of failure P_f, which is the complement of the probability of success P_s ($P_f = 1 - P_s$).

In the context of mineral exploration situations, the payoff criterion is not so widely useful as the efficiency criterion, particularly when we are interested in the optimization of individual stages of the exploration sequence. It is not rewarding to match the expected gross value of exploration success against the cost of failure of an airborne or ground geophysical survey or even a drilling program of average size, because of the imbalance between the orders of magnitude of the two terms of the comparison. However, if we pit the expected gross profit, say 20% of the gross value of the prize, instead of the whole gross value, against the expected cost of a whole exploration sequence, we obtain a much more realistic and useful criterion which has been used in the literature during the past two decades (see Chapter 1, Ref. 4, 17, 30, 41, and 42).

4.2.3. Strategies

After quantifying the corporate goals as criteria, we need set rules in order to enable us to select the optimal course of action among a variety of available options. The guiding rule is referred to as "strategy" in the terminology of statistical decision theory. In that sense, an optimizing strategy is not a value judgment; it is only a uniform rule of action which is established in accordance with prespecified corporate policies. A strategy is meant to guide a choice of action but is not a particular action at a given time.

The role played by strategies in the methodology of decision theory is as follows: each option available to the decision maker is assigned a numerical score which is the value taken by the criterion for that particular option. All options are then ranked in order of the magnitude of their scores for the decision maker to make his selection according to the prespecified strategy. There are three main categories of strategies: (1) those seeking extrema of one type of the other (maxima or minima) of the criterion without constraints, (2) those seeking extrema of one type or the other under constraints, and (3) hybrid strategies seeking to combine extrema of two types, such as the "minimax" strategy so favored by many Operations Research workers.[16]

Strategies calling for maximization will lead to the maximization of a criterion which implies favorable consequences, such as gross value, profit, precision, detection, occurrence, success, etc. The strategy requires simply the ranking of the various options based on their scores, i.e., the numerical values of the criterion, in order of decreasing magnitude, leading to the selection of the top-scoring option as the most desirable course of action. In the opposite case, the selected strategy will be that of minimization, based on a criterion which implies unfavorable circumstances, such as cost, loss, risk, error, etc. Again, the numerical values of the criterion will be used as scores for the various available options; the scores will be ranked in order of increasing magnitude, and the option with the smallest score will point to the most favored course of action.

However, most realistic business situations are much more complex than the ones described above and require more sophisticated strategies with built-in constraints. These are known as "optimization strategies." The two most favored kinds are (a) maximization of a "favorable" criterion under the constraint of cost or risk minimization, and (b), conversely, the minimization of an "unfavorable" criterion such as cost or risk, under a constraint requiring a specified acceptable level of reward. Both are very useful in mineral exploration situations and are the cornerstones of the optimization of the search for six types of ore deposits in North America.

4.3. METHODOLOGY OF OPTIMIZATION

4.3.1. Introduction

In the previous section, the general concepts of optimization were described in a qualitative manner outside any mathematical context. However, since the coming of age of the new science of "Operations Research" born from research into World War II resources allocation problems, analytical methods of systems optimization based on a mathematical apparatus have rapidly grown in use and sophistication. The remarkable development of computer capabilities in the past two decades has greatly assisted the inroad of the mathematical approach, because of the very large computational requirements of analytical solutions, particularly when dealing with probabilistic situations.

4.3.2. Fundamentals

4.3.2.1. Introduction of a Mathematical Model. The methodology of maximization of favorable criteria and minimization of unfavorable criteria can be translated mathematically into the study of the analytical behavior of an "objective function." The function is represented by a dependent variate z whose value is uniquely determined by a number of independent variates $(x_1, x_2, \ldots, x_i, \ldots, x_n)$. Optimization in a mathematical context is turned into the search for extrema of the objective function which represent the criterion. The extrema may be either maxima or minima, depending on the type of criterion and strategy chosen. Extrema may be of a "local nature," i.e., occurring within specified intervals of the value of the independent variates. They may be of a "global" nature if they occur within the full range of variation of all independent variates. Optimization is achieved when we find the numerical values of the independent variates for which the objective function reaches an extremum of the type required by the chosen strategy.

However, we should appreciate that in realistic business situations most systems operate under some kind of constraint, whether physical, monetary, or time-related. Therefore, our concern should be to translate the concept of constraint into mathematical language. Harbaugh[6] very aptly illustrates the quantification of the concept of constraint by means of a geometric analogy. (See Figure 4.2.) In order to be more general, he considers two types of objective functions z of two independent variates x_1 and x_2. One model is linear in nature, but the other one is quadratic.

The linear objective function is simply represented by the expression

$$z = ax_1 + bx_2 \quad \text{with } x_1, x_2 \geq 0 \tag{1}$$

The geometric representation is that of a plane of height z (Figure 4.2, upper left corner) which is represented by straight line contours restricted to the upper right quadrant because of the initial restrictions on the independent variates. If there are no other constraints, there will be no point on the plane for which the objective function will reach a maximum, because as x_1 and x_2 increase indefinitely to the right and upward, so will z. If we follow the convention of shading the permissible area, we find that all points of the surface yield feasible solutions because there are no constraints.

In order to be more realistic, we should consider adding constraints to the objective function. If the constraints are linear, they can be represented by straight lines which describe linear inequalities, as follows:

$$cx_1 + dx_2 \leq e \tag{2}$$

$$fx_1 + gx_2 \leq h \tag{3}$$

The three expressions (1), (2), and (3) are graphically represented in the middle-left portion of Figure 4.2 by the contoured plane and two oblique lines which define a permissible (shaded) region yielding all feasible solutions. Among them, the optimum solution, which is maximal in this example, lies at the intersection of the two inequality lines. At that point, z attains its maximum without going beyond the constraints. The reasoning would be the same if the inequalities are not linear (lower left portion of Figure 4.2).

In many instances, the objective function is of a quadratic nature and may

OPTIMIZING ORE DETECTION 83

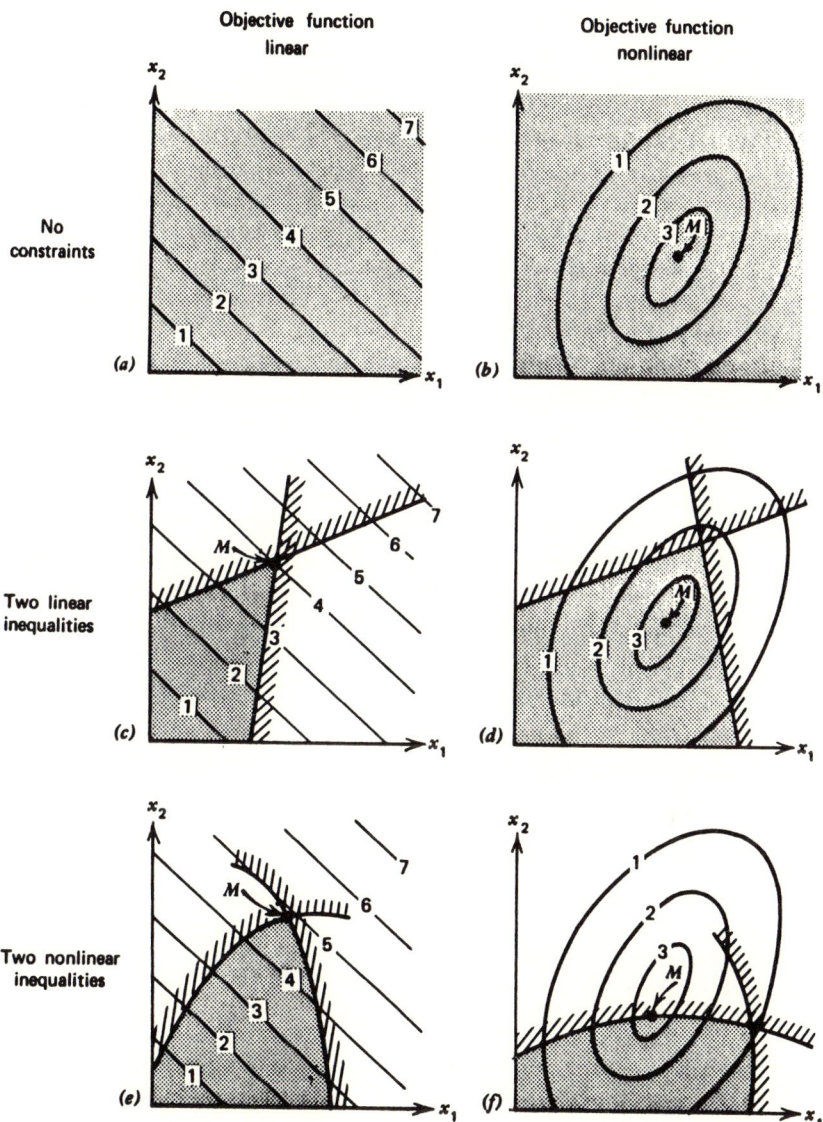

FIGURE 4.2
Geometric Representation of Optimization Under Constraints. [From Harbaugh, S. W., and Bonham-Carter, G. R.,[6] Figure 8.3, by permission of John Wiley, New York.]

be represented by the following expression, which is a general second-degree polynomial, where a,b,c,d,e are coefficients:

$$z = ax_1 + bx_2 + cx_1^2 + dx_1x_2 + ex_2^2 \qquad (4)$$

The geometric representation of the objective function is a parabolic surface shown in the upper right portion of Figure 4.2. In this case a local maximum exists at point M whose location is specified by appropriate values of the two independent variates x_1 and x_2. If we apply the constraints described by the two lines (2) and (3), we find that the permissible region does include the local maximum, which then becomes a unique solution for the optimization problem (center left portion of Figure 4.2).

It would not be difficult to imagine other cases where the coefficients c,d,e,f are such that the two constraint lines intersect below the local maximum M. In such a case, the actual maximum under specified constraints will lie at the intersection of the two lines, being lower in absolute value than the original local maximum, which is now out of bounds. In many complex problems, the constraints could themselves be represented by portions of circles, parabolas, ellipses, etc., instead of straight lines, and there could be more than two independent variables, which would preclude any geometric analogy.

4.3.2.2. Methods of Optimization. Optimization is now redefined as the search for extrema of an objective function under constraints and guidance of a strategy. Within this mathematical context, optimization may be conducted by a variety of methods which are reviewed in detail by Wilde[10,11] and aptly summarized by Harbaugh.[6] These authors distinguish two broad classes of optimizing methods, direct search and indirect search for extrema, in two different contexts: deterministic and probabilistic.

In the deterministic context, which is that of most engineering optimization situations, the independent variates can be determined in a unique manner through a specified functional model. In most general business situations and in mineral exploration in particular, the deterministic definition of the variates does not apply. The independent variates are represented by probability distributions or by expected values within confidence intervals. As a result, the dependent variate of the objective function and the extrema are similarly expressed in a stochastic manner.

4.3.3. Indirect Search for Extrema Based on Calculus

In its simplest form involving only one dependent variate $Z(x)$ and one independent variate x within the constraints $a \leq x \leq b$, the indirect search based on calculus calls for the setting of the first derivative $Z'(x)$ to zero. The next step consists of finding the roots of the equation $Z'(x) = 0$ which lie within the required interval (a,b). Among these, only real roots should be retained, the complex ones being discarded. Depending on the prespecified strategy, either the roots that result in the maximization of the objective function (maximal roots) or the minimal roots are of interest in the problem at hand. The screening is done by calculating the second derivative $Z''(x)$ and examining its sign for each extremal value of the independent variate: a negative sign indicates a maximal root and a positive sign a minimal one.

The method can be easily extended to the more complex and general n-dimensional case, where we have to deal with n independent variates rather than only one, such as $x_1, x_2, x_i, \ldots, x_n$. The problem is handled by setting the first partial derivative of Z with respect to each of the n independent variates to zero, one at a time. The result is a system of n simultaneous equations. The system is solved in order to determine the values of the independent variates which satisfy the constraints and generate extremal values of the objective function Z of the type required by the chosen strategy.

The calculus approach to optimization, straightforward as it may seem, is fraught with difficulties, many of which are listed by Bellman.[1] First, the form and structure of the objective function, even in the simpler uni- or bivariate cases, may be such as to involve products of transcendental functions which would make the derivation and finding of extremal roots an intractable computational problem, beyond the capability of even the most modern high-speed computers. A second difficulty pointed out by Bellman is the ambiguity which occurs when several extremal roots of the required type are found within the permissible area. Furthermore, we may have to contend with artificial extrema resulting from discontinuities of the objective function within the permissible area, or from the presence of boundaries. Finally, in the multidimensional case, there may be too many variates and too many constraints to make the calculus method workable at all.

4.3.4. Direct Search Approach

The direct search approach is heuristic and sequential in nature, leading to successively improved solutions by series of iterations determined by an algorithm. When no further improvement can be obtained, the solution reached, although an approximate one, is considered as the best solution to the problem at hand. Several types of algorithms are commonly used, including the "hill climbing," and the "elimination" algorithms, both leading to the extrema sought by direct ascent.

The direct search approach may provide the only feasible solution to optimization in two very common types of situations: (1) when the form of the objective function is too complex and proves intractable for the calculus approach, and (2) when the nature of the objective function is not sufficiently well known to be translatable into pure mathematical terms. The direct search approach can be successfully handled graphically when dealing with one or two independent variates and a similarly small number of constraints.

4.3.5. Indirect Search Approach Based on Mathematical Programming

The mathematical programming approach may be used to advantage in complex business and engineering situations, when dealing with n independent variates and m constraints, a type of situation which is beyond the capabilities of the indirect search based on calculus and the direct search approach. The mathematical programming approach is based on recursive algorithms, which are well suited for computerization. Readers should note that mathematical programming and computer programming are not to be confused; although the former requires the latter, the two terms have quite different meanings. The two best-known methods of mathematical programming are "linear programming" and "dynamic programming."

Linear programming is the most favored method when dealing with single stage optimization problems involving n independent variates and n linear constraints, such as are most of the complex allocation problems dealt with in engineering. The SIMPLEX algorithm, which is well suited for computer application, is one of the best known. Harbaugh[6] describes its structure and its application in some detail by means of a numerical example. If the objective function and the constraints are not linear in structure, other types of algorithms have to be used.

Dynamic Programming is a more recently developed approach designed to handle optimization problems quite different from that tackled by linear programming. Instead of dealing with one stage and numerous variates and constraints, we may wish to optimize multistage sequential situations involving only a few independent variates and constraints. Although lumped with linear programming as part and parcel of mathematical programming, dynamic programming is of a radically different nature. Dynamic programming is not an algorithmic method as are the direct search and linear programming approaches, but it should be described as a technique of problem formulation.

Dynamic programming parcels out an initially complex problem into a sequence of simpler and more tractable ones which can then be optimized either by the indirect calculus-backed approach or by the direct search method. It is a very versatile and flexible approach which may be applied to any type of problem, whether linear or nonlinear, deterministic or probabilistic, provided that the problem at hand can be reformulated as a multistage decision process. The recursive nature of the procedure allows the application of the results of the ith stage to facilitate the optimization of the $(i + 1)$th stage of the sequence.

The very flexibility and versatility of dynamic programming is just one aspect of a more general philosophy of optimization of complex systems, referred to as Operations Research, whose application is spreading into many fields of business and engineering.

4.3.6. Operations Research

The term "operations research" (OR) was coined during World War II by British scientists who pioneered it as a systematic approach to solve very complex decision problems involved in optimal allocation of scarce resources of war material and manpower in the context of war situations. OR has since been considerably broadened to encompass problems of optimal allocation in any type of business or engineering situation. A relatively new and very useful addition to OR is the field of sensitivity analysis. Sensitivity analysis is the study of likely effects of changes introduced to the structure or mode of operation of complex systems. Its purpose is to differentiate and isolate the most important factors which are responsible for the changes to the systems from the ones with the least or no effect. A simple example of sensitivity analysis has already been described in Section 2.5.6 of Chapter 2.

As a problem-solving approach, OR has to be regarded as both an art and a science. The science aspect pertains to the use of a mathematical apparatus and algorithms appropriate for the objective solving of complex decision problems. It is an applied science devoted to observing, analyzing, and predicting the behavior of man–machine systems based on "models." The art part of the definition of OR is justified because most complex decision problems include important elements which cannot be translated into mathematical terms, one of them being the human element, which is an integral part of every decision environment.

Much of the interest in OR originates from business and engineering circles. With the exception of mining engineering production optimization problems, OR has had only a small impact on the mineral industry. A survey of the available literature in the English language regarding the application of OR to the mineral industry was done by Coyle in 1969.[15] Out of a total of 75 papers reviewed, 60 dealt mainly with production optimization, mine design, particularly open pit optimization, and production scheduling, using the linear programming approach. Only 15 papers dealt with mineral exploration optimization, based on direct and indirect search approaches, but none on the dynamic programming method. A diligent search conducted by the writers during the past two years led to the conclusion that the imbalance between mining engineering and mineral exploration applications of OR is persisting, if not worsening.

4.4. APPLICATION OF OPTIMIZATION THEORY TO ORE DETECTION

4.4.1. Introduction

By bringing together detection probabilities and cost functions for each type of field program as the two key building blocks we can now construct a model for the optimization of ore detection which takes full account of the basic theory of optimization, as described in Sections 4.2 and 4.3 of the present chapter.

The "efficiency" model, already mentioned in Section 4.2.2, is the one selected for the study of ore search optimization, because it is well suited for the optimization of individual stages of the mineral exploration sequence. The diagram shown in Figure 4.3 illustrates the makeup of the model in terms of the probability of detection as a reward function, cost function, and scaling factors. It should be pointed out that we are interested in optimizing detection, i.e., technical success, only, not commercial success, as they are defined in Section 1.1.1 of Chapter 1, so that no consideration of economic worth of the expected target is introduced in the model.

4.4.2. Practical Application of the Efficiency Model

4.4.2.1 Introduction. Since we are dealing with the application of the theory of linear optimization, the objective function should be linear in structure with all terms expressed in terms of a common variate, in the general form

$$F(x) = KR(x) - MC(x)$$

where $R(x)$ and $C(x)$ are the reward function and the cost function, respectively, K and M are scaling factors, and x is the independent variate.

In the context of the application of the efficiency model to control grid optimization, the grid spacing s is the independent variate, the probability of detection P is the reward function, and the objective function is written as

$$F(s) = KP(s) - MC(s)$$

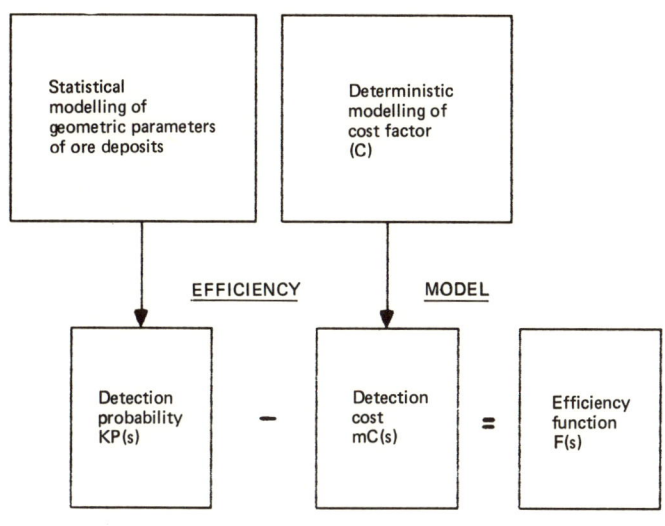

FIGURE 4.3
Construction of an Efficiency Model for the Optimization of Ore Detection.

On the one hand, the range of variation of the reward function $P(s)$ is by definition restricted to the 0–1 interval. On the other hand, that of the cost function $C(s)$ extends from 0 to infinity, theoretically at least. In order to deal with this disparity, we have to choose the values of the scaling factors K and M in such a way as to keep the terms $KP(s)$ and $MC(s)$ within compatible domains. This can be achieved by defining M as the reciprocal of the cost required to ensure a 0.95 probability of detection of a target with expected parameters L_t and R_t, and K as the reciprocal of 0.95, equal to 1.05. As a result, the efficiency function will start out in the negative region when s is small and reach a peak when s increases, heading again for negative values when the grid spacing becomes quite large, as illustrated by the graphs shown in Figure 4.4. If we wish to optimize the grid orientation or drilling angle, the independent variate is an angular interval instead of the spacing s.

The actual determination of optimal survey parameters, including grid spacing and orientation and detector orientation, results from the sequential application of either the indirect calculus-based or the direct search approach within the framework of dynamic programming. The two main types of surveys have to be handled separately because of the fundamental differences between con-

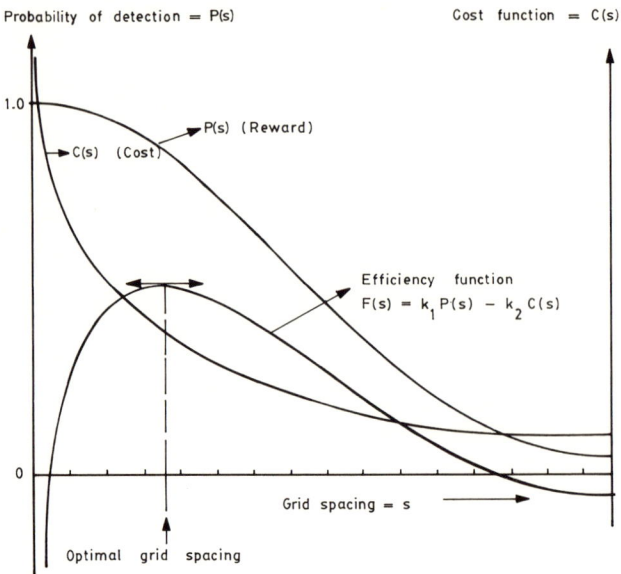

FIGURE 4.4
Determination of Optimal Grid Size Based on the Efficiency Criterion.

tinuous sampling (airborne surveys), on the one hand, and discrete sampling (ground surveys), on the other hand.

4.4.2.2. Optimization of Airborne Geophysical Surveys. Two main types of grids are used as control for airborne geophysical surveys for the purpose of detecting elliptical targets with long axis L_t and shape ratio $R_t = B_t/L_t$. The grids are of the parallel design with spacing s or of the orthogonal design with spacings s and t, t being greater than or equal to s.

If we use the parallel design, the probabilities of detection are initially maximized by orientating the grid at right angles with the expected direction of the long axis of the target (See Chapter 2, Section 2.4.4). The variate $U_t = L_t/s$ describes the relative grid–target geometry. As noted in Chapter 2, Section 2.4.3, the probability of detection (reward function) shows a discontinuity for $U_t = 1$, i.e., when the grid spacing equals the long axis of the target. When U_t lies in the 0–1 interval, the structure of the reward function is rather simple (See Table 2.4, Chapter 2). However, when U_t is greater than 1, the makeup of the function becomes more complex, involving radicals and inverse trigonometric terms which would unduly complicate the differentiation and the finding of extremal roots, if the indirect calculus approach is envisaged. Furthermore, the discontinuity for $U_t = 1$ mentioned above introduces an artificial maximum without practical value. Under these circumstances, the direct search approach is obviously a more rewarding avenue for the optimization of airborne surveys.

The computer is programmed to calculate the numerical value of the efficiency function in terms of the target parameters L_t and R_t, with the grid spacing s as variate. An iteration algorithm guides the search for the value of s which maximizes the efficiency function when s varies. The output of the program includes (a) the optimal value of the spacing s, (b) the corresponding coverage cost per unit of area, (c) the associated optimal detection probability, and finally, (d) a printed graph of the efficiency function.

If we consider the square design s by s, the structure of the efficiency function is a rather simple quadratic function of $1/s$, which may be easily differentiated, leading to the extraction of the maximal root by the calculus approach. When dealing with rectangular grids with spacing s and shape parameter $w = s/t$, we find from Chapter 2, Section 2.4.3.4 that the probability of detection may be calculated as a weighted average of the probabilities associated with the parallel and square designs, as follows:

$$\text{Probability (rectangular grid)} = w \cdot \text{Probability (square grid)} + (1 - w) \text{Probability (parallel grid)}$$

As the cost function may also be expressed in terms of s and w in a similar manner (See Chapter 3, Section 3.3.3), it is possible to calculate the optimal grid size parameters for the airborne detection of a target of specified parameters L_t and R_t.

4.4.2.3. Optimization of Ground Surveys and Drilling Programs. The probabilities of detection of an elliptical target with parameters L_t and R_t by discrete sampling on a square grid of size s by s are calculated in terms of L_t, R_t, and s by means of the OPTGRID program designed by the second writer (see Chapter 2, Section 2.5.3). The structure of the expression of the probability of detection (reward function) is a very complex combination of transcendental functions including radicals and inverse trigonometric terms. Not only is the expression difficult and cumbersome to differentiate in terms of s, but the search for extremal roots proves quite intractable, so that the indirect calculus approach has to be rejected in favor of the direct search approach, as was the case for the optimization of parallel grids for airborne surveys.

We are faced with four considerations: two of them involve the target orientation. These are random orientation and expected orientation within a confidence interval of ±10 degrees. The other two involve the relative configuration of target and detector. These are vertical detection by geophysical surveys and drilling, and angled detection by drilling.

The simplest case is that of vertical detection of a randomly oriented elliptical target L_t, R_t with a mean dip of a degrees. The optimization of the grid is a one-stage computerized procedure based on the iterative algorithm mentioned above. The output consists of the optimal spacing, corresponding coverage cost per unit of area, probability of optimal detection, and a graph printout of the efficiency function.

The most complex case involves the detection by drilling to a specified vertical depth at an angle of b degrees searching for an elliptical target with dimensional parameters L_t and R_t, mean dip a degrees, and with a preferred orientation within a stated confidence interval. The procedure is completed sequentially by the dynamic programming approach, as follows: (1) optimization of the orientation of the grid with respect to the expected target direction, (2) optimization of the spacing of the grid which has been previously optimally oriented, and (3) optimization of the drilling angle within the acceptable range of 35–90 degrees, when the holes are optimally spaced along a grid which has been, in turn, optimally oriented in stage (1).

We are now ready to consider the application of the methodology described above in Chapters 2–4 to the optimization of the search for six types of ore deposits in three regions of the North American continent, which will be covered in the following Chapters 5–11.

REFERENCES AND SELECTED READINGS

Methodology of Optimization

1. BELLMAN, R.E., and DREYFUS, S. E., 1962, *Applied Dynamic Programming,* Princeton University Press, Princeton, New Jersey.
2. BROOKS, S. H., 1959, Comparison of maximum-seeking methods, *Oper. Res.* **7,** 430–457.
3. DANTZIG, G., 1963, *Linear Programming and Extensions,* Princeton University Press, Princeton, New Jersey.
4. DIXON, L. C. W., 1972, *Non-linear Optimization,* Chap. 1, The English University Press, London.
5. DIXON, L. C. W. (Ed.), 1976, *Optimization in Action,* Proceedings of the Optimization Conference, Univ. of Bristol, January 1975, Chap. 1, Academic, London.
6. HARBAUGH, S. W., and BONHAM-CARTER, G. R., 1970, *Computer Simulation in Geology,* Chap. 8, Wiley, New York.

7. PETERSON, E. L., 1961, *Statistical Analysis and Optimization of Systems,* Wiley, New York.
8. ROSENBROCK, H. H., 1960, An automatic method for finding the greatest or least value of a function, *Comput. J.* **3,** 175–184.
9. SIVAZLIAN, B. D., and STANFEL, L. E., 1975, *Optimization Techniques in Operations Research,* Prentice-Hall, Englewood Cliffs, New Jersey.
10. WILDE, D. J., 1964, *Optimum-Seeking Methods,* Prentice-Hall, Englewood Cliffs, New Jersey.
11. WILDE, D. J., and BEIGHTLER, C. S., 1967, *Foundations of Optimization,* Prentice-Hall, Englewood Cliffs, New Jersey.
12. WILLIAMS, K. B., and HALEY, K. B., 1959, A practical application of Linear Programming in the mining industry, *Oper. Res. Q.* **10,** 131–137.

Operations Research

13. COBB, H., 1960, Operations Research: A tool in oil exploration, *Geophysics* **25,** 1009–1022.
14. COX, J. L., 1961, Cutting costs through operations research, *Min. Congr. J.* **47,** 45–46.
15. COYLE, R. G., 1969, Review of the literature on operations research in the mining industry, *Trans. Inst. Min. Metall. London* **78,** A1–A19.
16. DUCKWORTH, E., 1962, *A Guide to Operations Research,* Methuen, London.
17. HAZEN, S. W., Jr., 1968, Operations Research: A growing force in the mineral industries, *Min. Eng.* **20,** 88–90.
18. PRUSS, D. E., and FREEMAN, G. W., 1961, Mining exploration: An operations research and simulation approach, *Min. World* **14,** 42–43.
19. ROUBENS, M. (Ed.), 1977, *Advances in Operations Research,* North-Holland, Amsterdam.

Optimization of Ore Detection

20. BARNES, H. L., and LAVERY, N. G., 1977, Use of primary dispersion for exploration of Mississippi Valley-type deposits; *J. Geochem. Expl.* **8,** 105–115.
21. CELASUN, M., 1964, The allocation of funds to reconnaissance drilling programs; *Q. Col. Sch. Mines* **59,** 169–185.
22. DE GEOFFROY, J., and WU, S. M., 1970, Design of a sampling plan for regional geochemical surveys, *Econ. Geol.* **65,** 340–347.
23. DREW, L. J., 1967, Grid drilling exploration and its application to the search for petroleum, *Econ. Geol.* **62,** 698–710.
24. DREW, L. J., 1979, Pattern drilling exploration: Optimum pattern types and spacings when searching for elliptical targets, *J. Math. Geol.* **11,**(2), 223–254.
25. ELLIS, R. M., and BLACKWELL, J. H., 1959, Optimum prospecting plans in mineral exploration, *Geophysics* **24**(2), 344–358.
26. FAVINI, G., and ASSAD, R., 1979, An optimized decision model for area selection in massive sulfide exploration, *Can. Inst. Min. Metall. Bull.* **72**(804) 118–126.
27. GRIFFITHS, J. C., and DREW, L. J., 1966, Grid spacing and success rations in exploration, A.P.C.O.M. Symposium No 6, Vol. 1, pp. Q1–Q24.
28. HEWLETT, R. F., 1965, Design of drill-hole grid spacings for evaluating low-grade copper deposits, *U.S. Bur. Mines Rep. Invest.* No. 6634.
29. HODGSON, W. A., 1962, Optimum spacing for soil sample traverses, A.P.C.O.M. 10th Symposium, Johannesburg, South Africa, pp. 75–78.

30. KELLEY, J. C., and McMANUS, D. A., 1969, Optimizing sediment sampling plans, *Marine Geol.* **7,** 465–471.
31. SHURYGIN, A. M., 1976, The probability of finding deposits and some optimal search grids, *J. Math. Geol.* **8**(3), 323–330.
32. SLICHTER, L. B., 1955, Geophysics applied to prospecting for ore, *Econ. Geol. Jubilee Volume* **50,** 885–969.

CHAPTER FIVE

APPLICATION OF THE OPTIMIZATION METHODOLOGY TO THE SEARCH FOR SIX TYPES OF ORE DEPOSITS IN NORTH AMERICA

5.1. GENERAL STATEMENT

The four previous chapters described the methodology of optimization of ore detection without reference to any specific type of ore deposit or geological region of occurrence. The six following chapters will illustrate the application of the methodology to the search for six genetic types of ore deposits in three geological regions of North America. The necessary data were obtained from private sources and available publications and were assembled in a computerized data bank.

For the sake of practicality, each of the six following chapters is designed to be self-contained in order to avoid constant references to other chapters, despite the risk of being repetitious. The standard sequence for each chapter comprises five sections:

   (i) General geological background;
   (ii) Synopsis of field detection methodology;
   (iii) Statistical modeling of the geometric parameters required for the calculation of detection probabilities;
   (iv) Construction of detection probability tables;
   (v) Design of three detection strategies, including the optimization of search grids.

5.2. CONSTRUCTION OF A DATABASE

The initial stages of the project are (a) selection of ore deposit types and regions to be covered, (b) acquisition of the necessary data, and (c) preparation of the data for computerized storage and statistical processing.

Various interrelated considerations are involved in the construction of a database. A principal consideration is the availability of data. Because the main emphasis of the book is on deposit detection, the main endeavor was to secure the required geometric data, which are usually much more difficult to obtain than the economic ones. Fortunately, we were able to secure about 50% of the required data from private sources. The second half was culled from many of the references listed at the end of the chapter, and from numerous issues of various North American journals.

Another important consideration deals with practical computational and statistical requirements, such as the availability of space in the computer memory, allocation of computer time for computation and outputting, type of output, statistical quality of samples, etc.

A summary of the database is presented in Table 5.1. The study covers a total of 392 commercial and subeconomic ore deposits which belong to six main genetic types and four subtypes in three regions of North America. Five of the

TABLE 5.1.

Database for the Optimal Detection of Six Types of North American Ore Deposits.

Region	Type of deposit	Number of deposits
Cordillera Belt	Porphyry-Cu-Mo	57
	Contact-metasomatic	113
	Volcanogenic sulfides	29
	Sub-total:	199
Precambrian Shield	Volcanogenic sulfides	102
	Ni-Cu-ultramafic	25
	Vein-gold	52
	Sub-total:	179
Arctic Paleozoic Platform	Mississippi Valley Pb-Zn	14
	Grand total:	392

deposit types, namely, porphyry–Cu–Mo, contact metasomatic, Ni–Cu ultramafic, volcanogenic massive sulfides, and Mississippi Valley-type Pb–Zn are important sources of base metals, while the sixth is a main source of precious metals. The contact metasomatic type is divided into four subtypes and the volcanogenic type covers two regions; otherwise, the study deals with one type of deposit and one region at a time.

The three geological regions covered by the study are the North American Precambrian Shield, the North American Cordillera Belt, and the North American Paleozoic Platform. The fourth main geological unit of North America, the Appalachian Foldbelt, was not included because adequate geometric data could not be obtained for a sufficient number of its volcanogenic sulfide deposits.

The geometric parameters measured on all 392 deposits of the database are those required for the calculation of detection probabilities. The parameters belong to two types: dimensional and attitudinal. The first category includes the length and breadth of the horizontal section of the portion of ore deposits which lies within the average range of present day detectors (300 feet), and the shape ratio R = breadth/length. The second category includes the unorientated true dip of the deposits and the strike direction measured in degrees from true North in the right half circle. All five parameters were available for four types of deposits, but strike direction data were inadequate for the porphyry–Cu–Mo and Mississippi Valley types which were consequently assumed to be randomly orientated. The total number of numerical data bits covering the 392 deposits which were stored in the computerized data bank amounts to 2025.

5.3. DATA PROCESSING

5.3.1. Introduction

The statistical processing of the data was carried out in three steps: (1) preliminary processing, (2) construction of detection probability tables, and (3) design of optimal field surveys for the detection of specific types of ore deposits.

Two types of computers and two types of word processors available at the New South Wales Institute of Technology, Sydney, Australia, were used by the second writer to store, summarize, and process the 2025 data bits. They include a main frame Honeywell computer, level 66/60, for storage and computations, a Hewlett-Packard HP9816 computer and HP7580 plotter, an APPLE II for simpler computations and table printing, and a LISA for composition and printing of more complex synoptic tables.

5.3.2. Preliminary Processing

5.3.2.1. Sampling. The foundation of statistical processing lies in the requirement of adequate sampling of a homogeneous population of objects, in the present case all ore deposits of a specific genetic type, whether discovered or undiscovered within a specific region. Two aspects of sample specification, namely nature and size, are of crucial importance in judging the validity of the conclusions drawn from the samples.[11]

According to statistical theory, the samples should be statistically representative of the whole population and be randomly selected. These requirements create difficulties when dealing with populations of ore deposits, because the definition of ore (material which can be mined at a profit) immediately introduces a bias in the sampling. A further bias of a geographic nature has to be taken into account, because some portions of regions are better documented than others. Furthermore, restrictions in access to data can be expected to result in additional bias.

The sample size requirement is also quite important. In most practical cases, we find that a sample size of 50 is more than satisfactory to meet the most exacting requirements of the theory of Statistics. Such is the case in the present study for the porphyry–Cu–Mo deposits of the Cordillera, volcanogenic sulfide deposits, and vein gold deposits of the Shield. Satisfactory sample sizes ranging from 25 to 50 were obtained for the Ni–Cu ultramafic deposits of the Shield, volcanogenic sulfide deposits of the Cordillera, pyritic halos associated with porphyry–Cu–Mo of the Cordillera, and contact metasomatic deposits of the Pb–Zn–Cu–Ag and W–Mo types. A smaller but still adequate sample, in the 15–25 range, was collected for the contact metasomatic Cu–Fe type. A less than satisfactory sample, in the 10–15 range, had be be accepted for the Arctic Mississippi Valley-type of Pb–Zn deposits. Any sample size below 10 would not be acceptable for standard parametric statistical processing: one should turn to nonparametric statistical methods which are specially devised to handle small samples.

5.3.2.2. Statistical Summarization of Data. Once a population of ore deposits of a specific type has been sampled, however imperfectly, the range of measurements of each parameter is calculated (largest–smallest), and the measurements are grouped into classes and counted as frequencies. The frequency distributions are summarized graphically by means of histograms and numerically by means of statistics which represent the concepts of centrality, dispersion, and shape of distributions.

The concept of centrality is expressed numerically by the arithmetic mean (average) which is particularly suitable for the normal and other types of symmetrical distributions. The geometric mean is more suitable when dealing with asymmetric distributions such as the log-normal type. If the data are ordered by magnitude, another statistic called median may be used. The median divides the area under the histogram into two equal portions and is referred to as the 50th percentile of the distribution. By extending this idea, we can think of other statistics which divide the area under the histogram into four equal portions and are called "quartiles." The first quartile or 25th percentile corresponds to 25% of the area, the second quartile or 50th percentile is the median, and the third quartile covers 75% of the area. Finally, the mode is another expression of centrality which corresponds to the highest frequency, that is the peak of the histogram. The mode may not be unique, as is the case in multimodal distributions.

The concept of dispersion or spread of the data about the mean has to be introduced to describe the variability of the data. The range, already mentioned above, is a very simple expression of the data spread. A more sophisticated one is the variance or its square root, the standard deviation. In many instances, it is more convenient to express the dispersion by a dimensionless coefficient such as the ratio standard deviation/mean.

Finally, the concept of shape has to be considered in order to describe fully the nature of frequency distributions. Mean, mode, and median coincide in symmetrical distributions such as the normal model. Most distributions are asymmetrical, being skewed either on the left or the right side. The coefficient of skewness expressed as the ratio (mean − mode)/standard deviation is a very convenient way of describing numerically the concept of shape in a dimensionless manner. It can be seen that the skewness is equal to zero when the distributions are symmetrical since mean and mode coincide. Mathematical transformations of the observed data may be used for the "deskewing" of observed distributions into more symmetrical and more tractable ones. In the common case of lognormal distributions of observed data, a logarithmic transformation will result in a normal distribution for the logarithms of the data.

5.3.2.3. Model Fitting. The fitting of various theoretical statistical models to the frequency distributions of observed data is an important next step in the statistical processing. Among the many advantages derived from the fitting of normal and other associated models, the computation of confidence intervals for the mean of the fitted distribution counts as a major one. The limits of the fiducial interval are a function of standard deviation, sample size, and a Student-t statistic provided by tables: the larger the sample size and the smaller the standard deviation, the narrower the fiducial interval of the mean. Most commonly, 95%

confidence intervals are used to indicate that we are confident that the true mean of the population will fall within the interval in 95% of the cases, the probability of error, that is of the mean falling outside either limit, being 2.5%.

A "goodness of fit" test is used to test the degree of fit between the theoretical model and the actual distribution of observed data at a specified level of confidence, generally 0.05. The chi-squared test of a nonparametric nature, which is most commonly used, is easily calculated in terms of the difference between observed and expected frequencies for each class. If the calculated expression is less than the critical value provided by tables for the required level of confidence and degree of freedom, then we can accept the model as a good fit at the 0.05 confidence level. In the present study, lognormal models were successfully fitted to the distributions of all dimensional parameters of the six types of deposits. As far as the attitudinal parameters are concerned, normal and circular normal models were found to be the best fits for the distributions of dip angles and strike directions, respectively, for all but the contact metasomatic types of ore deposits.

5.4. COMPUTATION OF DETECTION PROBABILITIES

5.4.1. Summary

As indicated in Sections 2.2.3 and 2.2.4 of Chapter 2, the computation of detection probabilities for the six types of ore deposits covered by the present study involves the following geometric considerations:

(i) Intrinsic Target Geometry. As described by the dimensional parameters of the horizontal sections of the portions of deposits within a 300-foot detection range, the shape of the horizontal section of the six types of deposits is generally elliptical, varying from nearly linear (vein gold deposits) to subcircular (pyritic halos of porphyry deposits).

(ii) Relative Configuration of Detector and Target. The deposits covered by the study belong to several categories on the basis of the dip parameter. Two types of deposits including porphyry–Cu–Mo and Mississippi Valley Pb–Zn are either subhorizontal or subvertical so that the target dimensions for detection purposes are the same as that of the deposits. The other four types are dipping deposits for which the target breadth is obtained by adding a dip component to the breadth of the horizontal section of the deposit, as illustrated by Figures 2.7 and 2.8 of Chapter 2, according to the following expressions:

$$B_t = B_h + (D - H) \cdot \cot a \quad \text{for vertical detection}$$

and

$$B_t = B_h + (D - H)[(\cot a + \cot b)] \quad \text{for angled detection}$$

where B_t and B_h are the target and deposit horizontal breadths, D is the detection range taken as 300 feet, and H is the overburden thickness which is taken as 30 feet based on an average of 40 measurements in the Cordillera and 50 measurements in the Shield. Initially, we use 55 degrees as an "all-purpose" drilling angle. Later, as a part of an optimization study by dynamic programming, the drilling angle is optimized under cost constraints.

(iii) Grid Geometry and Relative Configuration of Grid and Target. Parallel and square grid designs are used for the airborne detection of all types of deposits, except vein gold and contact metasomatic (W–Mo), which are not detectable by airborne surveys. Square grids are used for the detection of all types of deposits by ground surveys and drilling programs. The relative configuration of grid and target involves a dimensional element, the ratio U_t = target length/grid spacing, and a directional element which is the orientation of the grid with respect to the expected strike direction of the target. Optimization studies show that the detection probabilities for airborne surveys on parallel grids are maximized when the grid is laid at 90 degrees to the expected target strike. The optimal ground grid orientation with respect to the expected strike direction lies in the 20–45 degree range, depending on the shape ratio of deposits. No optimal grid orientation is considered for the porphyry–Cu–Mo deposits and their associated pyritic halos and for the Mississippi Valley Pb–Zn deposits, which are assumed to be randomly orientated owing to the lack of adequate information.

5.4.2. Construction of Detection Probability Tables

5.4.2.1. Description of Tables. The probability tables dealing with the detection of six types of ore deposits in North America are of two designs. The first design provides the probabilities of single detection (at least one target intersection by the grid) and of confirmed detection (at least two intersections) for varying grid spacings. The second design, which should prove very useful for planning purposes, provides the grid spacings and associated coverage cost per unit of area required to obtain prespecified levels of probability of detection by various types of field programs.

Table 5.2 illustrates the first design, taking the detection of porphyry–Cu–Mo deposits of the Cordillera by ground geophysical surveys or vertical drilling as an example. The head of the table displays all pertinent infor-

TABLE 5.2.

Example of Probability Table for the Detection of Porphyry–Cu–Mo Deposits of the North American Cordillera Belt by Ground Geophysical Surveys or Vertical Drilling to a Depth of 300 Feet.

Survey design: square grid with spacings S by S feet. the expected shape of the ore deposits is elliptical with major axis = L in the 95% confidence interval: L.c.l. (geometric mean (U.c.l., and minor axis = B feet, R is defined as the ratio B/L with geometric mean = 0.46.

	Spacing S in feet	Probability of detection		
		L.c.l. L = 2820	G. mean 3110	U.c.l. 3820 ft
	800	1.000	1.000	1.000
	1000	0.986	1.000	1.000
mx	1200	0.914	0.991	1.000
	1400	0.788	0.942	0.999
	1600	0.660	0.852	0.973
	1800	0.547	0.745	0.925
	2000	0.455	0.643	0.847
lm	2200	0.383	0.552	0.761
	2400	0.324	0.475	0.677
	2600	0.277	0.411	0.600
	2800	0.239	0.358	0.530
	3000	0.208	0.313	0.470
	3200	0.183	0.276	0.419
	4000	0.117	0.176	0.273
	4800	0.081	0.123	0.189
	5600	0.060	0.090	0.139
	6000	0.052	0.078	0.121

N.B. Mx: Grid spacing range for 0.980 detection probability.
Lm: Grid spacing range for 0.500 detection probability.

mation required for the calculation of detection probabilities, including (a) grid design, (b) target mean shape ratio and mean length within its 95% confidence interval, as derived from model fitting and, (c) optimal grid orientation with respect to target strike within a ±10 degree confidence interval. The latter statement is omitted in Table 5.2, because porphyry–Cu–Mo deposits are considered as randomly orientated. The body of the table displays the numerical values of detection probabilities for grid spacings varied within a realistic range of 800–6000 feet in steps of 200–800 feet. For each spacing, the table provides

TABLE 5.3.

Example of Table Displaying Grid Sizes and Associated Coverage Cost per Mile Square for Specific Probability Levels in the Detection of Ni–Cu Ultramafic Deposits of the North American Shield by Vertical Drilling to a Depth of 300 Feet.

Survey Design: Square Grid With Spacings S by S Feet
Unit Cost = $6000 Per Diamond Drill-Hole

Probability Level Of Detection	Grid Size In Feet	Drilling Cost Per Mile Square In U.S.$. Thousands
0.05	1850	89.58
0.15	1150	188.25
0.25	900	283.73
0.35	750	388.81
0.45	700	438.91
0.55	600	577.42
0.65	550	675.43
0.75	500	803.19
0.85	450	974.35
0.95	300	2077.99

the 95% confidence interval for the detection probability based on the 95% fiducial interval for the target length.

Table 5.3 displays an example of the second design of probability tables which covers the detection of Ni–Cu ultramafic deposits of the Shield by vertical drilling to a depth of 300 feet. Since this type of table is always inserted between those of the first design, the target information is not repeated in the heading of the table. The specified levels of probability of detection are varied within a range of 0.05–0.95 with 0.05 increments.

5.4.2.2. Organization of Tables. Since in practical situations individual exploration programs deal with one specific type of deposit at a time, all detection probability tables are organized in a standard sequence which is repeated for each of the six types of deposits, resulting in a total of 118 tables interspersed within the next six chapters. The standard sequence is divided into three sections of three or four tables each. The first section covers airborne geophysical detection and includes four tables. Since there are no known airborne geophysical methods to assist the detection of contact metasomatic (W–Mo) type or vein-gold, the first section is omitted in the sequence of tables displayed in Chapters 7 and 11.

The second section deals with vertical ground detection by geophysical surveys or drilling, and the third section covers angled detection by drilling.

5.5. DESIGN OF THREE STRATEGIES FOR THE DETECTION OF SIX TYPES OF ORE DEPOSITS IN NORTH AMERICA

5.5.1. Introduction

An inspection of Tables 5.2 and 5.3 immediately shows that the probability of detection rapidly decreases to a fraction of percentage as the grid spacing increases and corresponding coverage cost decreases. Conversely, as the grid spacing decreases and the corresponding coverage cost rises very rapidly, the probability of detection reaches it maximum value of unity. It seems intuitive that three strategies may be considered, including two options covering each end of the range, and a "middle of the road" option bracketed by the other two. They are referred to as the "liminal" strategy for the lower end of the range, "maximal" strategy for the upper end, and "optimal" strategy as the compromise option.

5.5.2. Liminal Strategy

At the lower end of the range, one may seek to minimize the coverage cost provided that the probability of detection exceeds a minimum acceptable level, referred to as "threshold" level—hence the term *"liminal"* derived from the latin "limen" for threshold. The 0.500 probability level is an obvious choice for the threshold: below this value, the probability of failure of detection which is complementary to the probability of detection would be greater than the probability of success, thus removing any incentive to carry out the survey.

This point is illustrated in Table 5.2. The three grid spacings corresponding to the liminal probability level of 0.500 for the mean target length and its 95% confidence limits are shown by short lines; they are, respectively, 1900, 2325, and 2900 feet. It can be seen from the table that 1900 feet is the grid spacing resulting in the cheapest coverage cost which provides a probability of detection of at least 0.500 in 95% of the cases. Any smaller spacing, say 1800 feet, will produce a detection probability greater than 0.500 for a higher coverage cost; any larger spacing, say 2200 feet, will result in a cheaper coverage cost but the probability of detection may fall below the required threshold of 0.500, if the true target length falls at the fiducial limits or outside the fiducial interval. The probability that the target length is smaller than the lower fiducial limit of 2820 feet resulting in a probability of detection lower than 0.500 is one half of 5% = 2.5%. The same reasoning may be applied to the confirmed detection situation, resulting in a liminal grid spacing which guarantees a 0.500 level of confirmed detection at least, with a probability of error of only 2.5%.

We feel that the liminal strategy should prove attractive to explorationists faced with the problem of covering large areas of little known potential within strict budget constraints.

5.5.3. Maximal Strategy

At the other end of the range, explorationists may be interested in the coverage of small areas of very high potential in well-known mining districts. They feel justified in requiring a very high level of probability of detection, as close as practically possible to certain detection, say 0.980, despite the much higher coverage cost. Since the probability of detection nears its *maximum* value of unity, the second strategy is referred to as "maximal."

Based on the same Table 5.2, it is found that the three grid spacings corresponding to the required probability level of 0.980 for the mean target length and its 95% fiducial limits are, respectively, 1000, 1300, and 1500 feet. The 1000-foot spacing corresponding to the lower fiducial limit for the target length is the smallest grid spacing, resulting in the cheapest coverage cost which can guarantee a detection probability of at least 0.980 in 95% of the cases. A cheaper coverage resulting from a larger grid spacing, say 1200 feet, may not produce a detection probability of at least 0.980 in 95% of the cases, because the true length of the target may fall near the lower end of the fiducial interval; as a result, the probability of detection would be 0.914, instead of the required 0.980 level.

5.5.4. Optimal Strategy

5.5.4.1. Methodology. The optimal strategy is a compromise option which should suit the majority of exploration programs requiring the coverage of moderately large areas of fair to good potential within strict budget constraints. The strategy calls for the maximization of the probability of detection subject to the constraint of cost minimization. The methodology of field program optimization is based on a sequential procedure comprising several stages, each one requiring the application of the efficiency criterion, as described in Chapter 4 (Dynamic Programming).

In the most general case of oriented dipping deposits, the procedure is broken into three steps. When dealing with randomly oriented deposits such as porphyry–Cu–Mo or associated pyritic halos or Mississippi Valley Pb–Zn deposits, the first step is omitted. The initial step is the optimization of the orientation of survey grids with respect to the expected strike direction of the target, which is estimated from previous surveys, photogeological studies, or the results of statistical modelling. As mentioned in Section 5.4.1, the optimal orientation of parallel airborne grids is 90 degrees; for ground surveys, it ranges between 18 and 45 degrees, depending on the type of deposit.

The second step is the optimization of grid spacings for optimally oriented grids. The optimization is carried out analytically for airborne surveys on square grids as described in Section 4.3.3 of Chapter 4. In the case of parallel grids for airborne surveys, the analytical approach runs into difficulties because of a discontinuity of the efficiency function when the grid spacing equals the target length. When considering ground surveys and drilling programs, the expression of the efficiency function is very complex and its derivation intractable, ruling out the analytical approach. The direct search approach has to be used in the two latter cases. (See Section 4.3.4 of Chapter 4.)

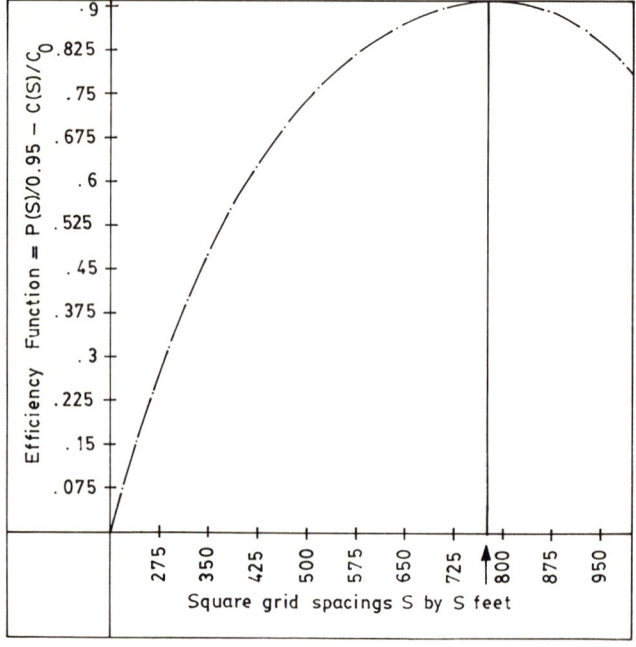

FIGURE 5.1.
Graph of Efficiency Function for the Detection of Volcanogenic Massive Sulfide Deposits of the North American Cordillera Belt by Optimally Angled (40 degrees) Drilling to a Vertical Depth of 300 Feet. The Optimal Grid Spacing is Indicated by an Arrow.

SEARCH FOR ORE DEPOSITS IN NORTH AMERICA

The third step is the optimization of the drilling angle which is omitted for subhorizontal deposits such as porphyry–Cu–Mo and Mississippi Valley Pb–Zn types. The optimization is based on the direct search approach using the efficiency criterion. This presupposes some knowledge of the likely direction of the dip, as may be obtained from previous surveys or photogeological studies, to avoid the risk of drilling down-dip. The graph shown in Figure 5.1 illustrates the three-stage optimization of drilling programs for the detection of volcanogenic massive sulfide deposits of the Cordillera.

5.5.4.2. Construction and Organization of Optimization Tables.
All tables are constructed on the basis of the statistical concept of 95% confidence interval for the mean target length, which is translated into 95% fiducial intervals for the parameters of optimized grids. The headings of tables provide additional pertinent information such as type of grid and unit costs. The parameters of optimized grids are (a) optimal grid spacing, corresponding coverage cost per unit of area, and associated optimal detection probability, and (b) optimal orientation of grid and optimal drilling angle when it applies. Altogether 30 such tables have been constructed to summarize the results of the optimization of field programs in the search for six types of deposits in North America. The appropriate optimization tables are inserted at the end of sections of detection probability tables for each type of deposit and each type of field program. For the reader's convenience, a total of 11 detection summary tables have been constructed to assemble the detection characteristics pertaining to the liminal, maximal, and optimal options.

REFERENCES AND SELECTED READINGS

1. ANNIS, R. C., CRANSTONE, D. A., and VALLEE, M., 1978, A survey of known mineral deposits in Canada that are not being mined, Mining Bulletin MR 181, Department of Energy, Mines, and Resources, Ottawa.
2. BATEMAN, A., 1942, *Economic Mineral Deposits*, Chap. 23, Wiley, New York.
3. BROBST, D. A., and, PRATT, W. P. (Ed.), United States mineral resources, *U.S. Geol. Surv. Prof. Pap.* No. 820.
4. *CANADIAN MINES HANDBOOK*, 1978–1979, The Northern Miner Press, Toronto.
5. CRANSTONE, D. A., and WHILLANS, R. T., 1979, Canadian reserves of copper, nickel, lead, zinc, molybdenum, silver, and gold as of January 1st, 1978, Mining Bulletin MR 185, Department of Energy, Mines, and Resources, Ottawa.
6. DERRY, D. R., 1980, *World Atlas of Geology and Mineral Deposits*, Mining Journal Books, London.
7. DOUGLAS, R. J. W., (Ed.), 1970, Geology and economic minerals of Canada, *Geol. Surv. Can. Econ. Geol. Rep.* No. 1, Ottawa.

8. GEOLOGICAL SERVICES, PROVINCE OF QUEBEC, 1967, Annotated bibliography on metallic mineralization of the region of Noranda, Mattagami, Val d'Or and Chibougamau, Special paper No. 2, Department of Resources, Quebec City.
9. GILBERT, C. (Ed.), 1957, Structural geology of Canadian ore deposits, Congress Volume, Canadian Institute of Mining and Minerals, Montreal.
10. GILL, J. E. (Ed.), 1972, Section 4, Mineral Deposits, 24th International Geology Congress, Ottawa, pp. 107–160.
11. GRIFFITHS, J. C., and ONDRICK, C. W., 1968, Sampling a geological population, Kansas Geological Survey Computer Contribution No 30.
12. LANG, A. H., 1970, Base metals, Canadian Shield, *Geol. Surv. Can. Econ. Geol. Rep.* No. 1, Ottawa.
13. LAZNICKA, P., 1972, The University of Manitoba File of World's non-ferrous metal deposits (MANIFILE), University of Manitoba, Department of Earth Science, Winnipeg, Manitoba.
14. LINDGREN, W., 1933, *Mineral Deposits*, McGraw-Hill, New York.
15. MARTIN, H. L., CRANSTONE, D. A., and ZWARTENDYK, J., 1976, Metal mining in Canada, 1976–2000, Mining Bulletin MR 167, Department of Energy, Mines, and Resources, Ottawa.
16. NIGGLI, P., 1954, *Rocks and Mineral Deposits*, W. H. Freeman, San Francisco.
17. NOBLE, J. A., 1955, The classification of ore deposits, *Econ. Geol. Jubilee Volume* **50**, 155–169.
18. PARK, C. F., and MACDIARMID, R. A., 1975, *Ore Deposits*, W. H. Freeman, San Francisco.
19. RIDGE, J. D. (Ed.), 1968, Ore deposits of the United States, 1935–1967, A.I.M.E., New York.
20. RIDGE, J. D., 1972, Annotated bibliographies on mineral deposits in the Western Hemisphere, *Geol. Soc. Am. Mem.* 131.
21. ROSS, C. S., 1935, Copper deposits of the Eastern United States; Copper Resources of the World; 16th International Geological Congress Vol. 1, pp. 158–162.
22. STOCKWELL, C. H., 1957, Geology and economic minerals of Canada, Geological Survey of Canada, *Econ. Geol. Ser.* No. 1.
23. THOMPSON, J. E., 1957, Metal Resources Circular No. 2; Ontario Department of Mines, Toronto.
24. WILSON, H. D. B., and LAZNICKA, P., 1972, Copper belts, lead belts and copper-lead lines of the World, 24th International Geological Congress, Montreal.
25. WOLF, K. H. (Ed.), 1976, *Handbook of Stratabound and Stratiform Ore Deposits*, Elsevier, Amsterdam.

CHAPTER SIX

DESIGNING OPTIMIZED FIELD PROGRAMS FOR THE DETECTION OF PORPHYRY–Cu–Mo DEPOSITS OF THE NORTH AMERICAN CORDILLERA BELT

6.1. GENERAL GEOLOGICAL BACKGROUND

6.1.1. Main Geological Features

The term "porphyry–copper–molybdenum" deposit was first coined some 70 years ago to describe a type of base metal occurrence in the North American Cordillera Belt. Since that time, geologists have applied the term to a broad variety of disseminated sulfide deposits associated with differentiated plutons that occur within well-defined orogenic belts. The deposits show many gradations between the "porphyry–copper," "molybdenum–stockwork" and "contact metamorphic" types. The latter is described in more detail in Section 7.5 of Chapter 7.

Intrusive rocks hosting the porphyry–copper–molybdenum deposits vary in composition from felsic to intermediate. Although the quartz monzonite type is the most common, there are many variations extending from syenite to diorite. The former class is known in the British Columbia section of the North American Cordillera, while the latter is found outside North America, mainly in the Southwest Pacific Island Arcs. A conspicuous development of porphyritic phases and breccia pipes is a common feature of plutons hosting copper–molybdenum deposits.

Lowell and Guilbert[18] describe the development of a roughly concentric pattern of hydrothermal alteration zones as another outstanding feature of the environment of porphyry–Cu–Mo deposits. Variations of the zoning pattern reflect mineralogic, lithologic, and structural features of the pre-ore environment, as well as the effect of post-ore geological events such as faulting and erosion.[10] In some orogenic belts such as the North American Cordillera, the zoning pattern is much better developed than in other regions and greatly assists field exploration.

6.1.2. Geographic and Chronologic Distribution

Since their first recognition in North America, porphyry–Cu–Mo deposits have been discovered along a limited number of orogenic belts of diverse ages. This peculiar geographic distribution has been recently explained in the context of global tectonics which emphasizes the role of zones of subduction along the edges of tectonic plates as a leading control for the occurrence of the deposits.[23] The model appears to explain rather well the distribution of deposits of the northern section of the Cordillera Belt, which extends over a length of some 2200 miles and width of 300 to 400 miles through British Columbia, the Yukon, and southern Alaska, along the structural zone separating the Pacific plate from the American plate. But it fails to account satisfactorily for the distribution of porphyry–Cu–Mo deposits within the U.S. section of the Cordillera Belt, which are scattered over a width of more than 1000 miles through the States of Montana, Utah, Colorado, New Mexico, and Arizona.

Recent measurements of absolute age by the potassium–argon dating method have shown that the age span of the porphyry–Cu–Mo deposits of the world extends over almost 400 million years, from the early Paleozoic times to late Miocene. In the Cordillera Belt of the Americas, the South American deposits are the more recent ones, with a modal age of 40–50 millions years. The North American age distribution is bimodal, with an early peak in the Permo–Triassic times, mainly in British Columbia, and a later peak in Cretaceous–Eocene times, chiefly in the U.S. section of the Cordillera.

6.1.3. Metallogenic and Economic Features

Two main types of porphyry deposits are covered in the present chapter: porphyry–copper and molybdenum stockwork. The first type may be subdivided into two categories: copper–molybdenum and copper–gold. The latter is characterized by unusually high ratios of precious metals/copper, generally greater than 0.01 oz of gold and silver per percentage point of copper. A frequent feature of the copper–gold category is an unusual abundance of magnetite, which assists the geophysical detection of the deposits from the air and on the ground.[25]

Both categories of porphyry deposits commonly exhibit a crudely concentric pattern of distribution of major and minor metals which is superimposed on the hydrothermal zoning pattern already mentioned. As many as four, but generally only three zones may be identified: they are termed as core, intermediate, and outer zones. Generally copper, molybdenum, gold, and silver are associated in the core zone, while zinc and lead appear in the intermediate zone. The outer zone features an association of lead, zinc, silver, and manganese.

The overall tonnage of porphyry deposits being presently mined in the North

American Cordillera Belt varies from a few million tons of high-grade ore as in some British Columbia deposits to several hundred million tons of low-grade material as in many U.S. deposits. The tonnage of deposits is controlled by physical and chemical factors. The former include (a) the volume of the intrusive, (b) attitude and lithology of wallrock, and (c) spacing and size of cracks shattering the host rock. The latter reflect the nature of the mineralizing solutions and that of the metal precipitating environment. Post-ore geological events such as uplifting and subsequent erosion may result in the removal of substantial tonnages of ore, while downfaulting may preserve large tonnages and promote upgrading through supergene alteration.

Current grades of porphyry–copper–molybdenum ore mined in the North American Cordillera Belt vary from 0.5% to 1.5% copper, accompanied by 0.02% to 0.06% molybdenite, with gold generally under 0.05 oz/ton and silver under 0.1 oz/ton. Conversely, in molybdenum–stockwork deposits, the molybdenite grade ranges from 0.1% to 0.3%, while the copper grade varies in the 0.15% to 0.25% only. Secondary enrichment of copper and gold in the upper portions of deposits is of little importance in the northern portion of the Cordillera Belt because of unfavorable climatic conditions. But in the southwestern U.S. section of the Cordillera, the extent and intensity of supergene enrichment have influenced production decisions in many cases, particularly in earlier days.

The increasing importance of porphyry-type base metal deposits as world sources of copper, molybdenum, and valuable minor metals such as gold, silver, selenium, tellurium, and rhenium justifies the prominent place given to the porphyry-target in present-day exploration planning. The large size of the porphyry deposits, their shallow occurrence, favorable attitude and even metal distribution make it possible to use large-scale, low-cost surface production methods that are amenable to computerized mining and scheduling, thus offsetting the low grade of the ore. However, porphyry projects are susceptible to economic and financial risks because of their long preproduction time requirements, high capital costs, and low profit margins resulting in long payback periods.

6.2. FIELD DETECTION METHODOLOGY

6.2.1. Detection Approaches

As indicated in Section 2.1 of Chapter 2, there are two approaches to the detection of any type of mineral deposits: direct and indirect methodologies. The first one is based on geological recognition either by visual means from the ground (unaided) or from airborne platform (assisted), or by mechanical means (drilling) from the surface. The second one relies on the recognition of the geophysical signature of mineral deposits on the one hand, or on the geophysical or geochemical detection of halos associated with the mineral deposits on the other hand. Ideally, the two approaches are used together, either simultaneously or sequentially, in order to boost their effectiveness.

6.2.2. Direct Geological Approach

During the past two decades, explorationists have developed an increasing interest in the use of LANDSAT satellite imagery to assist the detection of outcropping or near-outcropping mineral deposits. The combination of syn-ore alteration of a pyrometasomatic or hydrothermal nature with post-ore weathering may produce intense color anomalies. Multiple Spectral Sensing (MSS) techniques can be used to detect the color anomalies from satellite or aircraft platforms.

Abrams[1] describes the typical "red thumb" signatures of known porphyry–Cu–Mo deposits in the southwest Cordillera of the United States. Among the most conspicuous are the Bisbee, San Manuel, and Pima–Mission deposits. The color anomalies, reaching one square mile in extent, consist of brownish and greenish discoloration accompanied by extensive iron and manganese staining. The intensity and extent of the anomalies depend on diverse factors, including (a) lithology, (b) abundance of sulfides, (c) intensity of pyrometasomatism, and (d) intensity of weathering.

Unfortunately, the field of application of the remote sensing method is severely restricted to regions with thin vegetation or alluvial cover, or both. Success can be expected in regions of arid climate, whether subtropical or arctic. The heavy vegetation and alluvial cover prevailing in tropical or temperate regions preclude the effective use of the method. Detailed ground mapping of lithologic and alteration features remains the most effective direct method in most regions.

6.2.3. Indirect Geophysical Approach

The indirect geophysical approach to the detection of porphyry–Cu–Mo deposits may be handled in two ways. The first one calls for the airborne or ground detection of the geophysical signature of the deposits themselves, followed by the drill testing of the anomaly. The second one relies on the detection of associated halos to be followed by the systematic drilling investigation of the halos in the search for the deposits. (See Chapter 2, Section 2.6.)

As many as one-half of the known porphyry–Cu–Mo deposits of the North American Cordillera Belt exhibit well-developed pyritic halos containing 5% or more pyrite by volume disseminated throughout the intrusive and wallrock. Magnetite-rich halos commonly surround copper-bearing intrusives, particularly the gold-rich type.[25] Both kinds of halos are good targets for airborne geophysical detection. A combination of electromagnetic technique in the time domain (INPUT) and magnetometric technique appears to be the most effective for airborne halo detection.

The ground detection of porphyry–Cu–Mo deposits themselves by geophysical techniques is a rather difficult problem. The geophysical environment is not very favorable because of the relatively low abundance of sulfides (5%–15% by volume) which occur in a discrete manner within large volumes of intensely fractured rocks. The standard techniques available at present are not always suitable to attack the problem.

The standard electromagnetic techniques are at a severe disadvantage, except for the detection of massive sulfide pods within intrusive aureolae. The self-potential technique, so attractive because of its simplicity, rapidity, and low coverage cost has proved rather disappointing owing to the great complexity of the electrochemical phenomena prevailing in the porphyry environment. Mixed success is the record of the gravimetric technique in the southwestern Cordillera of the United States. Gravity lows are recorded over the zones of greatest alteration in the intrusive bodies, while gravity highs occur in the vicinity of massive sulfide bodies within the intrusive aureolae.

In most circumstances, the winning ground geophysical approach requires the combination of the fast and low-cost magnetometric technique with the slower and costlier induced polarization method, which is noted for its ability to detect disseminated sulfide material. A recently developed hybrid technique, known as magnetic induced polarization (M.I.P.), is described by Seigel.[22] M.I.P. appears to provide a capability of detection of both disseminated and massive sulfides, while avoiding the cumbersome features of the induced polarization technique.

6.2.4. Indirect Geochemical Approach

A frequently used indirect approach relies on the detection of the geochemical halos associated with porphyry–Cu–Mo deposits which are of two kinds. One type, referred to as primary halos, is syn-ore and occurs as a three-dimensional envelope in the bedrock surrounding the deposits. The second type, known as secondary halos, occurs in the surficial weathered environment of deposits and does not necessarily surround the deposits. Detection of porphyry–Cu–Mo deposits through their primary halos has gained increasing attention in recent times.

The success of the method relying on the detection of primary halos depends on the presence of sufficient rock exposure or the availability of drill core to supplement poor outcropping. The technique is keyed on the distribution of minor metal associations which are used as proximal indicators. For example, molybdenum–rhenium or copper–selenium associations denote the core zone of the mineralized environment; the cadmium–gallium–germanium association with zinc characterizes the outer zone of deposits.

Secondary halos are the product of the weathering and transport of material from the mineral deposits themselves or from their primary halos. Sampling media are residual topsoil, stream silt, plant matter, or ground water. Sillitoe[25] reports on the merits of using gold as a tracer element in residual soils because gold abundance in soils appears to reflect closely that of the hypogene gold in the underlying protore.

6.3. STATISTICAL MODELING OF THE GEOMETRIC PARAMETERS OF PORPHYRY–Cu–Mo DEPOSITS AND THEIR ASSOCIATED PYRITIC HALOS

6.3.1. Scope of Study

A sample of 57 commercial and subeconomic porphyry–Cu–Mo and molybdenum–stockwork deposits was assembled to represent statistically the whole population of known and undiscovered porphyry-type deposits of the North American Cordillera Belt. The deposits are listed by names and geographic regions in Table 6.1. The geographic breakdown is as follows: 25 deposits of the sample occur in the U.S. section of the Cordillera Belt and 32 were found in the British Columbia–Yukon section of the belt. Pyritic halo data of a quality adequate for statistical processing were available for 43 of the 57 deposits included in the data base.

The three geometric parameters required for the calculation of detection probabilities and the design of optimized field programs were measured for each of the 57 deposits. They consist of the length, breadth, and shape ratio (breadth/length) of the horizontal section of the portion of deposits which lie within the 300-foot maximum range of the commonly used types of field detectors. The parameter measurements were grouped into frequency distributions, to which statistical models were fitted.

TABLE 6.1
List of Porphyry–Cu–Mo Deposits of the North American Cordillera Included in the Database.

British Columbia (Canada)		United States	
Red Mountain	Gibraltar	Castle Dome	Questa
Boss Mountain	Brenda	Copper City	Butte
Sileurian Chieft.	Hudson Bay Mtn.	Mineral Park	San Manuel
Alwin	Schaft Creek	Esperanza	Morenci
Granisle	Lornex	Yerrington	Twin Butte
Cariboo Bell	Endako	Glacier Peak	Urad-Henderson
Polyanna	Gem	Silver Bell	Santa Rita
Newman Island	Island Copper	Tyrone	Climax
Ingerbelle	Valley Copper	Bagdad	Bingham
Similkameen	Casino	Safford	
B.C. Molybdenum	Serb Creek	Mission	
Huckleberry	Salal	Miami-Inspir.	
Stikine	Galaxy	Bisbee	
Adanac	Sheba	Ely	
Bethlehem	Trojan	Ray	
Highmont	Lytton	Ajo	

6.3.2. Statistical Modeling

The frequency distributions of parameter measurements are summarized by means of four statistics which are listed in the lower half of Table 6.2 for the deposits themselves and of Table 6.3 for the associated pyritic halos. The statistics are as follows: the arithmetic mean of the observed data, mode, standard deviation, and coefficient of skewness. Their makeup and significance have been described previously in Section 5.3.2 of Chapter 5.

The asymmetry of the distributions of deposit parameters is indicated by the magnitude of the skewness coefficient. The skewness is rather slight for the length and shape ratio (less than 0.01), but is rather strong for the breadth (greater than unity). The skewness of the distributions of all three parameters for the pyritic halos is moderate (0.1–1.0 range).

TABLE 6.2

Summary of Statistical Modeling of Geometric Parameters of Porphyry–Cu–Mo Deposits of the North American Cordillera Belt (Sample Size: 57 Deposits).

Statistic of fitted model	Length of horiz. section in feet	Breadth of horiz. section in feet	Shape ratio R = B/L
25%ile	2800	1140	0.35
95%L.c.l.	2820	1550	0.38
G. mean	3110	1640	0.46
95%U.c.l.	3830	1950	0.55
75%ile	6000	5300	0.78
Disp. coef.	0.06	0.05	0.94
Statistic of observed data			
Arith. mean	3510	3240	0.52
Mode	3500	2000	0.50
Stand. dev.	320	260	0.25
Skewness	0.03	4.8	0.08

TABLE 6.3

Summary of Statistical Modeling of Geometric Parameters of Pyritic Halos Associated with Porphyry–Cu–Mo Deposits of the North American Cordillera Belt (Sample Size: 43 of the 57 Deposits Included in Table 6.1).

Statistic of fitted model	Length of horiz. section in feet	Breadth of horiz. section in feet	Shape ratio R = B/L
25%ile	4591	2449	0.50
95%L.c.l.	4870	2920	0.54
G. mean	6374	3829	0.62
95% U.c.l.	8344	5022	0.71
75%ile	10,564	5000	0.77
Disp. coef.	0.39	0.51	0.21
Statistic of observed data			
Arith. mean	7660	4416	0.65
Mode	5450	4000	0.60
Stand. dev.	4355	2500	0.19
Skewness	0.51	0.17	0.26

DETECTION OF PORPHYRY DEPOSITS 117

Lognormal models were successfully fitted to the observed distributions of all three parameters for the deposits and their halos, as indicated by the results of χ^2 goodness of fit tests run at the 0.05 confidence level (see Chapter 5, Section 5.3.2). The statistics describing the fitted models are listed in the upper half of Table 6.2 for the deposits and Table 6.3 for the halos. They include (a) geometric mean, (b) the upper and lower limits of the 95% confidence interval of the mean, (c) the 25th and the 75th percentiles, and (d) the coefficient of dispersion. The latter indicates the degree of variability of the data about the mean of the fitted data in a dimensionless manner. A small variability (less than 0.1) affects the length and breadth parameters of the deposits, but a larger variability, close to 1.0, is indicated for the shape ratio. The variability of the three parameters is moderate (0.1–1.0 range) for the pyritic halos.

6.4. CONSTRUCTION AND ORGANIZATION OF DETECTION PROBABILITY TABLES

6.4.1. Introduction

For detection purposes, porphyry–Cu–Mo deposits are described as either tabular subhorizontal or pipelike subvertical bodies with horizontal cross sections of elliptical shape. The calculations of detection probabilities therefore apply to plane elliptical targets whose dimensions are those of the horizontal section of the deposits without the addition of dip components (see Section 2.2.3 of Chapter 2). Because of the lack of accurate orientation data, both porphyry deposits and their associated halos are considered as randomly oriented plane targets.

6.4.2. Description of Tables

There are three kinds of tables for both porphyry deposits and their associated halos. The first two cover the cases of single detection (at least one target intersection by the detector) and confirmed detection (at least two intersections) and are constructed in a similar manner. The third kind of table displays grid spacings and corresponding coverage costs per unit of area for specified levels of detection probability.

The heads of the first two types of tables display the statistics of the fitted models as obtained from Tables 6.2 and 6.3, including the mean length and its 95% confidence limits and the mean shape ratio. The probabilities of detection or confirmed detection are calculated for grid spacings increased in steps of 50–250 feet, as required by the type of field program under consideration, first for the mean length, and then for the upper and lower limits of the 95% confidence interval of the mean. As a result, the tables display the probability of detection or confirmed detection within its 95% confidence interval for each grid spacing.

6.4.3. Organization of Tables

Since most readers will have to consider only one specific type of field program at one time, the tables have been grouped into blocks, each covering one kind of program. The first block covers airborne geophysical surveys for the detection of porphyry deposits based on parallel and square grids (Tables 6.6–6.9). The second block deals with the detection of halos by airborne geophysical surveys also on parallel and square grids (Tables 6.11–6.14). The third block relates to ground geophysical surveys and vertical drilling programs to a depth of 300 feet (Tables 6.16–6.18). Ground detection tables for the pyritic halos were not considered because the halos are commonly detected from the air, following which the search for the deposits themselves is carried out on the ground within the halos. (See section 6.2.3 of this chapter.)

6.5. DESIGNING THREE STRATEGIES FOR THE DETECTION OF PORPHYRY–Cu–Mo DEPOSITS IN THE NORTH AMERICAN CORDILLERA BELT

6.5.1. Search Strategies

Three kinds of single-stage strategies aimed at the detection of porphyry deposits and their associated halos are described below. They are referred to as the liminal, maximal, and optimal options, which are designed to deal with three different kinds of exploration situations. The detection characteristics corresponding to the three options are listed in Table 6.4. for the porphyry deposits and in Table 6.5. for the pyritic halos. The listed parameters include probability of detection, corresponding grid spacing, and associated coverage cost per unit of area.

A two-stage variation on the optimal option is introduced to cover the sequential procedure described previously in Section 6.2.3. The first stage consists of the optimal detection of pyritic halos. The second one calls for the optimal detection of the deposit within the halos by means of systematic vertical drilling programs. The sequential optimal option is based on the dynamic programming approach described in Section 4.3.5. of Chapter 4.

TABLE 6.4
Summary of Detection Characteristics of Porphyry–Cu–Mo Deposits of the North American Cordillera Belt for Three Types of Detection Strategies.

Type of survey and grid geometry	Liminal detection		Optimal detection		Maximal detection	
	Coverage cost: US$/mile sq. Grid size: feet		Coverage cost US$/mile sq. Grid size: ft	Detection proba-bility	Coverage cost: US$/mile sq. Grid size: feet	
	Detection	Confirmed detection			Detection	Confirmed detection
Airborne geophysical parallel grid	$235 3900 ft	$375 1600 ft	$268 2900 ft	0.79	$560 930 ft	$585 850 ft
Airborne geophysical square grid	$258 6750 ft	–	$328 3960 ft	0.86	$412 2750 ft	–
Ground geophysical square grid	$540 1900 ft	$858 1225 ft	$595 1700 ft	0.80	$898 1100 ft	$1040 960 ft
Vertical drilling square grid	$86,000 1900 ft	$205,000 1225 ft	$102,000 1700 ft	0.80	$205,000 1100 ft	$262,500 960 ft

TABLE 6.5
Summary of Detection Characteristics of Pyritic Halos Associated with Porphyry–Cu–Mo Deposits of the North American Cordillera Belt for Three Types of Detection Strategies.

Type of survey and grid geometry	Liminal detection		Optimal detection		Maximal detection	
	Coverage cost: US$/mile sq. Grid size: feet		Coverage cost: US$/mile sq. Grid size: ft	Detection probability	Coverage cost: US$/mile sq. Grid size: feet	
	Detection	Confirmed detection			Detection	Confirmed detection
Airborne geophysical parallel grid	$185 8000 ft	$220 4600 ft	$200 6120 ft	0.86	$275 2700 ft	$371 1600 ft
Airborne geophysical	$152 12,000 ft	-	$242 7230 ft	0.93	$295 4800 ft	-

6.5.2. Liminal Option

The liminal option is based on a grid design which provides a single or confirmed detection probability of at least 0.500, as described in Section 5.5.2 of Chapter 5. The option is attractive only when explorationists face the problem of ensuring a reasonably effective coverage of large areas of little known potential within strict budgetary constraints. The detection parameters pertaining to the liminal option are listed in the leftmost portions of Table 6.4 for the porphyry deposits and Table 6.5 for the halos. A comparison of the two tables immediately shows that the coverage costs per unit of area required to ensure liminal detection are substantially lower for the halos than for the porphyry deposits, which offer much smaller targets for detection.

The cost multiplying factor required to ensure confirmation of detection instead of single detection is about 1.6 for the detection of porphyry–Cu–Mo deposits by airborne and ground geophysical surveys. However, for the associated pyritic halos, the cost required to obtain confirmation is only 1.2 times that of single detection; this emphasizes the benefit of initially requiring confirmed detection for halos rather than single detection. The cost multiplying factor escalates to 2.4 when confirmed detection of porphyry deposits by vertical drilling is considered.

Two kinds of airborne survey control grids are considered in the present study, the parallel design and the square design. A comparison of the merits of the two designs is a matter of interest for exploration planning. The coverage cost per unit of area provides a simple yardstick for the comparison. An inspection of the leftmost portions of Tables 6.4 and 6.5 shows that the selection of the square grid design is justified only for the search for halos because it leads to a reduction of 18% of the coverage cost. On the contrary, the selection of the square grid design is not recommended for porphyry deposit detection, because it would result in a 10% coverage cost increase.

6.5.3. Maximal Option

If, instead of minimizing coverage cost in return for a specified detection probability level (0.50), we wish to maximize the probability of detection to the near certainty level, say 0.980, we will then select the maximal option. Explorationists will find the maximal option quite attractive when considering the coverage of small areas of high potential in well prospected regions. The detection characteristics pertaining to the maximal option are listed in the rightmost portions of Tables 6.4 and 6.5 covering single and confirmed detection of porphyry deposits and their associated halos.

The cost multiplier required to obtain confirmation of detection of porphyry deposits by airborne and ground geophysical surveys is only 1.1 instead of 1.6 for the liminal option. In the case of halo detection, however, the cost multipliers are 1.4 for the maximal option against 1.2 for the liminal strategy. As a result, explorationists choosing the maximal option should be planning for the confirmed detection of porphyry deposits, while being satisfied with the single detection of the pyritic halos. When planning drilling programs for porphyry deposits, the confirmed detection approach of the maximal option is the most attractive one, because the cost multiplier is only 1.3 against 2.4 for the liminal option.

When faced with the choice between parallel and square grid designs for the airborne geophysical detection of porphyry deposits and associated halos, the coverage cost per unit of area provides a convenient criterion. One may see from Table 6.4 that the selection of the square grid design leads to a coverage cost saving of 26% for porphyry deposit detection under the maximal option. On the other hand, it can be seen from Table 6.5 that the square grid design is not recommended for halo detection because the corresponding coverage cost is 10% higher than that of the parallel design.

6.5.4. Optimal Option

The optimal option is based on a compromise between the low cost of the liminal option which guarantees at least a 0.500 detection probability and the high cost maximal option which provides nearly certain detection. The optimal option should prove most attractive in the majority of exploration situations which call for the coverage of moderately large areas of good potential within strict budgetary constraints.

The optimization of the field program designs was carried out by a computerized direct search method based on the efficiency criterion, as described in Section 4.4 of Chapter 4. Optimization tables are constructed for each of the three main types of field programs, namely, airborne geophysical surveys, ground geophysical surveys, and vertical drilling, in a manner described in Section 5.5.3. of Chapter 5. All tables provide the optimized grid parameters, including probability of optimal detection, optimal grid spacing, and corresponding coverage cost per unit of area, within 95% confidence limits which are based on the fiducial limits of the mean target length.

Tables 6.10. and 6.15 display the parameters of optimized grids for the detection of porphyry deposits and associated halos by airborne geophysical surveys on parallel and square grid designs. Optimization results for the detection of porphyry deposits by ground geophysical surveys and vertical drilling are listed in Tables 6.19a and 6.19b, respectively. No ground optimization results are provided for the halos because ground detection is not considered for them.

DETECTION OF PORPHYRY DEPOSITS

The detection characteristics pertaining to the optimal option are culled from the tables listed above and assembled into the central portions of Tables 6.4 (porphyry deposits) and 6.5 (halos). An inspection of both tables shows that detection parameters corresponding to the optimal option are bracketed by those associated with the liminal and maximal options, as could be expected. Table 6.4 shows that the optimal probabilities of detection of porphyry deposits by all types of geophysical surveys lie within a range of 0.79–0.86 with corresponding coverage cost per unit of area only marginally higher than of the liminal option. The optimal probability of detection by drilling programs (0.80) compares favorably with that of the maximal option, while the corresponding coverage cost is only one half of that of the maximal cost. The optimal probabilities of detection of halos by airborne surveys are close to that provided by the maximal option, while coverage costs are substantially lower (Table 6.5).

If we wish to compare the detection performances of parallel and square grids for airborne geophysical surveys searching for porphyry deposits or halos under the optimal option, we cannot rely on the coverage cost per unit of area as a criterion, because the probability of detection is not held at a specified level. The "expected loss" criterion combining coverage cost and detection probability can be used to advantage. It is written as a product as follows:

$$\text{Expected Loss} = (\text{Coverage cost}) \times (\text{Probability of failure})$$

The latter term is the complement of the optimal probability of detection written as $(1 - P_d)$. Obviously, the choice of design will be based on the minimization of the expected loss due to the failure of detection. We find from Table 6.4 that the expected loss per unit of area is $46 for the optimal square grid design against $56 for the optimal parallel design, which should be rejected. Similarly, the square grid design should be our choice for the optimal detection of pyritic halos by airborne surveys because the expected loss is only $17 as compared with $28 for the second design.

6.5.5. Sequential Optimal Strategy

After describing three single-stage strategies referred to as liminal, maximal, and optimal options, we should consider a two-stage optimal approach which illustrates the use of the dynamic programming method covered in Chapter 4. The purpose of the first stage of the sequence is to detect and outline pyritic halos associated with porphyry deposits by optimized airborne geophysical surveys, because they are much larger and more easily detectable targets than the deposits themselves. The following step is the optimized detection of the deposits by systematic drilling within the halos.

The theory of geometric probabilities provides that the probability of a randomly chosen sample point falling on a target of area a_t which is entirely included within a larger area A is equal to the ratio a_t/A. Taking the porphyry deposits of the North American Cordillera Belt and their pyritic halos as an example, we find that the probability mentioned above equals only 0.15, as derived from the ratio of the mean deposit area (2.85 million square feet) over the mean halo area (18.6 million square feet).

As seen in Section 1.1.3 of Chapter 1, the probability of a compound event is the product of the component probabilities, provided that the corresponding events are statistically independent. Based on the results listed in the central portions of Tables 6.4 and 6.5, we find that the compound optimal probability of detection of porphyry deposits of the Cordillera Belt by optimized airborne geophysical surveys on parallel grids followed by optimized drilling is equal to $0.86 \times 0.80 = 0.69$. The corresponding result would be equal to $0.93 \times 0.80 = 0.74$, if we choose the optimal square grid design instead of the parallel one at the initial stage of the program. These results are equal to nearly five times the probability of detecting a prophyry–Cu–Mo deposit by random sampling within its halo.

6.5.6. Detection Probability and Optimization Tables

The reader will find in the following section a total of ten tables that display the probabilities of detection and optimal designs in the search for porphyry–Cu–Mo deposits of the Cordillera Belt and their associated pyritic halos of airborne geophysical surveys (pages 125–134). A further five tables display the probabilities of detection and optimal designs pertaining to the search for porphyry–Cu–Mo deposits by ground geophysical surveys and by vertical drilling (pages 135–138).

TABLE 6.6
Probabilities of Detection of Porphyry–Cu–Mo Deposits of the North American Cordillera Belt by Airborne Geophysical Surveys (Continuous Readings) on Parallel Grids.

Survey design: parallel lines with spacing S feet,
Randomly orientated elliptical targets with expected major axis = L feet
in the confidence interval: l.c.l. (geom. mean (u.c.l.
and minor axis = B feet
R is defined as the ratio B/L with geometric mean = 0.46

	Probability of detection		
Grid spacing S feet	l.c.l. L=2820	geom. mean 3110	u.c.l. 3840 feet
800	1.000	1.000	1.000
1000	0.920	1.000	1.000
1200	0.903	0.922	1.000
1400	0.886	0.908	1.000
1600	0.869	0.895	0.916
1800	0.851	0.881	0.905
2000	0.832	0.866	0.894
2200	0.811	0.851	0.883
2400	0.789	0.836	0.871
2600	0.752	0.820	0.859
2800	0.699	0.803	0.847
3000	0.652	0.784	0.835
3200	0.611	0.751	0.822
3400	0.575	0.707	0.808
3600	0.543	0.668	0.794
3800	0.515	0.632	0.778
4000	0.489	0.601	0.747
4200	0.466	0.572	0.711
4400	0.445	0.546	0.679
4600	0.425	0.522	0.649
4800	0.408	0.501	0.622
5000	0.391	0.481	0.597
5200	0.376	0.462	0.574
5400	0.362	0.445	0.553
5600	0.349	0.429	0.533
5000	0.337	0.414	0.515
6000	0.326	0.401	0.498

TABLE 6.7
Determination of Grid Size for Specified Probability Levels in the Detection of Porphyry–Cu–Mo Deposits of the North American Cordillera Belt by Airborne Geophysical Surveys on Parallel Grids.

Survey design: parallel lines with spacing S feet
Unit cost =US$70 / line mile

Specified detection probability level	Required grid spacing S in feet	Corresponding cost in $ per mile square
0.10	20100	$ 158.39
0.15	14200	$ 166.03
0.20	11000	$ 173.60
0.25	9000	$ 181.07
0.30	7600	$ 188.63
0.35	6500	$ 196.86
0.40	5800	$ 203.72
0.45	5200	$ 211.08
0.50	4700	$ 218.64
0.55	4300	$ 225.95
0.60	3900	$ 234.77
0.65	3600	$ 242.67
0.70	3400	$ 248.71
0.75	3200	$ 255.50
0.80	2600	$ 282.15
0.85	2000	$ 324.80
0.90	1300	$ 424.31
0.95	500	$ 879.20

TABLE 6.8
Probabilities of Confirmed Detection of Porphyry–Cu–Mo Deposits of the North American Cordillera Belt by Airborne Geophysical Surveys (Continuous Readings) on Parallel Grids.

Survey design: parallel lines with spacing S feet,
Randomly orientated elliptical targets with expected major axis = L feet
in the confidence interval: l.c.l. (geom. mean (u.c.l.
and minor axis = B feet
R is defined as the ratio B/L with geometric mean = 0.46

	Probability of confirmed ratio		
Grid spacing S feet	l.c.l. L=2820	geom. mean 3110	u.c.l. 3840 feet
800	1.000	1.000	1.000
1000	0.838	1.000	1.000
1200	0.685	0.860	1.000
1400	0.581	0.725	0.923
1600	0.507	0.627	0.795
1800	0.455	0.554	0.698
2000	0.417	0.499	0.623
2200	0.391	0.457	0.564
2400	0.375	0.425	0.516
2600	0.000	0.401	0.478
2800	0.000	0.384	0.447
3000	0.000	0.373	0.423
3200	0.000	0.000	0.404
3400	0.000	0.000	0.389
3600	0.000	0.000	0.378
3800	0.000	0.000	0.371
4000	0.000	0.000	0.000

TABLE 6.9
Probabilities of Detection of Porphyry–Cu–Mo Deposits of the North American Cordillera Belt by Airborne Geophysical Surveys (Continuous Readings) on Square Grids.

```
Survey design:  square grid with spacings S by S feet,
Randomly orientated elliptical targets with expected major axis = L feet
in the confidence interval:   l.c.l. ( geom. mean ( u.c.l.
and minor axis = B feet
R is defined as the ratio B/L with geometric mean = 0.46
```

	Probability of detection		
Grid spacing S feet	l.c.l. L=2820	geom. mean 3110	u.c.l. 3840 feet
2500	1.000	1.000	1.000
3000	0.896	1.000	1.000
3500	0.818	0.921	1.000
4000	0.749	0.855	0.958
4500	0.688	0.794	0.905
5000	0.636	0.740	0.852
5500	0.590	0.691	0.803
6000	0.550	0.647	0.758
6500	0.515	0.608	0.716
7000	0.484	0.574	0.679
7500	0.456	0.542	0.644
8000	0.432	0.514	0.613
8500	0.409	0.489	0.584
9000	0.389	0.466	0.558
9500	0.371	0.445	0.534
10000	0.355	0.425	0.512

TABLE 6.10
Optimal Design of Airborne Geophysical Surveys (Continuous Readings) for the Detection of Porphyry–Cu–Mo Deposits of the North American Cordillera Belt.

Confidence interval: all results are reported as 95% confidence intervals respectively			
Expected target length in feet:	2820	3110	3830
Expected shape ratio R = 0.46			
Unit cost: $70/l.ml.			
(i) Grid Design: Parallel Grid			
Optimal grid spacing in feet:	2400	2900	3600
Cost in US$ per mile square:	294	267	243
Probability of detection:	0.79	0.79	0.79
(ii) Grid Design: Square Grid			
Optimal grid spacing in feet:	3510	3960	4490
Cost in US$ per mile square:	351	327	305
Probability of detection:	0.82	0.86	0.91

TABLE 6.11
Probabilities of Detection of Pyritic Halos Associated with Porphyry–Cu–Mo Deposits of the North American Cordillera Belt by Airborne Geophysical Surveys (Continuous Readings) on Parallel Grids.

Survey design: parallel lines with spacing S feet,
Randomly orientated elliptical targets with expected major axis = L feet
in the confidence interval: l.c.l. (geom. mean (u.c.l.
and minor axis = B feet
R is defined as the ratio B/L with geometric mean =0.62

Grid spacing S feet	Probability of detection		
	l.c.l. L=4870	geom. mean 6374	u.c.l. 8344 feet
2500	1.000	1.000	1.000
3000	0.920	1.000	1.000
3500	0.905	1.000	1.000
4000	0.889	0.918	1.000
4500	0.871	0.906	1.000
5000	0.810	0.894	0.921
5500	0.737	0.881	0.913
6000	0.675	0.866	0.904
6500	0.623	0.812	0.894
7000	0.579	0.754	0.884
7500	0.540	0.704	0.874
8000	0.506	0.660	0.859
8500	0.477	0.621	0.808
9000	0.450	0.586	0.764
9500	0.427	0.555	0.723
10000	0.405	0.528	0.687

TABLE 6.12
Determination of Grid Size for Specified Levels of Probability in the Detection of Pyritic Halos Associated with Porphyry–Cu–Mo Deposits of the North American Cordillera Belt by Airborne Geophysical Surveys on Parallel Grids.

```
Survey design:  parallel lines with spacing S feet
Unit cost = US$70 / line mile
```

Specified detection probability level	Required grid spacing S in feet	Corresponding cost in $ per mile square
0.10	45800	$ 148.07
0.15	32400	$ 151.41
0.20	25000	$ 154.78
0.25	20400	$ 158.12
0.30	17200	$ 161.49
0.35	14900	$ 164.81
0.40	13100	$ 168.21
0.45	11700	$ 171.59
0.50	10600	$ 174.87
0.55	9700	$ 178.10
0.60	8900	$ 181.53
0.65	8200	$ 185.07
0.70	7700	$ 188.00
0.75	7200	$ 191.33
0.80	6700	$ 195.16
0.85	6200	$ 199.61
0.90	4100	$ 230.15
0.95	1600	$ 371.00

TABLE 6.13
Probabilities of Confirmed Detection of Pyritic Halos Associated with Porphyry–Cu–Mo Deposits of the North American Cordillera Belt by Airborne Geophysical Surveys (Continuous Readings) on Parallel Grids.

Survey design: parallel lines with spacing S feet,
Randomly orientated elliptical targets with expected major axis = L feet in the confidence interval: l.c.l. (geom. mean (u.c.l.
and minor axis = B feet
R is defined as the ratio B/L with geometric mean = 0.62

Grid spacing S feet	Probability of confirmed detection		
	l.c.l. L=4870	geom. mean 6374	u.c.l. 8344 feet
1500	1.000	1.000	1.000
2000	0.861	1.000	1.000
2500	0.761	0.883	1.000
3000	0.699	0.794	0.929
3500	0.660	0.734	0.845
4000	0.636	0.693	0.784
4500	0.622	0.663	0.739
5000	0.000	0.642	0.704
5500	0.000	0.628	0.678
6000	0.000	0.621	0.658
6500	0.000	0.000	0.642
7000	0.000	0.000	0.631
7500	0.000	0.000	0.624
8000	0.000	0.000	0.000

TABLE 6.14
Probabilities of Detection of Pyritic Halos Associated with Porphyry–Cu–Mo Deposits of the North American Cordillera Belt by Airborne Geophysical Surveys (Continuous Readings) on Square Grids.

```
Survey design:  square grid with spacings S by S feet
Randomly orientated elliptical targets with expected major axis = L feet
in the confidence interval:  l.c.l. ( geom. mean ( u.c.l.
and minor axis = B feet
R is defined as the ratio B/L with geometric mean = 0.62
```

	Probability of detection		
Grid spacing S feet	l.c.l. L=4870	geom. mean 6374	u.c.l. 8344 feet
4500	1.000	1.000	1.000
5000	0.963	1.000	1.000
5500	0.927	1.000	1.000
6000	0.890	1.000	1.000
6500	0.852	0.963	1.000
7000	0.816	0.936	1.000
7500	0.782	0.908	1.000
8000	0.749	0.879	0.980
8500	0.719	0.850	0.962
9000	0.691	0.823	0.941
9500	0.664	0.796	0.920
10000	0.639	0.770	0.898

TABLE 6.15
Optimal Design of Airborne Geophysical Surveys (Continuous Readings) for the Detection of Pyritic Halos Associated with Porphyry–Cu–Mo Deposits of the North American Cordillera Belt.

Confidence interval: all results are reported as 95% confidence intervals respectively			
Expected target length in feet:	4870	6374	8344
Expected shape ratio R = 0.62			
Unit Cost: $70/l.ml.			
(i) Grid Design: Parallel Grid			
Optimal grid spacing in feet:	4700	6120	7970
Cost in US$ per mile square:	219	200	186
Probability of detection:	0.86	0.86	0.86
(ii) Grid Design: Square Grid			
Optimal grid spacing in feet:	5830	7230	9100
Cost in US$ per mile square:	267	242	221
Probability of detection:	0.91	0.93	0.94

TABLE 6.16
Probabilities of Detection of Porphyry–Cu–Mo Deposits of the North American Cordillera Belt by Ground Geophysical Surveys (Discrete Readings) or Vertical Drilling to a Depth of 300 Feet.

```
Survey design: square grid with spacings S by S feet
The expected shape of the ore deposits is elliptical
with major axis = L in the 95% confidence interval:
L.c.l. ( geometric mean ( U.c.l., and minor axis = B feet
R is defined as the ratio B/L with geometric mean = 0.46
```

Spacing S in feet	Probability of detection		
	L.c.l. L = 2820	G. mean 3110	U.c.l. 3820ft
800	1.000	1.000	1.000
1000	0.986	1.000	1.000
1200	0.914	0.991	1.000
1400	0.788	0.942	0.999
1600	0.660	0.852	0.973
1800	0.547	0.745	0.925
2000	0.455	0.643	0.847
2200	0.383	0.552	0.761
2400	0.324	0.475	0.677
2600	0.277	0.411	0.600
2800	0.239	0.358	0.530
3000	0.208	0.313	0.470
3200	0.183	0.276	0.419
4000	0.117	0.176	0.273
4800	0.081	0.123	0.189
5600	0.060	0.090	0.139
6000	0.052	0.078	0.121

TABLE 6.17
Determination of Grid Size for Specified Probability Levels in the Detection of Porphyry–Cu–Mo Deposits of the North American Cordillera Belt by Vertical Drilling to a Depth of 300 Feet.

Survey Design: Square Grid With Spacings S by S Feet
Unit Cost = $6000 Per Diamond Drill-Hole

Probability Level Of Detection	Grid Size In Feet	Drilling Cost Per Mile Square In U.S.$. Thousands
0.05	6900	18.91
0.15	4300	30.05
0.25	3300	40.87
0.35	2900	48.08
0.45	2500	58.48
0.55	2300	65.56
0.65	2000	79.93
0.75	1800	93.30
0.85	1600	111.46
0.95	1400	137.17

TABLE 6.18
Probabilities of Confirmed Detection of Porphyry–Cu–Mo Deposits of the North American Cordillera Belt by Ground Geophysical Surveys or Vertical Drilling to a Depth of 300 Feet.

Survey design: square grid with spacings S by S feet.
Randomly oriented elliptical ore bodies
with major axis = L feet in the 95% confidence interval:
l.c.l (geometric mean (u.c.l., minor axis = B feet
R is defined as ratio B/L with geometric mean = 0.46

	Probability of confirmed detection		
Spacing S feet	L.c.l. L = 2820	G. mean 3110	U.c.l. 3820ft
900	0.951	1.000	1.000
1000	0.926	0.995	1.000
1100	0.793	0.982	1.000
1200	0.383	0.943	0.998
1300	0.253	0.873	0.995
1400	0.166	0.570	0.989
1500	0.109	0.352	0.938
1600	0.071	0.251	0.884
1700	0.046	0.178	0.739
1800	0.030	0.126	0.418
1900	0.019	0.089	0.319
2000	0.012	0.063	0.243
2200	0.004	0.031	0.140
2400	0.000	0.015	0.080
2600	0.000	0.006	0.045
2800	0.000	0.002	0.026
3000	0.000	0.000	0.014

TABLE 6.19(a)
Optimal Design of Ground Geophysical Surveys on Square Grids for the Detection of Porphyry–Cu–Mo Deposits of the North American Cordillera Belt.

```
Survey Design:   Square Grid With Spacing S by S Feet.
Confidence Interval:  All results are reported as 95% confidence intervals
respectively.
Expected Target Length in Feet:         2820      3110      3830
Expected Shape Ratio R = 0.46
Unit Cost:  $6/Station
```

Optimal Grid Spacing in Feet:	1350	1700	2100
Cost in $ Per Mile Square	734	594	496
Probability of Detection	0.82	0.80	0.88

TABLE 6.19(b)
Optimal Design for the Detection of Porphyry–Cu–Mo Deposits of the North American Cordillera Belt by Vertical Drilling to a Depth of 300 Feet.

```
Survey design:  square grid with spacing S by S feet
Confidence interval:  all results are reported as 95% confidence intervals
respectively
Expected target length in feet:         2820      3110      3830
```

(i) Percussion drilling: Unit Cost = $2000/hole			
Optimal grid spacing in feet:	1350	1700	2100
Cost in $thou. per mile square:	48	34	25
Probability of detection:	0.82	0.80	0.88
(ii) Diamond drilling: Unit Cost = $6000/hole			
Optimal grid spacing in feet:	1350	1700	2100
Cost in $thou. per mile square:	146	102	75
Probability of detection:	0.82	0.80	0.88

REFERENCES AND SELECTED READINGS

1. ABRAMS, M. J., BROWN, D., and LEPLEY, L., 1983, Remote sensing for porphyry–copper deposits in Southern Arizona, *Econ. Geol.* **78,** 591–604.
2. CLARK, K. F., 1972, Stockwork molybdenum deposits in the Western Cordillera, *Econ. Geol.* **67,** 731–758.
3. CREASEY, S. C., 1959, Some relations in the hydrothermally altered rocks of porphyry–copper deposits, *Econ. Geol.* **54,** 354–373.
4. DE GEOFFROY, J., and WIGNALL, T. K., 1972, A statistical study of the geological characteristics of the porphyry–copper–molybdenum deposits of North and South America, *Econ. Geol.* **67,** 656–668.
5. DE GEOFFROY, J. and WIGNALL, T. K., 1973, Statistical models for porphyry–copper–molybdenum deposits of the Cordilleran Belt of North and South America, *Can. Inst. Min. Metall. Bull.* **66**(735), 84–90.
6. ENERGY, MINES, AND RESOURCES DEPARTMENT OF CANADA, 1981, Canadian mineral deposits not being mined in 1980, Report M.R. I. 80–7.
7. FOUNTAIN, D. K., 1968, Geophysics applied to the exploration and development of Cu and Mo deposits of British Columbia, *Can. Inst. Min. Metall. Bull.* **61**(676), 1199–1206.
8. FOUNTAIN, D. K., 1972, Geophysical case histories of disseminated sulfide deposits in British Columbia, *Geophysics* **37,** 142–159.
9. GODWIN, C. I., 1976, Casino porphyry–Cu–Mo deposit; Porphyry deposits of the Canadian Cordillera, *Can. Inst. Min. Metall. Special Volume* **15,** 344–354.
10. GUILBERT, J. M., and LOWELL, J. D., 1974, Variations in zoning patterns in porphyry ore deposits, *Can. Inst. Min. Metall. Bull.* **67,** 99–109.
11. HOLLISTER, V. F., 1973, Characteristics of porphyry copper deposits of South America, *Min. Eng.* **25,** 51–56.
12. HOLLISTER, V. F., POTTER, R. R., and BARKER, A. L., 1974, Porphyry-type deposits of the Appalachian Orogen, *Econ. Geol.* **69,** 618–630.
13. HOLLISTER, V. F., ANZALONE, S. A., and PRICHTER, D. H., 1975, Porphyry–copper deposits of Southern Alaska and contiguous Yukon Territories, *Can. Inst. Min. Metall. Bull.* **68**(755), 104–111.
14. HOLLISTER, V. F., 1978, *Geology of the Porphyry–Copper Deposits of the Western Hemisphere,* A.I.M.E., New York.
15. JAMES, A. H., 1971, Hypothetical diagrams of several porphyry–copper deposits, *Econ. Geol.* **66,** 43–47.
16. KESLER, S. E., 1973, Copper, molybdenum and gold abundance in porphyry–copper deposits, *Econ. Geol.* **68,** 106–112.
17. LIVINGSTON, D. E., 1973, A plate tectonic hypothesis for the genesis of porphyry–copper deposits of the Southern Basin and Range Province, U.S.A., *Earth Planet. Sci. Lett.* **20,** 171–176.
18. LOWELL, J. D., and GUILBERT, J. M., 1970, Lateral and vertical alteration-mineralization zoning in porphyry deposits, *Econ. Geol.* **65,** 373–408.
19. ROSE, A. W., 1970, Zonal relations of wallrock alteration and sulfide distribution at the porphyry–copper deposits, *Econ. Geol.* **65,** 910–936.
20. ROWE, R. B., 1973, Porphyry deposits of the Canadian Cordillera, *Can. Min. J.* **94**(11), 35–38; **94**(12), 37–41.
21. SAWYER, J. P. B., and DICKINSON, R. A., 1976, Mount Nansen porphyry deposit; Porphyry deposits of the Canadian Cordillera, *Can. Inst. Min. Metall. Special Volume* **15,** 336–343.

22. SEIGEL, H. O., 1974, The magnetic induced polarization (M.I.P.) method, *Geophysics* **39**, 321–339.
23. SILLITOE, R. H., 1972, A plate tectonic model for the origin of porphyry–copper deposits, *Econ. Geol.* **67**, 184–197.
24. SILLITOE, R. H., 1973, The tops and bottoms of porphyry–copper deposits, *Econ. Geol.* **68**, 799–815.
25. SILLILTOE, R. H., 1979, Some thoughts on gold-rich porphyry–copper deposits, *Miner. Deposita* **14**, 161–174.
26. STRINGHAM, B., 1960, Differences between barren and productive intrusives, *Econ. Geol.* **55** 1622–1630.
27. SUMNER, J. S., 1967, Geophysical aspects of porphyry–copper deposits, *Geol. Surv. Can. Rep.* No. 26, 322–335.
28. SUTHERLAND-BROWN, A., 1969, Mineralization in British Columbia and the copper–molybdenum deposits, *Can. Inst. Min. Metall. Bull.* **62**(681), 26–32.
29. TITLEY, S. R., and HICKS, C. L. (Eds.), 1966, *Geology of Porphyry–Copper Deposits of Southwestern North America,* University of Arizona Press, Tucson.
30. TITLEY, S. R. (Ed.), 1982, *Advances in Geology of the Porphyry–Copper Deposits,* University of Arizona Press, Tucson.

CHAPTER SEVEN

OPTIMIZED SEARCH FOR FOUR TYPES OF CONTACT METASOMATIC DEPOSITS OF THE NORTH AMERICAN CORDILLERA BELT

7.1. GENERAL GEOLOGICAL BACKGROUND

7.1.1. Introduction and Terminology

7.1.1.1. Contact Metamorphism. The peculiar effect of intrusive bodies of acid to intermediate composition upon adjacent rock formations has been investigated during the past 100 years and has long been favored as a ready explanation for ore genesis. Granitic intrusions are commonly surrounded by halos of gradually fading metamorphism which may extend for a few hundreds of feet up to two miles, but averages half a mile in width. In contrast, effusive rocks and basic intrusive bodies have only a moderate to strong baking action over a few tens of feet at the most.

On the mineralogic standpoint, contact metamorphism consists essentially in a recrystallization of minerals accompanied by a transfer of volatiles such as H_2O and CO_2. The nature of the mineralogic transformation reflects mostly the original composition of the countryrock and the intensity of metamorphism, with a lesser influence of the nature of the intrusive.[10,22]

Generally, clastic sediments or effusive rocks of silicic composition or metasediments show little change, unless they contain calcareous or pelitic impurities. Pelitic sediments (shale, slate, silstone, marl) are converted into "hornfels" which are made up of assemblages of biotite, hornblend, plagioclase, and quartz, with aggregates of cordierite, staurolite, and andalusite. Carbonate sediments (limestone, dolomite) undergo conspicuous recrystallization as marbles, and being generally more reactive than other rocks, are subjected to metasomatic alteration.

7.1.1.2. Metasomatism and the Formation of Skarn. In contrast to contact metamorphism, metasomatism involves some marked changes in the proportion of nonvolatile elements such as Fe, Mg, and Si, in conditions well above the critical pressure and temperature of water. The final product, an assemblage of new lime-rich, iron-rich minerals is called "skarn." The term was first coined by Medieval Swedish miners and referred to the lime–silicate gangue of certain types of iron or sulfide ores occurring in the Scandinavian Precambrian Shield. Most skarn are coarsely crystallized and exhibit crude zonal assemblages which reflect sedimentary features of the country rock or the geometry of the intrusive-wall rock contact.[5]

Einaudi[5] and Smirnov[22] distinguish two main types of skarns, including "endoskarn" and "exoskarn." The latter replaces country rock outside the intrusive bodies over widths of a few hundreds to more than one thousand feet. The former extending inward, on the intrusive side, replaces the intrusive rock over widths of a few tens of feet to a few hundreds of feet. Exoskarns are, in turn, divided into "calcic" skarns, the most common type, made up of lime-rich silicates, hydrosilicates, oxides, and sulfides, and "magnesian" skarns, less common, which are characterized by different types of hydrosilicates such as phlogopite, tremolite, and talc.

7.1.1.3. Ore Skarn. Ore skarn occurs within the reaction skarn when shrinkage cracks resulting from metasomatism are filled up with iron and base metal sulfides, scheelite, barite, fluorite, and secondary quartz. Ore skarn forms most commonly in the exoskarn zone. In a minority of cases, the ore skarn straddles the contact and extends into the endoskarn zone (see Figure 7.1).

Ore skarns are made up of early silicates of Ca,Mg,Fe,Al, derived partly from the wall rock, partly from the intrusive body. The silicates are followed in the paragenetic sequence by scheelite and oxides such as cassiterite and magnetite. The sequence ends with sulfides including pyrite and pyrrhotite and base metal sulfides (Cu,Zn,Pb,Mo).

Ore skarns often exhibit zoning patterns at differing scales. At the local scale, crude banding results from rhythmic metasomatism, while at a larger scale the zoning reflects stratigraphic features of the wall rock. Finally, at a semiregional scale, the distribution of the mineralization shows a zonal arrangement within the metamorphic aureola of the intrusive: scheelite mineralization is narrowly restricted to the immediate contact, while copper spreads outward for some distance, and lead and zinc, farther away.

7.1.1.4. Classification of Contact-Associated Ore Deposits. Most genetic classifications of ore deposits include a category of deposits which may be labeled as "contact-associated," including such terms as "contact metamorphic" or "pyrometamorphic" or "pyrometasomatic" deposits. In this study, we have chosen the term "contact metasomatic" to emphasize the joint role of contact and metasomatism in the formation of the deposits.

DETECTION OF CONTACT METASOMATIC DEPOSITS

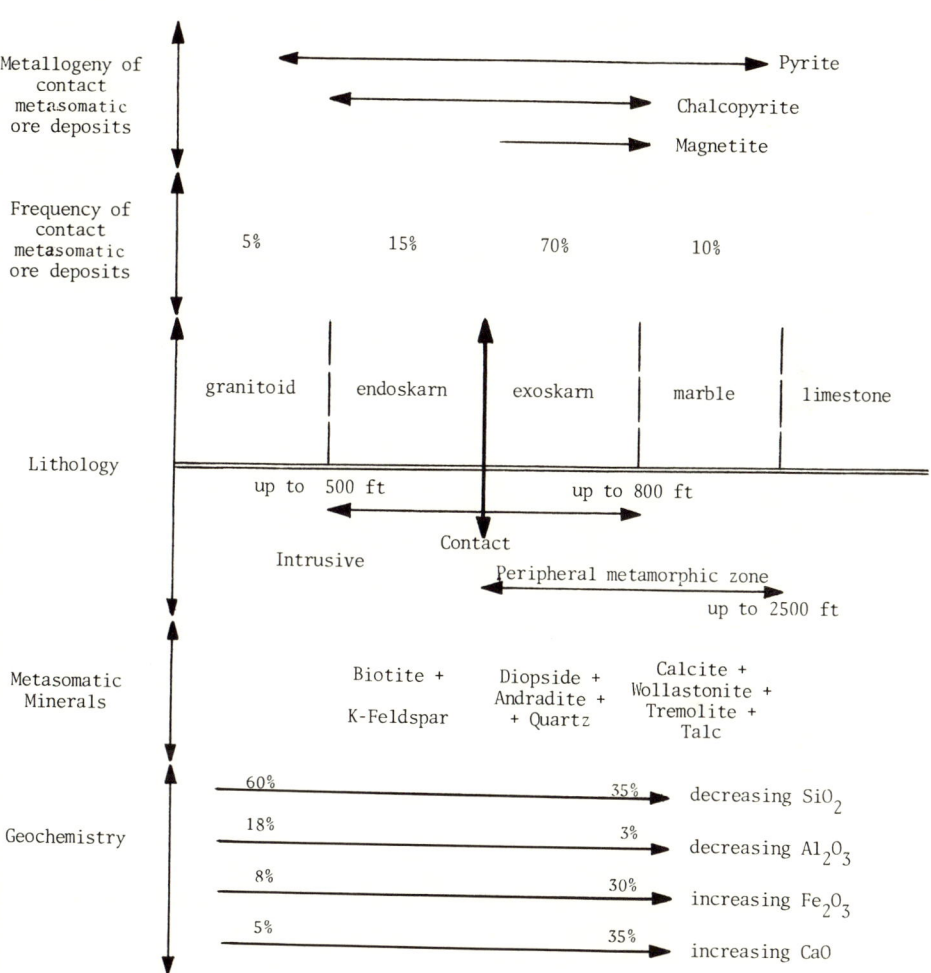

FIGURE 7.1
Environment of Contact Metasomatic Ore Deposits.

There are many types and subtypes of contact metasomatic deposits, and their classification generally emphasizes the chief minerals of economic importance as labels for the categories. In practice, there are innumerable gradations between four common types which are covered in this chapter. They include (a) Fe–Cu–Au deposits, (b) complex Pb–Zn–Cu–Ag deposits, sometimes related to small intrusives, sometimes without obvious relationship with intrusives, (c) Cu–Mo–Au deposits which are generally associated with differentiated plutons of the porphyry–copper-bearing type, and (d) W–Mo deposits.

7.1.2. Characteristics of Contact Metasomatic Deposits

7.1.2.1. Geometry of Contact Metasomatic Deposits.

Smirnov[22] classifies the shape of contact metasomatic deposits into three main types of increasing complexity. The least complex are the stratified deposits. Generally tabular in gross shape, they follow favorable beds along the strike for up to 8000 feet, down-dip for up to 2000 feet, with varying thickness up to 400 feet. The ore material terminates abruptly across the strike of the host bed and gradually pinches out down-dip.

The second type consists of steeply inclined pipes and stocks whose dimensions range from a few hundreds of feet to 1000 feet. The mineralized material often crosses the strike of the host formation and terminates abruptly. The third type, the most complex one, consists of intricately ramified bodies grading into stockworks, whose dimensions are generally much smaller than those of the two previous types, ranging from a few tens of feet up to a few hundreds.

7.1.2.2. Control of Contact Metasomatic Ore Deposits.

The formation of contact metasomatic ore deposits is believed to be promoted by the conjunction of two main controlling influences: (1) a physicochemical factor reflecting the lithology of both wall rock and intrusive and conditions of emplacement, and (2) a structural factor reflecting the geometry of the intrusive–wall rock contact. The role of the wall rock lithology has been abundantly documented on a worldwide basis. An important control is exerted by stratigraphic discontinuities, which appear to act as "ore traps." The role of the intrusive lithology is less commonly stressed, though it may be quite significant. In the Cordilleran Belt of North America, granitic intrusives, commonly exhibiting conspicuous contact metamorphic aureolae, do not generate contact metasomatic ore deposits. Most commonly the deposits are genetically related to intrusives of intermediate composition (quartz monzonite, monzonite, granodiorite), more rarely to basic intrusives (diorite, gabbro).

DETECTION OF CONTACT METASOMATIC DEPOSITS 145

Smirnov[22] cites three main types of structural control for contact metasomatic deposits, including (a) the geometry of the intrusive–wallrock interface, (b) geometry of the bedding and jointing of the wall rock, and (c) the geometry of the fractures affecting both wall rock and intrusive. In the early stages of ore deposition, (a) and (b) appear to have a dominating influence, while (c) prevails in the later stages.

The influence of the geometry of the intrusive–wall rock contact as control for the deposition of tungsten mineralization is well documented in the North American Cordillera.[10] Cooling creates a corrugated contact surface by spalling, generating troughs and grooves which act as ore traps. Kerr[10] also shows that the concordance of the contact surface with bedding planes of the wall rock is far less favorable for ore deposition than discordance. Finally, the influence of fracturing is illustrated by the accumulation or mineralization in interstrata crush zones produced during cooling as a result of the difference in competency between wall rock formations. Furthermore, cross-cutting and oblique faults are believed to act as feeder channels for ore-bearing solutions.

7.1.3. Geographic Distribution of Contact Metasomatic Deposits

Occurrence of contact metasomatic deposits is as widespread geographically as it is chronologically. Such deposits are well known in the late Precambrian carbonate formations of the Grenville Province in the eastern portion of the North American Shield and in the Paleozoic fold belt of the Appalachian. They are particularly numerous in Mesozoic foldbelts because of the favorable combination of extensive Paleozoic carbonate formations and numerous intrusions of intermediate composition and Mesozoic age, as can be found in the North American Cordillera Belt.

In the northern half of the Cordillera, the deposits are concentrated along two main belts. The westernmost belt consists mainly of Fe–Cu skarn deposits and extends through the coastal and insular regions of Southeast Alaska and British Columbia. The Intermontane Belt extends through eastern British Columbia and the Yukon and exhibits a regional zoning from Cu–Fe skarn type on the west side to complex Pb–Zn–Ag type in the central part of the belt, and to W–Mo skarn type along the east and northeast edge of the belt.

In the U.S. section of the Cordillera, Fe-rich skarn deposits are quite common in the States of Utah, Nevada, California, and New Mexico. Complex Pb–Zn–Ag metasomatic deposits are mainly concentrated in the central portion of the Cordillera, in the States of Idaho, Utah, Nevada, and Colorado. The Cu–Mo–Au type, mainly associated with copper-bearing differentiated plutons is well represented in the southeastern Cordillera, in the States of Utah, Nevada, Arizona, and New Mexico, as well as in northern Mexico. Finally the W-rich skarn deposits occur mainly in the southwestern portion of the Cordillera, in the States of Utah, California, and Nevada.

7.2. FIELD DETECTION METHODOLOGY

7.2.1. Introduction

In contrast with the porphyry–Cu–Mo deposits, most contact metasomatic deposits offer targets of modest size for detection purposes. This rules out direct detection by remote sensing techniques from satellite platforms. Other direct detection methods including photogeology from aircraft platforms, ground mapping, and drilling have proved most useful in the detection of contact metasomatic deposits. The indirect detection approach based on the recording of the geophysical signature of deposits from aircraft platforms or on the ground has proved more useful than indirect geochemical methods based on the detection of halos.

7.2.2. Direct Geological Detection

Direct geological detection has proved very valuable in the search for contact metasomatic deposits in two cases: typical contact-associated deposits, and complex metasomatic deposits that do not show any close relationship with intrusive contacts. In the first case, photogeological mapping of the contact followed by ground mapping is the obvious method of delineating the potentially productive areas, based on lithologic and mineralogic clues, to be followed by geophysical coverage and drill testing of areas of highest merit. The second case is vastly more difficult to handle, because the complex metasomatic deposits are generally strata-bound and have no visible surface expression. The discovery of the Burgin No. 1 and No. 2 deposits in the Tintic District of Utah under a thick capping of recent lava flows, in the mid-1950s, is one of the best-known cases of successful detection of these elusive targets by direct geological methods.[16,20] The targeting relied on the conjunction of stratigraphic, structural, mineralogic, and geochemical clues and was followed by drill testing.

7.2.3. Indirect Geophysical Approach

The slow-flying and ground-hugging helicopter platform is to be preferred to the cheaper fixed-wing aircraft platform for the indirect detection of contact metasomatic deposits by airborne geophysical surveys, because of the relatively small size of the deposits and the ruggedness of the Cordillera topography. Airborne magnetic surveys are quite suitable for the indirect detection of Fe–Cu and Cu–Mo types of deposits, but provide little assistance in pinpointing small and weakly magnetized W–Mo deposits. Electromagnetic surveys in the frequency domain or in the time domain (INPUT) may be of assistance in the detection of "blind" massive sulfide deposits of the complex Pb–Zn–Cu–Ag type or Cu–Mo type, providing that the capping is less than 200 feet in thickness.

On the ground, magnetometric surveys are a very effective and low-cost method of detection of Cu–Fe skarn deposits, while other more expensive techniques such as electromagnetic and induced polarization surveys did not prove themselves. The combination of magnetics and electromagnetics proved quite successful in the search for massive sulfide deposits in skarn zones in New Mexico and Arizona. Success was also reported in the Tintic District of Utah in the search for blind complex Pb–Zn–Cu deposits by the "down the hole" resistivity method from drill holes and underground workings.[20] The indirect geophysical search for W–Mo skarn deposits is most troublesome because of the small size of the targets and the weakness of the magnetization generated by disseminations of magnetite and pyrrhotite which surround the deposits. Although disseminated pyrite is widespread in the immediate environment of the W–Mo deposits, there are no records of the successful use of the induced polarization technique for their detection in the Cordillera. Klichnikov[11] reports success in locating W–Mo skarn deposits in Central Asia, using a combination of gravimetry to locate the buried apices of small intrusives, magnetics to delineate contact aurolae, and induced polarization to locate the pyritic halos of W–Mo deposits within the aurolae.

7.2.4. Indirect Geochemical Approach

The detection of primary halos of complex Pb–Zn–Cu metasomatic deposits based on the pioneering work of Lovering[17] in the Tintic District of Utah is a good example of the successful application of the indirect geochemical approach in the search for contact metasomatic deposits. Later on, an even more sophisticated approach was devised in the Tintic District,[20] relying on the delineation of "ghost" halos caused by leakages from primary halos through thick barren lava capping.

Detection of secondary halos by residual soil sampling proved successful in Central Asia in the search for W–Mo–skarn deposits.[11] However, very little success was met with in the search for Cu–Fe skarn deposits in central British Columbia based on soil sampling or geobotanical sampling.

7.3. PROBABILITY OF DETECTION OF Cu–Fe CONTACT METASOMATIC DEPOSITS

7.3.1. Geological Synopsis

7.3.1.1. Introduction. The Cu–Fe contact metasomatic deposits are commonly tabular or lenticular in overall shape. They crudely follow the stratification of favorable beds, but often terminate abruptly across the bedding. Ore deposits consist of an envelope of iron-rich and lime-rich silicates enclosing pods and lenses of massive magnetite with subordinate specularite, with minor sulfides including pyrite, pyrrhotite, and chalcopyrite occurring along the edges of the deposits. In some cases (Alaska, British Columbia), chalcopyrite is much more abundant and becomes the principal economic mineral. The paragenesis is as follows: after the formation of the early silicates, pyrite magnetite and specularite develop, replacing some the silicates. The sequence ends with the replacement of silicates, magnetite, and calcite by chalcopyrite.

7.3.1.2. British Columbia and Alaska Section of the Cordillera. Cu–Fe skarn deposits have been mined for many years in the coastal and insular portion of British Columbia and Alaska.[21] Good bessemer-grade iron ore and copper ore of moderate grade have been gained from the Kasan Peninsula, Prince of Wales Island, Southeast Alaska.[24] The ore lies in skarnified limestone intercalations in Triassic volcanics intruded by Jurassic quartz diorite stocks. Similar deposits occur in the Queen Charlotte Islands and in Vancouver Island, off the coast of British Columbia. A second belt of Cu–Fe deposits lies in the Intramontane zone of mainland British Columbia. An important metallogenic feature of the region is the crude zoning of deposit types from the west where Fe predominates to the east where Cu prevails (example: Craigmont deposit).

7.3.1.3. U.S. Section of the Cordillera. The Cu–Fe skarn deposits of the U.S. section of the Cordillera occur in the same general geological setting as those of British Columbia. However, there is a gap comprising the states of Washington, Oregon, and Idaho between the Cu–Fe skarn belts of British Columbia and those of the U.S. which stretch through the states of Utah, Nevada, California, and New Mexico.

The mineralization is chiefly magnetite and specularite with very little or no copper. The larger deposits occur in Utah and mostly in California, with the principal deposit, the Eagle Mountain mine, totaling some 80 million tons of good grade iron ore, low in phosphorus and sulfur. The deposit consists of a string of eight lenses along a six-mile-long horizon of early Paleozoic dolostone folded to vertical during the Laramide orogeny and intruded by a stock of

DETECTION OF CONTACT METASOMATIC DEPOSITS

Cretaceous porphyritic quartz monzonite. Local contamination of the intrusive by the dolostone country rock is indicated by the presence of many xenoliths and the change in composition from monzonite to quartz diorite.

7.3.2. Statistical Modeling of Geometric Parameters of Cu–Fe Contact Metasomatic Deposits

A total of 23 deposits were included in the database to represent the population of Cu–Fe deposits of the Cordillera. The deposits are listed by name and region of occurrence in Table 7.1. The geographic breakdown is as follows: 13 deposits in British Columbia, 3 in Alaska, and 7 deposits in the rest of the United States.

The geometry of deposits is quantitatively described by means of dimensional and attitudinal parameters. The former include the length, breadth, and shape ratio of the horizontal section of the portions of the deposits which lie

TABLE 7.1

List of Contact Metasomatic (Cu–Fe) Deposits of the North American Cordillera Belt Included in the Database.

B.C. Islands (Canada)	S.W. Alaska (U.S.A.)
Quatsino	Mamie
F.L. mine	Poor Man
Iron Hill	Mt. Andrew
Prescott	Utah (U.S.A.)
Paxton	Iron Spring
Yellow Kid	McCahill
Lake	Nevada (U.S.A.)
Ford	Dayton
Ridge	California (U.S.A.)
Jessie	Eagle Mountain
B.C. Mainland (Canada)	Dale
Craigmont	Iron Mountain
Merry Widow	New Mexico (U.S.A.)
King Fisher	Hanover

within the average range of detection (300 feet). The latter include the unoriented dip and the strike direction of deposits referred to true North and measured in degrees within the right half circle. The range of measurements of the parameters for the 23 deposits of the sample is from 500 to 5300 feet for length, 35 to 520 feet for breadth, and 25 to 85 degrees for dip.

The measurements are grouped into classes and assembled into frequency distributions which are summarized by means of statistics. The statistics include arithmetic mean, mode, standard deviation, and coefficient of skewness and are listed in the lower half of Table 7.2. The coefficient of skewness gives an indication of the asymmetry of the distributions of observed data; the skewness is moderately high (0.35) for the lengths, but is quite high for the four other parameters (0.60–0.90).

The observed distributions of the dimensional parameters were fitted by lognormal models at the 0.05 confidence level. The statistics of the fitted models, including geometric means, and their 95% fiducial limits, first and third quartiles, and coefficients of dispersion are listed in the upper half of Table 7.2. The

TABLE 7.2

Statistical Modeling Summary for Contact Metasomatic (Cu–Fe) Deposits of the North American Cordillera Belt (Sample Size: 23 Deposits).

Statistic of fitted model	Length of horiz. sect. in feet	Breadth of horiz. sect. in feet	Shape ratio R=B/L	Strike dir. from T. North in degrees	Dip in degrees
25%ile	800	150	0.12	12	0
95% L.C.L.	982	183	0.14	33	23
Mean	1273	230	0.18	52	36
95% U.c.l.	1650	288	0.23	72	48
75%ile	1700	400	0.25	95	70
Disp. coef.	0.21	0.09	0.11	0.38	0.12
Statistic of observed data					
Arithm. mean	1667	266	0.23	56	42
Mode	1150	350	0.12	15	15
Stand. dev.	1477	128	0.18	48	34
Skewness	0.35	0.65	0.92	0.85	0.83

DETECTION OF CONTACT METASOMATIC DEPOSITS 151

dispersion coefficient which gives an indication of the variability of the data is rather small (0.1) for the breadths, and is moderate (0.2) for the lengths and shape ratios. The observed distributions of dip and strike direction could not be fitted by normal or circular normal models at the required 0.05 confidence level.

7.3.3. Construction and Organization of Detection Probability Tables

For detection purposes, the Cu–Fe contact metasomatic deposits are considered as orientated dipping slabs with a horizontal section of elliptical shape. The addition of a dip component to the horizontal dimensions of ore deposits is required in order to obtain the dimensions of the targets for detection purposes. Since strike orientation measurements were available for all deposits of the sample, the grid orientation factor was introduced in the calculation of probabilities of detection. It was found that the probability of detection by airborne surveys is maximized when the grid is laid at 90 degrees to the expected strike of the target. Optimal grid orientation for vertical detection by geophysical surveys or drilling is 30 degrees ± 10, from target strike.

Two types of tables are constructed for the display of the probability of detection of Cu–Fe deposits. One type covers the calculation of the probabilities of single detection (at least one target intersection by the grid) and confirmed detection (at least two intersections) by three types of field programs for grid spacings varied between 100 and 6000 feet in steps of 100–200 feet. In both cases, the probabilities are provided within 95% confidence intervals, which are based on the 95% fiducial intervals for the expected target length.

The second type of table is provided for planning purposes. The tables display grid spacings and corresponding coverage cost per unit of area for specified levels of probability of detection for each main type of field program. The range of probability levels is from 0.10 to 0.95, and steps of 0.05 are used.

The detection probability tables are organized into three blocks, each covering one type of field program. The first block covers airborne geophysical surveys on parallel and square grids (Tables 7.3–7.6). The second block deals with vertical ground detection by geophysical surveys or drilling to a depth of 300 feet (Tables 7.7–7.11). The third block refers to angled detection by drilling to a vertical depth of 300 feet (Tables 7.10–7.12).

7.3.4. Detection Probability Tables

Four tables displaying the probabilities of detection of contact metasomatic (Cu–Fe) deposits are to be found in pages 152–155. Six more tables cover the probabilities of detection of these deposits by ground geophysical surveys and drilling programs (pages 156–161).

TABLE 7.3.
Probabilities of Detection of Contact Metasomatic (Cu–Fe) Deposits of the North American Cordillera Belt by Airborne Geophysical Surveys on Parallel Grids.

```
Survey design:  parallel lines with spacing S feet
Elliptical targets with expected major axis = L  feet
in the confidence interval:  l.c.l. ( geom. mean ( u.c.l.
and minor axis = B feet
R is defined as the ratio B/L with geometric  mean = 0.25
Orientation of flight lines:  90 degrees with strike-line.
```

Grid spacing	Probability of detection		
S feet	l.c.l. L= 982	geom. mean 1273	u.c.l. 1650 feet
200	1.000	1.000	1.000
400	0.901	0.924	1.000
600	0.849	0.885	0.912
800	0.792	0.844	0.882
1000	0.753	0.800	0.850
1200	0.627	0.750	0.817
1400	0.538	0.697	0.781
1600	0.470	0.610	0.740
1800	0.418	0.542	0.703
2000	0.376	0.488	0.632
2200	0.342	0.443	0.575
2400	0.314	0.407	0.527
2600	0.289	0.375	0.486
2800	0.269	0.348	0.452
3000	0.251	0.325	0.422
3200	0.235	0.305	0.395
3400	0.221	0.287	0.372
3600	0.209	0.271	0.351
3800	0.198	0.257	0.333
4000	0.188	0.244	0.316
4200	0.179	0.232	0.301
4400	0.171	0.222	0.287
4600	0.164	0.212	0.275
4800	0.157	0.203	0.263
5000	0.151	0.195	0.253
5200	0.145	0.188	0.243

TABLE 7.4
Determination of Grid Size for Specified Levels of Probability in the Detection of Contact Metasomatic (Cu-Fe) Deposits of the North American Cordillera Belt by Airborne Geophysical Surveys on Parallel Grids.

Survey design: parallel lines with spacing S feet
Unit Cost = US$70 per line-mile

Specified detection probability level	Required grid spacing S in feet	Corresponding cost in $ per mile square
0.10	8180	$ 185.18
0.15	5620	$ 205.77
0.20	4280	$ 226.36
0.25	3460	$ 246.82
0.30	2900	$ 267.45
0.35	2500	$ 287.84
0.40	2200	$ 308.00
0.45	1960	$ 328.57
0.50	1780	$ 347.64
0.55	1620	$ 368.15
0.60	1480	$ 389.73
0.65	1380	$ 407.83
0.70	1280	$ 428.75
0.75	1180	$ 453.22
0.80	960	$ 525.00
0.85	740	$ 639.46
0.90	480	$ 910.00
0.95	220	$1820.00

TABLE 7.5
Probabilities of Confirmed Detection of Contact Metasomatic (Cu–Fe) Deposits of the North American Cordillera Belt by Airborne Geophysical Surveys on Parallel Grids.

```
Survey design:  parallel lines with spacing S feet,
Randomly orientated elliptical targets with expected major axis = L feet
in the confidence interval:  l.c.l. ( geom. mean ( u.c.l.
and minor axis = B feet
R is defined as the ratio B/L with geometric mean = 0.25
```

	Probability of confirmed detection		
Grid spacing S feet	l.c.l. L= 982	geom. mean 1273	u.c.l. 1650 feet
150	1.000	1.000	1.000
300	0.926	1.000	1.000
450	0.653	0.915	0.934
600	0.432	0.628	0.901
750	0.319	0.456	0.661
900	0.262	0.353	0.509
1050	0.000	0.290	0.408
1200	0.000	0.257	0.339
1350	0.000	0.000	0.293
1500	0.000	0.000	0.264
1650	0.000	0.000	0.250
1800	0.000	0.000	0.000

DETECTION OF CONTACT METASOMATIC DEPOSITS

TABLE 7.6
Probabilities of Detection of Contact Metasomatic (Cu-Fe) Deposits of the North American Cordillera Belt by Geophysical Surveys on Square Grids.

Survey design: square grid with spacings S by S feet
Randomly orientated elliptical targets with expected major axis = L feet in the confidence interval: l.c.l. (geom. mean (u.c.l.
and minor axis = B feet
R is defined as the ratio B/L with geometric mean = 0.25

Grid spacing S feet	Probability of detection		
	l.c.l. L= 982	geom. mean 1273	u.c.l. 1650 feet
800	1.000	1.000	1.000
1000	0.957	1.000	1.000
1200	0.863	1.000	1.000
1400	0.780	0.919	1.000
1600	0.709	0.848	1.000
1800	0.648	0.785	0.923
2000	0.597	0.728	0.868
2200	0.552	0.678	0.816
2400	0.513	0.634	0.769
2600	0.480	0.595	0.726
2800	0.450	0.560	0.688
3000	0.424	0.529	0.652
3200	0.400	0.501	0.620
3400	0.379	0.476	0.591
3600	0.361	0.453	0.564
3800	0.343	0.433	0.540
4000	0.328	0.414	0.517
4200	0.313	0.396	0.496
4400	0.300	0.380	0.477
4600	0.288	0.365	0.459
4800	0.277	0.351	0.442
5000	0.267	0.339	0.427
5200	0.257	0.327	0.412

TABLE 7.7
Probabilities of Detection of Contact Metasomatic (Cu–Fe) Deposits of the North American Cordillera Belt by Ground Geophysical Surveys or Vertical Drilling to a Depth of 300 Feet.

Survey design: square grid with spacings S by S feet.
The expected shape of the ore deposits is elliptical
with major axis = L in the 95% confidence interval:
L.c.l. (geometric mean (U.c.l., and minor axis = B feet.
R is defined as the ratio B/L with geometric mean = 0.25
grid orientation 30 degrees + or - 10 from the expected
strike line.

Spacing S in feet	Probability of detection		
	L.c.l. L = 982	G. mean 1273	U.c.l. 1650ft
300	1.000	1.000	1.000
400	0.964	1.000	1.000
500	0.728	0.981	1.000
600	0.525	0.820	0.997
700	0.386	0.639	0.934
800	0.296	0.497	0.786
1000	0.189	0.318	0.533
1200	0.131	0.221	0.371
1400	0.097	0.162	0.273
1600	0.074	0.124	0.209
1800	0.058	0.098	0.165
2000	0.047	0.080	0.134
2400	0.033	0.055	0.093
2800	0.024	0.041	0.068
3200	0.018	0.031	0.052
3600	0.015	0.025	0.041
4000	0.012	0.020	0.033

TABLE 7.8
Determination of Grid Size for Specified Probability Levels in the Detection of Contact Metasomatic (Cu–Fe) Deposits of the North American Cordillera Belt by Vertical Drilling to a Depth of 300 Feet.

Survey Design: Square Grid With Spacings S by S Feet.
Unit Cost = $6000 Per Diamond Drill-Hole.

Probability Level Of Detection	Grid Size In Feet	Drilling Cost Per Mile Square In U.S.$. Thousands
0.05	2350	63.64
0.15	1450	129.81
0.25	1150	188.25
0.35	950	258.82
0.45	850	312.92
0.55	800	347.47
0.65	700	438.91
0.75	650	500.48
0.85	600	577.42
0.95	550	675.43

TABLE 7.9
Probabilities of Confirmed Detection of Contact Metasomatic (Cu–Fe) Deposits of the North American Cordillera Belt by Ground Geophysical Surveys or Vertical Drilling to a Depth of 300 Feet.

Survey design: square grid with spacings S by S feet. the expected shape of the ore deposits is elliptical with major axis = L in the 95% confidence interval: L.c.l. (geometric mean (U.c.l., and minor axis = B feet. R is defined as the ratio B/L with geometric mean = 0.25 grid orientation 0 degrees + or - 10 from the expected strike line.

Spacing S in feet	Probability of confirmed detection		
	L.c.l. L = 982	G. mean 1273	U.c.l. 1650ft
200	1.000	1.000	1.000
250	0.981	1.000	1.000
300	0.858	1.000	1.000
350	0.619	0.943	1.000
400	0.422	0.829	1.000
450	0.293	0.639	0.947
500	0.194	0.467	0.866
550	0.124	0.355	0.749
600	0.078	0.265	0.570
700	0.028	0.136	0.375
800	0.008	0.066	0.238
900	0.001	0.030	0.142
1000	0.000	0.012	0.081
1200	0.000	0.001	0.024

TABLE 7.10
Probabilities of Detection of Contact Metasomatic (Cu–Fe) Deposits of the North American Cordillera Belt by 55-Degree Angled Drilling to a Vertical Depth of 300 Feet.

```
Survey design: square grid with spacings S by S feet.
the expected shape of the ore deposits is elliptical
with major axis = L in the 95% confidence interval:
L.c.l. ( geometric mean ( U.c.l., and minor axis = B feet,
R is defined as the ratio B/L with geometric mean = 0.40
grid orientation 30 degrees + or - 10 from the expected
strike line.
```

	Probability of detection		
Spacing S in feet	L.c.l. L = 982	G. mean 1273	U.c.l. 1650ft
400	1.000	1.000	1.000
500	0.963	1.000	1.000
600	0.801	0.994	1.000
700	0.616	0.908	1.000
800	0.473	0.768	0.986
1000	0.303	0.509	0.810
1200	0.210	0.354	0.593
1400	0.155	0.260	0.436
1600	0.118	0.199	0.334
1800	0.094	0.157	0.264
2000	0.076	0.127	0.214
2400	0.053	0.088	0.148
2800	0.039	0.065	0.109
3200	0.030	0.050	0.084
3600	0.023	0.039	0.066
4000	0.019	0.032	0.053

TABLE 7.11
Determination of Grid Size for Specified Probability Levels in the Detection of Contact Metasomatic (Cu–Fe) Deposits of the North American Cordillera Belt by 55-Degree Angled Drilling to a Vertical Depth of 300 Feet.

Survey Design: Square Grid With Spacings S by S Feet.
Unit Cost = $7325 Per Diamond Drill-Hole.

Probability Level Of Detection	Grid Size In Feet	Drilling Cost Per Mile Square In U.S.$. Thousands
0.15	1800	114.29
0.25	1400	167.46
0.35	1200	214.39
0.45	1100	248.11
0.55	1000	289.81
0.65	900	346.39
0.75	850	382.03
0.85	750	476.30
0.95	700	535.83

TABLE 7.12
Probabilities of Confirmed Detection of Contact Metasomatic (Cu–Fe) Deposits of the North American Cordillera Belt by 55-Degree Angled Drilling to a Vertical Depth of 300 Feet.

Survey design: square grid with spacings S by S feet. The expected shape of the ore deposits is elliptical with major axis = L in the 95% confidence interval: L.c.l. (geometric mean (U.c.l., and minor axis = B feet. R is defined as the ratio B/L with geometric mean = 0.40 grid orientation 0 degrees + or - 10 from the expected strike line.

Spacing S in feet	Probability of confirmed detection		
	L.c.l. L = 982	G. mean 1273	U.c.l. 1650ft
350	1.000	1.000	1.000
400	0.865	1.000	1.000
450	0.575	1.000	1.000
500	0.399	0.955	1.000
600	0.183	0.525	1.000
700	0.075	0.294	0.747
800	0.023	0.159	0.477
900	0.003	0.080	0.304
1000	0.000	0.034	0.191
1200	0.000	0.002	0.065

7.4. PROBABILITY OF DETECTION OF Pb–Zn–Cu–Ag CONTACT METASOMATIC DEPOSITS

7.4.1. Geological Synopsis

7.4.1.1. Introduction. The Pb–Zn–Cu–Ag type of contact metasomatic deposits of the Cordilleran Belt exhibits marked differences from the other three types covered by the present chapter. Here, the emphasis is on metasomatism rather than contact association. The proximal environment of the deposits is intensely altered and skarnified, though the mineralogic context is quite different from that of typical contact-skarn deposits. The newly formed silicates accompanying the complex Pb–Zn–Cu–Ag mineralization are mainly amphiboles, epidote, and chlorite with a general absence of garnets which are so conspicuous in the contact skarns. Finally, there is often no obvious spatial relationship between the deposits and their metasomatic halos on the one hand, and intrusive bodies, if any in the vicinity, on the other hand.

The complex Pb–Zn–Cu–Ag deposits are not so common nor so widely distributed throughout the Cordillera as the Fe–Cu skarn deposits. In the Canadian section of the Cordillera, these deposits occur in two regions including the Yukon and the southcentral portion of British Columbia. The deposits of the central portion of the U.S. section of the Cordillera are larger, richer, and more numerous than in the Canadian section, and are the models of the type.

7.4.1.2. British Columbia and Yukon Section of the Cordillera. At the northern end of the Canadian Cordillera, in the Yukon, six Cu–Au–Ag metasomatic deposits occur in a skarnified Cambrian limestone horizon over a strike length of about 10 miles, near the town of Whitehorse. While copper, silver, gold mineralizations are present throughout, magnetite, and specularite are abundant only in the central portion of the mineralized belt, and molybdenite only at both ends.

In the Salmo area of southcentral British Columbia, two deposits of the Pb–Zn–Ag metasomatic type, the Emerald and the Jersey, occur within a structural trough of metasomatized Cambrian limestone and dolomite, between two small granitic stocks of Jurassic age.[19] Farther south and west, metasomatic deposits unusually rich in gold and showing only low copper values occur near the town of Hedley, at the Nickel Plate mine, in the general vicinity of basic intrusive stocks ranging in composition from diorite to gabbro. At the near-by and smaller French mine, the ore is mostly copper with minor gold and silver values.

7.4.1.3. U.S. Section of the Cordillera. Complex Pb–Zn–Cu–Ag metasomatic deposits are more common in the U.S. section of the Cordillera than in the Canadian section, particularly in the States of Idaho, Utah, and Colorado. In the Mackay district of Idaho, metasomatic deposits carrying mainly copper with minor lead–zinc–silver values occur in large blocks of skarnified limestone engulfed by a monzonitic intrusive. Similar mineralizations have been mined in the past in the San Francisco Mountains, in Utah,[2] as well as in Nevada and in Arizona.

However, the richest and most complex metasomatic base metal sulfide deposits of the whole North American Cordillera are found in the Tintic district of Utah and the Leadville district of Colorado.[20] The Tintic district had produced

DETECTION OF CONTACT METASOMATIC DEPOSITS

prior to the mid-1950s some 15 million tons of high-grade silver–lead–zinc ore with minor copper. The deposits occur as massive sulfide replacements in the highly metasomatized, folded, and faulted Ophir limestone formation of Cambrian age, which is cut across by innumerable dykes and sills of porphyritic quartz monzonite. The district was considered exhausted, until some geological rethinking in the mid-1950s led to the discovery of the high-grade Burgin No. 1 and large, low-grade Burgin No. 2 deposits, under a thick capping of Mississippian clastics and Eocene lava flows. This success spurred an intensified search for similar deposits in the presently revitalized district.

Farther to the east, in the central portion of the State of Colorado, the discovery of blanket or "manto"-type metasomatic deposits in the Leadville district made history because of the very high grade of the silver mineralization accompanied by moderate to low grades in base metals (lead, zinc, copper). The mineralization is found in two limestone horizons, the Blue Limestone and the White Limestone, separated by the barren Parting Quartzites, and cut across by the White Porphyry and the younger Grey Porphyry. The ore extends over considerable distances along the bedding of the Blue Limestone, either immediately below the White Porphyry–Blue Limestone contact, or between tongues of Grey Porphyry. In the lower White Limestone member, the sulfide mineralization extends between tongues of Grey Porphyry, or along the Parting Quartzite contact.

7.4.2. Statistical Modeling of Geometric Parameters of Pb–Zn–Cu–Ag Metasomatic Deposits

A total of 41 deposits were selected and included in the database to represent the whole population of complex base metal sulfide metasomatic deposits of the Cordillera. The deposits are listed by name and region of occurrence in Table 7.13. The geographic breakdown is as follows: Canadian section of the Cordillera: 5 deposits; Tintic district of Utah: 12 deposits; Leadville district of Colorado: 14 deposits. The remaining 10 deposits are scattered across the States of Nevada, Idaho, and Arizona.

The geometry of the deposits is described by means of dimensional and attitudinal parameters. The former include length, breadth, and shape ratio of the horizontal section of deposits, and the latter, dip and strike direction of the deposits. The range of measurements obtained from the 41 deposits is from 450 to 7500 feet for length, 15 to 420 feet for breadth, and 0 to 75 degrees for dip. The measurements are grouped into classes and assembled into frequency distributions whose summarizing statistics are listed in the lower half of Table 7.14.

TABLE 7.13
List of Contact Metasomatic (Pb–Zn–Cu) Deposits of the North American Cordillera Belt Included in the Database.

Utah (U.S.A.)	Colorado (U.S.A.)
Horn Silver	Iron Hill
Cactus	Henriett-Maid
Red Warrior	Cord
Tintic Standard	Minnie-Helen
North Lily	Sellers
Eureka Lily	Imes
Burgin	Giltedge
Gemini	Comstock
Mammoth Chief	Ibex
Plutus	Rickard-Stone
Iron Blossom	Golden Eagle
Godiva	Mahala
	Tucson
Nevada (U.S.A.)	Carbonate Hill
Simon	**Idaho (U.S.A.)**
Blue Stone	
Mason Valley	Empire
Casting Copper	
Ludwig	**B.C. & Yukon (Canada)**
Arizona (U.S.A.)	Jersey
	Emerald
Johnson	Nickel Plate
Copper World	French
Hardshell	Whitehorse Copper
Mowry	

7.4.3. Construction and Organization of Detection Probability Tables

As is the case for the Cu–Fe skarn deposits of Section 7.3, the complex Pb–Zn–Cu–Ag metasomatic deposits are considered as oriented dipping slabs with a horizontal section of elliptical shape, for detection purposes. One type of table is constructed to provide probabilities of detection or confirmed detection of the complex metasomatic deposits by means of three types of field programs for varying grid spacings. A second type of table displays the grid spacings and

DETECTION OF CONTACT METASOMATIC DEPOSITS 165

TABLE 7.14
Statistical Modeling Summary for Contact Metasomatic (Pb–Zn–Cu) Deposits of the North American Cordillera Belt (Sample Size: 41 Deposits).

Statistic of fitted model	Length of horiz.sect. in feet	Breadth of horiz. sect. in feet	Shape ratio R=B/L	Strike dir. from T.North in degrees	Dip in degrees
25%ile	900	85	0.06	12	0
95%L.c.l.	949	85	0.06	35	31
Mean	1218	111	0.09	55	42
95%U.c.l.	1562	146	0.14	74	53
75%ile	1600	220	0.18	110	70
Disp. coef.	0.11	0.18	0.54	0.38	0.18

Statistic of observed data					
Arithm.mean	1784	155	0.17	62	42
Mode	1100	210	0.18	10	0
Stand.dev.	2018	180	0.18	62	34
Skewness	0.62	0.30	0.56	0.83	1.21

corresponding coverage costs per unit of area which are required to obtain prespecified levels of detection probabilities.

The tables are organized into three blocks, one for each type of field program. Tables 7.15–7.18 cover airborne geophysical surveys; Tables 7.19–7.21 deal with vertical ground detection by geophysical surveys or drilling to a depth of 300 feet. Finally, Tables 7.22–7.24 refer to angled drilling to a vertical depth of 300 feet.

7.4.4. Detection Probability Tables

Pages 166–169 cover the probabilities of detection of contact metasomatic (Pb–Zn–Cu) of the Cordillera Belt by airborne geophysical surveys. The following pages, 170–175, include six tables of probabilities of detection of these deposits by ground geophysical surveys and drilling programs.

TABLE 7.15
Probabilities of Detection of Contact Metasomatic (Pb–Zn–Cu) of the North American Cordillera Belt by Airborne Geophysical Surveys on Parallel Grids.

```
Survey design:  parallel lines with spacing S feet
Elliptical targets with expected major axis = L feet
in the confidence interval:  l.c.l. ( geom. mean ( u.c.l.
and minor axis = B feet
R is defined as the ratio B/L with geometric mean = 0.17
Orientation of flight lines:   90 degrees with strike-line.
```

	Probability of detection		
Grid spacing S feet	l.c.l. L= 949	geom. mean 1217	u.c.l. 1562 feet
200	0.944	1.000	1.000
400	0.887	0.912	0.932
600	0.827	0.867	0.897
800	0.760	0.819	0.861
1000	0.691	0.767	0.824
1200	0.576	0.705	0.785
1400	0.493	0.633	0.741
1600	0.432	0.554	0.710
1800	0.384	0.492	0.632
2000	0.345	0.443	0.568
2200	0.314	0.403	0.517
2400	0.288	0.369	0.474
2600	0.266	0.341	0.437
2800	0.247	0.316	0.406
3000	0.230	0.295	0.379
3200	0.216	0.277	0.355
3400	0.203	0.260	0.334
3600	0.192	0.246	0.316
3800	0.182	0.233	0.299
4000	0.173	0.221	0.284
4200	0.164	0.211	0.271
4400	0.157	0.201	0.258
4600	0.150	0.193	0.247
4800	0.144	0.185	0.237
5000	0.138	0.177	0.227
5200	0.133	0.170	0.219

TABLE 7.16
Determination of Grid Size for Specified Probability Levels in the Detection of Contact Metasomatic (Pb–Zn–Cu) Deposits of the North American Cordillera Belt by Airborne Geophysical Surveys on Parallel Grids.

```
Survey design:  parallel lines with spacing S feet
Unit Cost = US$70 per line-mile
```

Specified detection probability level	Required grid spacing S in feet	Corresponding cost in $ per mile square
0.10	8160	$ 185.29
0.15	5600	$ 206.00
0.20	4280	$ 226.36
0.25	3460	$ 246.82
0.30	2900	$ 267.45
0.35	2500	$ 287.84
0.40	2200	$ 308.00
0.45	1960	$ 328.57
0.50	1760	$ 350.00
0.55	1600	$ 371.00
0.60	1480	$ 389.73
0.65	1360	$ 411.76
0.70	1200	$ 448.00
0.75	1040	$ 495.38
0.80	840	$ 580.00
0.85	640	$ 717.50
0.90	420	$1020.00
0.95	200	$1988.00

TABLE 7.17
Probabilities of Confirmed Detection of Contact Metasomatic (Pb–Zn–Cu) Deposits of the North American Cordillera Belt by Airborne Geophysical Surveys on Parallel Grids.

```
Survey design:  parallel lines with spacing S feet,
Randomly orientated elliptical targets with expected major axis = L feet
in the confidence interval:  l.c.l. ( geom. mean ( u.c.l.
and minor axis = B feet
R is defined as the ratio B/L with geometric mean = 0.17
```

	Probability of confirmed detection		
Grid spacing S feet	l.c.l. L= 949	geom. mean 1217	u.c.l. 1562 feet
500	0.486	0.736	1.000
600	0.349	0.545	0.818
700	0.262	0.415	0.640
800	0.207	0.325	0.510
900	0.176	0.261	0.414
1000	0.000	0.216	0.341
1100	0.000	0.187	0.286
1200	0.000	0.171	0.243
1300	0.000	0.000	0.211
1400	0.000	0.000	0.189
1500	0.000	0.000	0.174

DETECTION OF CONTACT METASOMATIC DEPOSITS

TABLE 7.18
Probabilities of Detection of Contact Metasomatic (Pb–Zn–Cu) Deposits of the North American Cordillera Belt by Airborne Geophysical Surveys on Square Grids.

```
Survey design:  square grid with spacings S by S feet
Randomly orientated elliptical targets with expected major axis = L feet
in the confidence interval:  l.c.l. ( geom. mean ( u.c.l.
and minor axis = B feet
R is defined as the ratio B/L with geometric mean = 0.17
```

	Probability of detection		
Grid spacing S feet	l.c.l. L= 949	geom. mean 1217	u.c.l. 1562 feet
800	1.000	1.000	1.000
1000	0.935	1.000	1.000
1200	0.833	1.000	1.000
1400	0.747	0.886	1.000
1600	0.676	0.811	0.950
1800	0.616	0.746	0.885
2000	0.565	0.689	0.826
2200	0.522	0.640	0.773
2400	0.484	0.597	0.725
2600	0.452	0.559	0.682
2800	0.424	0.525	0.644
3000	0.398	0.495	0.610
3200	0.376	0.468	0.578
3400	0.356	0.444	0.550
3600	0.338	0.423	0.524
3800	0.322	0.403	0.501
4000	0.307	0.385	0.479
4200	0.293	0.368	0.459
4400	0.281	0.353	0.441
4600	0.270	0.339	0.424
4800	0.259	0.326	0.409
5000	0.249	0.314	0.394
5200	0.240	0.303	0.380

TABLE 7.19
Probabilities of Detection of Contact Metasomatic (Pb–Zn–Cu) Deposits of the North American Cordillera Belt by Ground Geophysical Surveys or Vertical Drilling to a Depth of 300 Feet.

Survey design: square grid with spacings S by S feet. The expected shape of the ore deposits is elliptical with major axis = L in the 95% confidence interval: L.c.l. (geometric mean (U.c.l., and minor axis = B feet. R is defined as the ratio B/L with geometric mean = 0.17 grid orientation 30 degrees + or - 10 from the expected strike line.

Spacing S in feet	Probability of detection		
	L.c.l. L = 949	G. mean 1218	U.c.l. 1562 ft
200	1.000	1.000	1.000
300	0.970	1.000	1.000
400	0.717	0.952	1.000
500	0.478	0.747	0.965
600	0.334	0.542	0.821
700	0.245	0.403	0.646
800	0.188	0.309	0.504
1000	0.120	0.198	0.326
1200	0.084	0.137	0.226
1400	0.061	0.161	0.166
1600	0.047	0.077	0.127
2000	0.030	0.049	0.081

TABLE 7.20
Determination of Grid Size for Specified Levels of Probability in the Detection of Contact Metasomatic (Pb–Zn–Cu) Deposits of the North American Cordillera Belt by Vertical Drilling to a Depth of 300 Feet.

Survey Design: Square Grid With Spacings S by S Feet.
Unit Cost = $6000 Per Diamond Drill-Hole.

Probability Level Of Detection	Grid Size In Feet	Drilling Cost Per Mile Square In U.S.$. Thousands
0.05	1850	89.58
0.15	1150	188.25
0.25	900	283.73
0.35	750	388.81
0.45	700	438.91
0.65	650	500.48
0.55	600	577.42
0.65	550	675.43
0.75	500	803.19
0.85	450	974.35
0.95	400	1211.54

TABLE 7.21

Probabilities of Confirmed Detection of Contact Metasomatic (Pb–Zn–Cu) Deposits of the North American Cordillera Belt by Ground Geophysical Surveys or Vertical Drilling to a Depth of 300 Feet.

Survey design: square grid with spacings S by S feet. The expected shape of the ore deposits is elliptical with major axis = L in the 95% confidence interval: L.c.l. (geometric mean (U.c.l., and minor axis = B feet. R is defined as the ratio B/L with geometric mean = 0.17 grid orientation 0 degrees + or - 10 from the expected strike line.

Spacing S in feet	Probability of confirmed detection		
	L.c.l. L = 949	G. mean 1218	U.c.l. 1562 ft
150	1.000	1.000	1.000
200	0.932	1.000	1.000
250	0.792	0.943	1.000
300	0.469	0.844	0.968
350	0.304	0.698	0.902
400	0.197	0.422	0.816
450	0.124	0.302	0.696
500	0.077	0.215	0.454
550	0.047	0.151	0.349
600	0.029	0.105	0.269
700	0.010	0.049	0.156
800	0.003	0.023	0.088

TABLE 7.22
Probabilities of Detection of Contact Metasomatic (Pb–Zn–Cu) Deposits of the North American Cordillera Belt by 55-Degree Angled Drilling to a Vertical Depth of 300 Feet.

Survey design: square grid with spacings S by S feet. The expected shape of the ore deposits is elliptical with major axis = L in the 95% confidence interval: L.c.l. (geometric mean (U.c.l., and minor axis = B feet. R is defined as the ratio B/L with geometric mean = 0.33 grid orientation 30 degrees + or - 10 from the expected strike line.

Spacing S in feet	Probability of detection		
	L.c.l. L = 949	G. mean 1218	U.c.l. 1562ft
300	1.000	1.000	1.000
400	0.998	1.000	1.000
500	0.853	1.000	1.000
600	0.643	0.918	1.000
700	0.476	0.754	0.983
800	0.365	0.598	0.882
1000	0.233	0.384	0.629
1200	0.162	0.267	0.439
1400	0.119	0.196	0.323
1600	0.091	0.150	0.247
2000	0.058	0.096	0.158
2400	0.041	0.067	0.110

TABLE 7.23
Determination of Grid Size for Specified Probability Levels in the Detection of Contact Metasomatic (Pb–Zn–Cu) Deposits of the North American Cordillera Belt by 55-Degree Angled Drilling to a Vertical Depth of 300 Feet.

Survey Design: Square Grid With Spacings S by S Feet.
Unit Cost = $7325 Per Diamond Drill-Hole.

Probability Level Of Detection	Grid Size In Feet	Drilling Cost Per Mile Square In U.S.$. Thousands
0.15	1550	142.87
0.25	1250	200.67
0.35	1050	267.10
0.45	950	315.98
0.55	850	382.03
0.65	800	424.21
0.75	700	535.83
0.85	650	611.00
0.95	600	704.93

DETECTION OF CONTACT METASOMATIC DEPOSITS 175

TABLE 7.24.
Probabilities of Confirmed Detection of Contact Metasomatic (Pb–Zn–Cu) Deposits of the North American Cordillera Belt by 55-Degree Angled Drilling to a Vertical Depth of 300 Feet.

Survey design: square grid with spacings S by S feet.
the expected shape of the ore deposits is elliptical
with major axis = L in the 95% confidence interval:
L.c.l. (geometric mean (U.c.l., and minor axis = B feet.
R is defined as the ratio B/L with geometric mean = 0.33
grid orientation 0 degrees + or - 10 from the expected
strike line.

Spacing S in feet	Probability of confirmed detection		
	L.c.l. L = 949	G. mean 1218	U.c.l. 1562ft
300	1.000	1.000	1.000
350	0.841	1.000	1.000
400	0.563	1.000	1.000
450	0.396	0.839	1.000
500	0.265	0.520	1.000
600	0.111	0.345	0.780
700	0.039	0.180	0.473
800	0.010	0.088	0.297
900	0.001	0.038	0.179
1000	0.000	0.014	0.103
1100	0.000	0.003	0.056
1200	0.000	0.000	0.027

7.5. PROBABILITY OF DETECTION OF Cu–Mo–Au CONTACT METASOMATIC DEPOSITS

7.5.1. *Geological Synopsis*

7.5.1.1. Introduction. Differentiated plutons, some of them of the copper-bearing porphyry type, are known to produce intense contact metasomatism in carbonate and other kinds of "reactive" wall rock formations, resulting in the formation of typical garnetiferous skarns. Many examples have been described by Einaudi[5] in the southwest United States Cordillera, and by others in the southwest Pacific Island Arcs, and in the USSR.[22]

7.5.1.2. Canadian Cordillera. At the northern end of the Canadian Cordillera, in the Yukon, the large low-grade W–Mo deposit referred to as Lotung, consists of scheelite mineralization in garnet–diopside skarn formed by the metasomatic alteration of Pennsylvanian argillite and limestone by a differentiated quartz monzonite stock of Cretaceous age. In central British Columbia, pyrite–chalcopyrite mineralization with or without magnetite and specularite occur in Paleozoic limestone formations intruded by differentiated grano–diorite stocks of Mesozoic age.

At the Red Mountain deposit, near Rossland, in southern British Columbia, pyrite–pyrrhotite–molybdenite mineralization is hosted by "skarn-tuff" horizons in Triassic series of argillite and andesite tuffs intruded by a Mesozoic grano–diorite stock. In the Princeton region, Similkameen district, the Ingerbelle deposit is classified as a hybrid skarn–porphyry type similar to those described by Einaudi[5] in the southwest U.S. The Ingerbelle mineralization consists of disseminations of pyrite and chalcopyrite in intensely skarnified and propylitized andesitic tuffs and fragmentals of the Nicola series of Triassic age.

7.5.1.3. U.S. Cordillera. Einaudi[5] stresses the marked differences between contact metasomatic deposits associated with differentiated copper-bearing porphyry-type plutons of the southwest Cordillera and those associated with other types of intrusives. The porphyry-associated skarns are more massive and finer-grained and characterized by a greater abundance of andradite and diopside than the nonporphyry skarns. Furthermore, in the porphyry-associated skarns, pyrite prevails over pyrrhotite and specularite over magnetite, and retrograde alteration to clay minerals is much more intense than in the nonporphyry environment.

Einaudi describes the following paragenetic sequence: (a) initial "contact metamorphic" phase altering carbonates into marble and pelitic series into hornfels, (b) "metasomatic" phase, altering marbles into garnetiferous skarns and introducing actinolite and biotite in the hornfels, (c) introduction of sulfides during the main phase of the potassic–silicic alteration of the intrusive, and (d) retrograde alteration which may be so intense that intrusive and skarn alike are reduced to quartz–carbonate–clay assemblages.

The predominant economic minerals are chalcopyrite and molybdenite, with subordinate amounts of zinc and lead sulfides and minor gold and silver values. Generally, the Cu/Mo and Zn/Cu ratios and protore Cu grades are higher in the skarn than in the intrusive itself. In some cases, such as the Ely, Nevada, deposit, the skarns are mineralized with W and Mo and very minor copper. Very little supergene enrichment in copper and gold takes place because of the buffering influence exerted by the carbonate formations on the ground water acidified by the oxidation of pyrite.

7.5.2. Statistical Modeling of Geometric Parameters of Cu–Mo–Au Contact Metasomatic Deposits

A total of 15 deposits were included in the database to represent the population of Cu–Mo–Au contact metasomatic deposits of the Cordillera. The deposits are listed by name and region of occurrence in Table 7.25. The geographic breakdown is as follows: Canadian section of the Cordillera: 3 deposits; U.S. section: 11 deposits; and Mexico section: 1 deposit.

The geometry of the deposits is described by means of dimensional and attitudinal parameters. The former include length, breadth, and shape ratio of the horizontal section of deposits, and the latter, dip and strike direction of the deposits. The range of measurements obtained from the 15 deposits of the sample is as follows: 180–4850 feet for length, 135–1250 feet for breadth, and 0–60 degrees for dip. The measurements are grouped into classes and assembled into frequency distributions which are summarized by means of statistics, as listed in the lower half of Table 7.26.

Lognormal models were successfully fitted to the observed distributions of dimensional parameters at the 0.05 confidence level. However, the observed distributions of attitudinal parameters could not be satisfactorily fitted by normal

TABLE 7.25

List of Contact Metasomatic (Cu–Mo–Au) Deposits of North American Cordillera Belt Included in the Database.

Arizona (U.S.A.)	New Mexico (U.S.A.)
Christmas	Santa Rita
Twin Buttes	Continental
Mission	
Bisbee	Mexico
Lake Shore	Cananea
Silver Bell	
Morenci	British Columbia (Canada)
Utah (U.S.A.)	Red Mountain
Bingham	Similkameen
Nevada (U.S.A.)	Yukon (Canada)
Ely	Logtung

TABLE 7.26
Statistical Modeling Summary for Contact Metasomatic (Cu–Mo–Au) Deposits of the North American Cordillera Belt (Sample Size: 15 Deposits).

Statistic of fitted model	Length of horiz. sect. in feet	Breadth of horiz. sect. in feet	Shape ratio R=B/L	Strike dir. from T. North in degrees	Dip in degrees
25%ile	2200	300	0.12	0	0
95%L.c.l.	2206	330	0.13	46	3
Mean	2648	480	0.18	82	22
95%U.c.l.	3180	698	0.26	105	41
75%ile	4000	1000	0.50	115	65
Disp. coef.	0.05	0.12	0.38	0.41	0.12
Statistic of observed data					
Arithm. mean	2796	617	0.23	75	22
Mode	3000	450	0.25	110	0
Stand. Dev.	907	480	0.18	57	37
Skewness	0.01	0.37	0.08	0.61	0.62

or circular normal models at the required 0.05 confidence level. The statistics summarizing the fitted models are listed in the upper half of Table 7.26.

7.5.3. Construction and Organization of Detection Probability Tables

The Cu–Mo–Au contact metasomatic deposits are considered as orientated dipping slabs with elliptical horizontal sections, for detection purposes. One kind of probability table is constructed to provide the probabilities of detection and confirmed detection of the deposits by means of three types of field programs for varying grid spacings. A second kind of table displays the grid spacing and corresponding coverage cost per unit of area which are required to obtain prespecified levels of probability of detection for each type of field program.

The tables are organized into three blocks, one for each type of field program. The first block covers airborne geophysical surveys on parallel and square grids (Tables 7.27–7.30). The second block deals with ground vertical detection by geophysical surveys or drilling to a depth of 300 feet (Tables 7.31–7.33). The third block refers to angled drilling detection (Tables 7.34–7.36).

7.5.4. Detection Probability Tables

Four tables to be found in pages 179–182 display probabilities of detection by airborne geophysical surveys. Six additional probability tables cover the ground detection of the deposits by geophysical surveys and drilling programs (pages 183–188).

TABLE 7.27

Probabilities of Detection of Contact Metasomatic (Cu–Mo–Au) Deposits of the North American Cordillera Belt by Airborne Geophysical Surveys on Parallel Grids.

```
Survey design:   parallel lines with spacing S feet
Elliptical targets with expected major axis = L feet
in the confidence interval:  l.c.l. ( geom. mean ( u.c.l.
and minor axis = B feet
R is defined as the ratio B/L with geometric mean = 0.19
Orientation of flight lines:  90 degrees with strike-line.
```

Grid spacing S feet	Probability of detection		
	l.c.l. L=2206	geom. mean 1273	u.c.l. 3180 feet
400	1.000	1.000	1.000
600	0.929	0.941	0.951
800	0.905	0.921	0.934
1000	0.881	0.901	0.917
1200	0.856	0.881	0.901
1400	0.830	0.860	0.883
1600	0.803	0.839	0.866
1800	0.774	0.817	0.848
2000	0.743	0.794	0.830
2200	0.707	0.770	0.811
2400	0.678	0.743	0.792
2600	0.626	0.714	0.771
2800	0.581	0.698	0.750
3000	0.542	0.651	0.726
3200	0.509	0.610	0.726
3400	0.479	0.575	0.683
3600	0.452	0.543	0.645
3800	0.428	0.514	0.611
4000	0.407	0.488	0.581
4200	0.387	0.465	0.553
4400	0.370	0.444	0.528
4600	0.354	0.425	0.505
4800	0.339	0.407	0.484
5000	0.325	0.391	0.465
5200	0.313	0.376	0.447

TABLE 7.28
Determination of Grid Size for Specified Probability Levels in the Detection of Contact Metasomatic (Cu–Mo–Au) Deposits of the North American Cordillera Belt by Airborne Geophysical Surveys on Parallel Grids.

```
Survey design:   parallel lines with spacing S feet
Unit Cost = US$70 per line-mile
```

Specified detection probability level	Required grid spacing S in feet	Corresponding cost in $ per mile square
0.10	17520	$ 161.10
0.15	12060	$ 170.65
0.20	9180	$ 180.26
0.25	7420	$ 189.81
0.30	6220	$ 199.42
0.35	5360	$ 208.96
0.40	4700	$ 218.64
0.45	4200	$ 228.00
0.50	3780	$ 237.78
0.55	3460	$ 246.82
0.60	3160	$ 256.96
0.65	2920	$ 266.58
0.70	2640	$ 280.00
0.75	2280	$ 302.11
0.80	1880	$ 336.60
0.85	1420	$ 400.28
0.90	920	$ 541.74
0.95	420	$1020.00

TABLE 7.29
Probabilities of Confirmed Detection of Contact Metasomatic (Cu–Mo–Au) Deposits of the North American Cordillera Belt by Airborne Geophysical Surveys on Parallel Grids.

```
Survey design:  parallel lines with spacing S feet,
Randomly oriented elliptical targets with expected major axis = L feet
in the confidence interval:  l.c.l. ( geom. mean ( u.c.l.
and minor axis = B feet
R is defined as the ratio B/L with geometric mean = 0.19
```

	Probability of confirmed detection		
Grid Spacing S feet	l.c.l. L=2206	geom. mean 2648	u.c.l. 3180 feet
350	1.000	1.000	1.000
500	0.941	1.000	1.000
650	0.923	0.936	0.947
800	0.897	0.921	0.935
950	0.690	0.906	0.922
1100	0.546	0.730	0.910
1250	0.440	0.597	0.795
1400	0.362	0.495	0.667
1550	0.304	0.417	0.566
1700	0.260	0.355	0.486
1850	0.228	0.307	0.421
2000	0.205	0.269	0.368
2150	0.192	0.239	0.325
2300	0.000	0.217	0.289
2450	0.000	0.201	0.260
2600	0.000	0.191	0.237
2750	0.000	0.000	0.218
2900	0.000	0.000	0.204
3050	0.000	0.000	0.194
3200	0.000	0.000	0.000

TABLE 7.30
Probabilities of Detection of Contact Metasomatic (Cu–Mo–Au) Deposits of the North American Cordillera Belt by Airborne Geophysical Surveys on Square Grids.

Survey design: square grid with spacings S by S feet
Randomly orientated elliptical targets with expected major axis = L feet
in the confidence interval: l.c.l. (geom. mean (u.c.l.
and minor axis = B feet
R is defined as the ratio B/L with geometric mean = 0.19

Grid spacing S feet	Probability of detection		
	l.c.l. L=2206	geom. mean 2648	u.c.l. 3180 feet
2200	1.000	1.000	1.000
2400	0.919	1.000	1.000
2600	0.875	1.000	1.000
2800	0.834	0.934	1.000
3000	0.796	0.897	1.000
3200	0.760	0.861	0.960
3400	0.727	0.828	0.928
3600	0.697	0.796	0.897
3800	0.668	0.766	0.867
4000	0.642	0.738	0.839
4200	0.618	0.712	0.812
4400	0.595	0.687	0.786
4600	0.574	0.664	0.762
4800	0.554	0.642	0.738
5000	0.536	0.622	0.716
5200	0.518	0.603	0.696

TABLE 7.31
Probabilities of Detection of Contact Metasomatic (Cu–Mo–Au) Deposits of the North American Cordillera Belt by Ground Geophysical Surveys or Vertical Drilling to a Depth of 300 Feet.

```
Survey design: square grid with spacings S by S feet.
The expected shape of the ore deposits is elliptical
with  major axis = L in the 95% confidence interval:
L.c.l. ( geometric mean ( U.c.l., and minor axis = B feet,
R is defined as the ratio B/L with geometric mean = 0.19
grid orientation 30 degrees + or - 10 from the expected
strike line.
```

	Probability of detection		
Spacing S in feet	L.c.l. L = 2206	G. mean 2648	U.c.l. 3180ft
600	1.000	1.000	1.000
700	0.989	1.000	1.000
800	0.933	0.997	1.000
900	0.824	0.969	1.000
1000	0.700	0.900	0.991
1200	0.501	0.700	0.901
1400	0.371	0.529	0.735
1600	0.284	0.408	0.580
1800	0.224	0.323	0.464
2000	0.182	0.262	0.377
2400	0.126	0.182	0.262
2800	0.093	0.133	0.192
3200	0.071	0.102	0.147
3600	0.056	0.081	0.116
4000	0.045	0.065	0.094

TABLE 7.32
Determination of Grid Size for Specified Probability Levels in the Detection of Contact Metasomatic (Cu–Mo–Au) Deposits of the North American Cordillera Belt by Vertical Drilling to a Depth of 300 Feet.

Survey Design: Square Grid With Spacings S by S Feet.
Unit Cost = $6000 Per Diamond Drill-Hole.

Probability Level Of Detection	Grid Size In Feet	Drilling Cost Per Mile Square In U.S.$. Thousands
0.05	4250	30.44
0.15	2650	54.09
0.25	2050	77.14
0.35	1750	97.31
0.45	1550	117.03
0.55	1450	129.81
0.65	1250	164.37
0.75	1150	188.25
0.85	1050	218.79
0.95	950	258.82

TABLE 7.33

Probabilities of Confirmed Detection of Contact Metasomatic (Cu–Mo–Au) Deposits of the North American Cordillera Belt by Ground Geophysical Surveys or Vertical Drilling to a Depth of 300 Feet.

```
Survey design: square grid with spacings S by S feet.
The expected shape of the ore deposits is elliptical
with major axis = L in the 95% confidence interval:
L.c.l. ( geometric mean ( U.c.l., and minor axis = B feet,
R is defined as the ratio B/L with geometric mean = 0.19
grid orientation 0 degrees + or - 10 from the expected
strike line.
```

	Probability of confirmed detection		
Spacing S in feet	L.c.l. L = 2206	G. mean 2648	U.C.l. 3180ft
400	1.000	1.000	1.000
450	0.978	1.000	1.000
500	0.937	0.999	1.000
550	0.884	0.972	1.000
600	0.820	0.937	0.999
700	0.549	0.843	0.950
800	0.385	0.698	0.879
1000	0.187	0.343	0.562
1200	0.082	0.187	0.344
1400	0.035	0.095	0.209
1600	0.014	0.047	0.120

TABLE 7.34
Probabilities of Detection of Contact Metasomatic (Cu–Mo–Au) of the North American Cordillera Belt by 55-Degree Angled Drilling to a Vertical Depth of 300 Feet.

Survey design: square grid with spacings S by S feet. The expected shape of the ore deposits is elliptical with major axis = L in the 95% confidence interval: L.c.l. (geometric mean (U.c.l., and minor axis = B feet. R is defined as the ratio B/L with geometric mean = 0.26 grid orientation 30 degrees + or - 10 from the expected strike line.

Spacing S in feet	Probability of detection		
	L.c.l. L = 2206	G. mean 2648	U.c.l. 3180 ft
700	1.000	1.000	1.000
800	0.999	1.000	1.000
900	0.975	1.000	1.000
1000	0.886	0.996	1.000
1200	0.675	0.887	0.996
1400	0.507	0.708	0.918
1600	0.388	0.557	0.766
1800	0.307	0.442	0.629
2000	0.248	0.358	0.516
2400	0.173	0.249	0.359
2800	0.127	0.183	0.263
3200	0.097	0.140	0.202
3600	0.077	0.110	0.159
4000	0.062	0.089	0.129

TABLE 7.35
Determination of Grid Size for Specified Probability Levels in the Detection of Contact Metasomatic (Cu–Mo–Au) Deposits of the North American Cordillera Belt by 55-Degree Angled Drilling to a Vertical Depth of 300 Feet.

Survey Design: Square Grid With Spacings S by S Feet.
Unit Cost = $7325 Per Diamond Drill-Hole.

Probability Level Of Detection	Grid Size In Feet	Drilling Cost Per Mile Square In U.S.$. Thousands
0.15	3050	55.04
0.25	2350	77.69
0.35	2050	94.17
0.45	1850	109.37
0.55	1650	129.83
0.65	1550	142.87
0.75	1350	177.39
0.85	1250	200.67
0.95	1150	229.82

TABLE 7.36
Probabilities of Confirmed Detection of Contact Metasomatic Deposits (Cu–Mo–Au) of the North American Cordillera Belt by 55-Degree Angled Drilling to a Vertical Depth of 300 Feet.

```
Survey design:   square grid with spacings S by S feet
The expected shape of the ore deposits is elliptical
with major axis = L in the 95% confidence interval:
L.c.l. ( geometric mean( U.c.l., and minor axis = B feet
R is defined as the ratio B/L with geometric mean = 0.26
grid orientation 0 degrees + or - 10 from the expected
strike line.
```

	Probability	of confirmed	detection
Spacing S in feet	L.c.l. L = 2206	G. mean 2648	U.c.l. 3180 feet
600	1.000	1.000	1.000
700	0.847	1.000	1.000
800	0.633	0.895	1.000
900	0.445	0.757	0.943
1000	0.323	0.545	0.856
1200	0.154	0.324	0.546
1400	0.067	0.176	0.356
1600	0.027	0.090	0.219
1800	0.009	0.043	0.127
2000	0.002	0.109	0.071
2200	0.000	0.007	0.038
2500	0.000	0.001	0.013

7.6. PROBABILITY OF DETECTION OF W–Mo CONTACT METASOMATIC DEPOSITS

7.6.1. Geological Synopsis

7.6.1.1. Introduction. Tungsten-bearing metasomatic deposits occur within a number of well-defined metallogenic districts of the world. The earliest known are associated with Paleozoic foldbelts, but Mesozoic fold belts are the most prolific setting for these deposits, and tungsten-rich skarn deposits are well represented in the North American Cordillera.

Scheelite is the principal tungsten-bearing mineral occurring in contact metasomatic deposits, whereas wolframite is the most common W-rich mineral in pegmatitic deposits. In the contact metasomatic deposits, scheelite is generally restricted to very fine-grained exoskarn referred to as "tactites." The tactites form around the apex of small stocks or apophyses of larger stocks of intermediate composition (grano–diorite, monzonite) resulting from the contamination of the more acidic intrusive phases by carbonate country rocks.

The silicate gangue of the ore is made up of a very fine-grained admixture of Ca-rich pyroxene and garnets, with secondary quartz and minor fluorite. The economic mineralization consists of scheelite with minor wolframite and sulfides. The latter include pyrite, chalcopyrite, molybdenite, galena, and sphalerite. The oxide phase, magnetite and specularite, is usually present, but is far less abundant than in other contact metasomatic deposits.

7.6.1.2. Canadian Section of the Cordillera. The northern end of the Canadian Cordillera, in the Yukon Territory, is well endowed with tungsten–skarn deposits.[3] Cathro lists a total of 22 scheelite–skarn occurrences in the Selwyn Tungsten Belt, which extends over a length of 400 miles and a width of 50 miles, northwest of the town of Whitehorse. The scheelite mineralization occurs in dark green skarns made up of tremolite, actinolite, epidote, and garnet. The skarn alteration affects Cambrian limestone formations, while overlying Silurian to Mississippian pelitic to clastic formations are altered into hornfels in the vicinity of small discordant stocks of Cretaceous quartz monzonite. The scheelite deposits are generally confined to within half a mile from intrusive contacts, and are of high grade but small size. However, the Tungsten Belt has yielded so far two major deposits. One of them, the Cantung deposit, was discovered in 1962;[23] it includes two scheeelite-rich horizons, one of which was amenable to open-pitting. The second one, the Mactung deposit, some 100 miles away, was discovered more recently. It is made up of four skarn horizons mineralized with scheelite, pyrrhotite, and chalcopyrite, separated by pyritized hornfels.

Much farther south, lies a smaller scheelite–skarn district, in the Salmo area of south central British Columbia, which was the main source of tungsten ore for the Canadian war effort of the 1940–1945 period.[1,19] The Emerald, Dodger, and Feeney W-ore deposits occur along a mile-long trough occupied by Cambrian limestone resting on Precambrian quartzites and overlain by Silurian argillite, which dip easterly against the subvertical greizenized contact of a Jurassic granitic stock. Some sections of the trough are also mineralized with lead–zinc–silver ore (Jersey and Emerald zones), as already mentioned in Section 7.4.1.2.

7.6.1.3. U.S. Cordillera Section. There are as many as 30 scheelite mineralized districts in the U.S. section of the Cordillera, which are scattered across nearly all western states but Oregon. The geographic breakdown is as follows: California: 11 districts; Nevada: 8 districts; Utah: 6 districts; the remaining five being shared between the states of Idaho and Washington.[10] The most productive belt extends through western California and Nevada over a length of 700 miles and a width of 200 miles. The main commercial districts of the tungsten belt are those of Atolia, Bishop, Fresno, and Kern in California[12] and the Mill City district in Nevada.

The main geological setting for the scheelite deposits is that of skarnified carbonate formations included in roof pendants which were engulfed by the large quartz monzonite batholiths of the Sierra Nevada. Generally, the intrusive is contaminated into quartz diorite in the vicinity of the pendants, and the contacts are greizenized in the vicinity of mineralized skarn beds. The scheelite-bearing skarns exhibit assemblages of andradite, diopside, and epidote with sprinklings of pyrite, chalcopyrite, galena, and sphalerite. Minor molybdenite is known to occur among nests of fibrous amphiboles, very close to the intrusive contact.

The Pine Creek deposits, among the largest in the world, are located in the Kern district.[20] They produced a total of 7 million tons of ore from six ore zones stretching through the "Tungsten Hills" of central-western California. The Mill City district of Nevada was also an important producer during the war years. Farther northeast, in Idaho, lies the important district of Yellow Pine, which includes 20 known occurrences of scheelite-bearing skarns within an area 200 miles long and 100 miles wide.

7.6.2. Statistical Modeling of Geometric Parameters of W–Mo Contact Metasomatic Deposits

A total of 37 deposits were selected and included in the database to represent the whole population of W–Mo contact metasomatic deposits of the Cordillera. The deposits are listed by name and by region of occurrence in Table 7.37. The geographic breakdown is as follows: Idaho: 1 deposit; Utah: 7 deposits; California: 7 deposits; Nevada: 16; and British Columbia–Yukon: 6 deposits.

The geometry of the deposits is described by means of dimensional and attitudinal parameters. The former include length, breadth, and shape ratio of the horizontal section of deposits, and the latter, dip and strike direction of the deposits. The parameters are measured on all 37 deposits of the sample, and the measurements are grouped into classes and assembled into frequency distributions. The distributions are summarized by means of statistics which are listed in the lower half of Table 7.38.

Lognormal models were fitted to the observed distributions of the dimensional parameters at the 0.05 confidence level, but the distributions of attitudinal

TABLE 7.37
List of Contact Metasomatic (W–Mo) Deposits of the North American Cordillera Belt Included in the Database.

Utah (U.S.A.)	Nevada (Cont'd)
Old Hickory	Sutton
Garnet	Humbolt
Strategic Metals	Oreana
Cupric	Victory
Copper Ranch	T.N.T.
Copper King	Riley
Daily Metal	Kirby
	Valley View
Idaho (U.S.A.)	Tip Top
Yellow Pine	Granite Creek
	Pacific
California (U.S.A.)	Alpine
Darwin Hill	Tempiute
St. Charles	
Durham	British Columbia
Little Sister	& Yukon (Canada)
Lookout	Mactung
Round Valley	Cantung
Pine Creek	Emerald
	Dodger
Nevada (U.S.A.)	Feeney
St. Anthony	Swakum Mountain
Rose Creek	
Friedman	

parameters could not be fitted satisfactorily by normal or circular normal models at the required 0.05 confidence level. The statistics summarizing the fitted distributions are listed in the upper half of Table 7.38.

7.6.3. Construction and Organization of Detection Probability Tables

As is the case for the other three types of contact metasomatic deposits, the W–Mo contact metasomatic deposits of the Cordillera are considered as oriented dipping slabs with elliptical horizontal section for detection purposes. One kind of table is constructed to provide probabilities of detection or confirmed detection of the W–Mo deposits by means of two types of ground field programs. No tables are provided for airborne geophysical detection since no such methods

TABLE 7.38
Statistical Modeling Summary for Contact Metasomatic (W–Mo) Deposits of the North American Cordillera Belt (Sample Size: 37 Deposits).

Statistic of fitted model	Length of horiz. sect. in feet	Breadth of horiz. sect. in feet	Shape ratio R=B/L	Strike dir. from T. North in degrees	Dip in in degrees
25%ile	500	25	0.04	0	25
95% L.c.l.	598	33	0.05	27	42
Mean	724	47	0.06	36	49
95% U.c.l.	875	66	0.09	55	57
75%ile	1050	67	0.13	70	65
Disp. coef.	0.93	0.31	0.21	0.15	0.12
Statistic of observed data					
Arithm. mean	880	93	0.12	41	49
Mode	800	40	0.04	0	60
Stand. dev.	633	90	0.06	45	25
Skewness	0.13	0.42	1.33	0.91	0.43

are available in the present state of the technology. A second kind of table displays the grid spacings and corresponding coverage cost per unit of area which are required to obtain prespecified levels of detection probabilities.

The tables are organized into two blocks. The first one covers ground vertical detection by geophysical surveys and drilling to a depth of 300 feet (Tables 7.39–7.41) and the second block deals with angled drilling detection (Tables 7.42–7.44).

7.6.4. Detection Probability Tables

The reader will find a series of six tables displaying the probabilities of detection of contact metasomatic (W–Mo) deposits of the Cordillera by ground geophysical surveys and drilling programs in pages 193–197.

TABLE 7.39

Probabilities of Detection of Contact Metasomatic (W–Mo) Deposits of the North American Cordillera Belt by Ground Geophysical Surveys or Vertical Drilling to a Depth of 300 Feet.

Survey design: square grid with spacings S by S feet.
The expected shape of the ore deposits is elliptical
with major axis = L in the 95% confidence interval:
L.c.l. (geometric mean (U.c.l., and minor axis = B feet,
R is defined as the ratio B/L with geometric mean = 0.27
grid orientation 30 degrees + or - 10 from the expected
strike line.

	Probability of detection		
Spacing S in feet	L.C.l. L = 598	G. mean 724	U.c.l. 875ft
200	1.000	1.000	1.000
300	0.792	0.978	1.000
400	0.474	0.679	0.897
500	0.303	0.445	0.640
600	0.211	0.309	0.451
700	0.155	0.227	0.331
800	0.118	0.174	0.254
1000	0.076	0.111	0.162
1200	0.053	0.077	0.113
1400	0.039	0.057	0.083
1600	0.030	0.043	0.063
2000	0.019	0.028	0.041

TABLE 7.40
Determination of Grid Size for Specified Probability Levels in the Detection of Contact Metasomatic (W–Mo) Deposits of the North American Cordillera Belt by Vertical Drilling to a Depth of 300 Feet.

Survey Design: Square Grid With Spacings S by S Feet.
unit Cost = $6000 Per Diamond Drill-Hole.

Probability Level Of Detection	Grid Size In Feet	Drilling Cost Per Mile Square In U.S.$. Thousands
0.05	1400	137.17
0.15	900	283.73
0.25	700	438.91
0.35	600	577.42
0.45	500	803.19
0.55	450	974.35
0.65	410	1157.27
0.75	390	1269.95
0.85	350	1554.43
0.95	310	1953.14

TABLE 7.41
Probabilities of Confirmed Detection of Contact Metasomatic (W–Mo) Deposits of the North American Cordillera Belt by Ground Geophysical Surveys or Vertical Drilling to a Depth of 300 Feet.

Survey design: square grid with spacings S by S feet.
The expected shape of the ore deposits is elliptical
with major axis = L in the 95% confidence interval:
L.c.l. (geometric mean (U.c.l., and minor axis = B feet,
R is defined as the ratio B/L with geometric mean = 0.27
grid orientation 0 degrees + or - 10 from the expected
strike line.

Spacing S in feet	Probability of confirmed detection		
	L.c.l. L = 598	G. mean 724	U.c.l. 875 ft
150	1.000	1.000	1.000
200	0.813	1.000	1.000
250	0.438	0.771	0.955
300	0.236	0.449	0.779
350	0.115	0.273	0.494
400	0.052	0.154	0.333
450	0.022	0.083	0.212
500	0.007	0.042	0.130
550	0.001	0.050	0.077
600	0.000	0.008	0.044
800	0.000	0.000	0.002

TABLE 7.42
Probabilities of Detection of Contact Metasomatic (W–Mo) Deposits of the North American Cordillera Belt by 55-Degree Angled Drilling to a Vertical Depth of 300 Feet.

Survey design: square grid with spacings S by S feet. The expected shape of the ore deposits is elliptical with major axis = L in the 95% confidence interval: L.c.l. (geometric mean (U.c.l., and minor axis = B feet, R is defined as the ratio B/L with geometric mean = 0.56 grid orientation 30 degrees + or - 10 from the expected strike line.

	Probability of detection		
Spacing S in feet	L.c.l. L = 598	G. mean 724	U.c.l. 875 ft
300	1.000	1.000	1.000
400	0.900	1.000	1.000
500	0.628	0.863	1.000
600	0.437	0.638	0.871
700	0.321	0.470	0.681
800	0.246	0.360	0.526
1000	0.157	0.231	0.337
1200	0.109	0.160	0.234
1400	0.080	0.118	0.172
1600	0.061	0.090	0.132
2000	0.039	0.058	0.084

TABLE 7.43

Determination of Grid Size for Specified Probability Levels in the Detection of Contact Metasomatic (W–Mo) Deposits of the North American Cordillera Belt by 55-Degree Angled Drilling to a Vertical Depth of 300 Feet.

Survey Design: Square Grid With Spacings S by S feet.
Unit Cost = $7325 Per Diamond Drill-Hole.

Probability Level Of Detection	Grid Size In Feet	Drilling Cost Per Mile Square In U.S.$. Thousands
0.15	1300	188.40
0.25	1000	289.81
0.35	900	346.39
0.45	800	424.21
0.55	750	480.02
0.65	700	535.83
0.75	650	611.00
0.85	600	704.93
0.95	500	980.56

TABLE 7.44

Probabilities of Confirmed Detection of Contact Metasomatic (W–Mo) Deposits of the North American Cordillera Belt by 55-Degree Angled Drilling to a Vertical Depth of 300 Feet.

Survey design: square grid with spacings S by S feet. The expected shape of the ore deposits is elliptical with major axis = L in the 95% confidence interval: L.c.l. (geometric mean (U.c.l., and minor axis = B feet, R is defined as the ratio B/L with geometric mean = 0.56 grid orientation 0 degrees + or - 10 from the expected strike line.

Spacing S in feet	Probability of confirmed detection		
	L.c.l. L = 598	G. mean 724	U.c.l. 875 ft
250	1.000	1.000	1.000
300	0.961	1.000	1.000
350	0.384	1.000	1.000
400	0.192	0.570	1.000
450	0.091	0.285	0.898
500	0.035	0.159	0.432
550	0.007	0.085	0.269
600	0.000	0.039	0.166
800	0.000	0.000	0.008

7.7. DESIGNING THREE STRATEGIES FOR THE DETECTION OF FOUR TYPES OF CONTACT METASOMATIC DEPOSITS OF THE NORTH AMERICAN CORDILLERA

7.7.1. Introduction

Three strategies are considered for the detection of four types of contact metasomatic deposits in the North American Cordillera, namely, the liminal, maximal, and optimal options, as described previously in Section 5.5 of Chapter 5.

The liminal strategy is designed to ensure a minimum level of 0.500 of detection probability at a relatively low cost for the coverage of large areas of little known potential. The maximal strategy ensures a near certain detection (probability level of 0.980); but, because of its high cost, the maximal option should be considered only for the coverage of small areas of very high potential. The optimal strategy, a compromise option bracketed by the first two, should be most attractive in the majority of exploration situations, when dealing with areas of moderate size and fair to good potential.

It is most useful for planning purposes to compare the detection performances of various types of field programs or grid designs. When the probability of detection is held at a specified level, as is the case for the first two options, the coverage cost per unit of area is a convenient yardstick for the comparisons. When dealing with the optimal option, however, we have to use the expected loss criterion, as described in Section 6.5.4 of Chapter 6, because the level of optimal probability of detection varies from one type of field program to another.

The detection characteristics associated with the three strategies are listed in Tables 7.45, 7.46 and 7.47, covering the four types of contact metasomatic deposits of the Cordillera.

7.7.2. Liminal Strategy

7.7.2.1. Cost Multiplier for Confirmed Detection. For planning purposes, we need to know what additional cost is required to obtain confirmed detection instead of single detection. For airborne geophysical surveys, the cost multiplier is the highest (2.3) for the smaller deposits (Pb–Zn–Cu–Ag type) and the lowest (1.7) for the larger deposits (Cu–Mo–Au type). The same range of variation of the cost multiplier applies when dealing with ground geophysical surveys. The cost multiplier for vertical and angled drilling programs varies from 2.2 for the larger deposits (Cu–Mo–Au type) to 2.8 for the smaller ones (W–Mo type).

TABLE 7.45
Liminal Detection Strategy: Summary of Detection Characteristics for Four Types of Contact Metasomatic Deposits of the North American Cordillera Belt.

Type of survey and grid geometry	Cu-Fe type Coverage cost: US$/mile sq. Grid size: feet		Pb-Zn-Cu type Coverage cost: US$/mile sq. Grid size: feet		Cu-Mo type Coverage cost: US$/mile sq. Grid size: feet		W-Mo type Coverage cost: US$/mile sq. Grid size: feet	
	Detection	Confirmed detection	Detection	Confirmed detection	Detection	Confirmed detection	Detection	Confirmed detection
Airborne geophysical parallel grid	$388 1500 ft	$830 550 ft	$425 1300 ft	$985 460 ft	$255 3200 ft	$448 1200 ft	–	–
Airborne geophysical square grid	$436 2500 ft	–	$470 2300 ft	–	$280 5400 ft	–	–	–
Ground geophysical square grid	$1675 630 ft	$3200 375 ft	$2500 450 ft	$4950 275 ft	$825 1200 ft	$1408 725 ft	$3198 375 ft	$6690 230 ft
Vertical drilling square grid	$538,000 630 ft	$1,380,000 375 ft	$974,000 450 ft	$1,815,000 275 ft	$175,000 1200 ft	$414,000 725 ft	$1,382,000 375 ft	$3,723,000 230 ft
55-degree angled drilling square grid	$460,000 770 ft	$1,110,000 450 ft	$634,000 650 ft	$1,480,000 400 ft	$166,400 1400 ft	$357,000 900 ft	$898,750 550 ft	$2,480,000 325 ft

TABLE 7.46
Maximal Detection Strategy: Summary of Detection Characteristics for Four Types of Contact Metasomatic Deposits of the North American Cordillera Belt.

Type of survey and grid geometry	Cu-Fe type		Pb-Zn-Cu type		Cu-Mo type		W-Mo type	
	Coverage cost: US$/mile sq. Grid size: ft		Coverage cost: US$/mile sq. Grid size: ft		Coverage cost: US$/mile sq. Grid size: ft		Coverage cost: US$/mile sq. Grid size: ft	
	Detection	Confirmed detection	Detection	Confirmed detection	Detection	Confirmed detection	Detection	Confirmed detection
Airborne geophysical parallel grid	$1525 300 ft	-	$1990 200 ft	-	$910 500 ft	-	-	-
Airborne geophysical square grid	$880 1000 ft	-	$975 900 ft	-	$465 2300 ft	-	-	-
Ground geophysical square grid	$2915 400 ft	$5500 260 ft	$5590 250 ft	$10,000 175 ft	$1355 750 ft	$2345 475 ft	$6690 230 ft	$11,105 160 ft
Vertical drilling square grid	$1,211,000 400 ft	$2,910,000 260 ft	$2,940,000 250 ft	$6,123,000 175 ft	$388,800 750 ft	$888,000 475 ft	$3,723,000 230 ft	$7,800,000 160 ft
55-degree angled drilling square grid	$900,000 520 ft	$2,480,000 350 ft	$1,286,000 450 ft	$3,480,000 300 ft	$323,000 925 ft	$705,000 600 ft	$1,520,000 420 ft	$3,680,000 275 ft

DETECTION OF CONTACT METASOMATIC DEPOSITS 201

7.7.2.2. Comparison of Airborne Grid Designs. A comparison of the detection performance of the parallel and square designs for the control grids of airborne geophysical surveys is of interest for planning purposes. In all three cases (Cu–Fe, Pb–Zn–Cu, Cu–Mo–Au types) there is no incentive in switching from the parallel design to the square design, because choosing the latter would lead to an average increase of coverage cost of 10%.

7.7.2.3. Comparison of Drilling Programs. When dealing with dipping targets such as the four types of contact metasomatic deposits, it is most appropriate to compare the merits of vertical drilling programs against angled drilling to the same vertical depth, because of the very high costs involved in detection by drilling. In all four cases, the 55-degree angled detection should be preferred to vertical drilling, because it leads to a substantial saving in coverage costs. The saving varies from 10% for the larger deposits (Cu–Mo–Au) to as much as 36% for the smaller deposits (W–Mo type).

7.7.3. Maximal Strategy

7.7.3.1. Cost Multiplier for Confirmed Detection. The cost multipliers required to obtain confirmation of single detection in the context of the maximal strategy are nearly the same as the ones listed above for the liminal option for the first three types of deposits. But in the case of the smaller deposits (W–Mo type), all maximal cost multipliers are lower than their liminal counterparts by some 20%–30%, thus making the confirmation requirement more attractive in the maximal situation than in the liminal one.

7.7.3.2. Comparison of Airborne Grid Designs. Contrary to the liminal situation, it is advantageous in all three cases (Cu–Fe, Pb–Zn–Cu, and W–Mo–Au types) to shift from the parallel to the square grid design, as underscored by the very substantial reduction in coverage cost per unit of area (average 46%) attained when using the square design.

7.7.3.3. Comparison of Drilling Programs. For all four types of contact metasomatic deposits, good planning will require the selection of angled over vertical drilling programs because of the substantial reduction in coverage costs incurred. The reduction is only modest (17%) for the larger deposits (Cu–Mo–Au type), but is very attractive (60%) for the detection of the smaller deposits (W–Mo type).

7.7.4. Optimal Strategy

7.7.4.1. Introduction. Design optimization based on the efficiency criterion was carried out for each type of field program for the detection of the

TABLE 7.47
Optimal Detection Strategy: Summary of Detection Characteristics for Four Types of Contact Metasomatic Deposits of North American Cordillera Belt.

Type of survey and grid geometry	Cu-Fe type		Pb-Zn-Cu type		Cu-Mo type		W-Mo type	
	Coverage cost: US$/mile sq. Grid size: ft	Detection probability	Coverage cost: US$/mile sq. Grid size: ft	Detection probability	Coverage cost: US$/mile sq. Grid size: ft	Detection probability	Coverage cost: US$/mile sq. Grid size: ft	Detection probability
Airborne geophysical parallel grid	$551 900 ft	0.82	$602 800 ft	0.82	$345 1800 ft	0.82	—	—
Airborne geophysical square grid	$599 1610 ft	0.84	$617 1550 ft	0.83	$428 2560 ft	0.98	—	—
Ground geophysical square grid	$1946 550 ft	0.92	$2190 500 ft	0.75	$990 1000 ft	0.90	$3483 350 ft	0.84
Vertical drilling square grid	$675,000 550 ft	0.92	$803,000 500 ft	0.75	$237,000 1000 ft	0.90	$1,554,000 350 ft	0.84
55-degree angled drilling square grid	$536,000 700 ft	0.92	$705,000 600 ft	0.93	$247,000 1100 ft	0.98	$981,000 500 ft	0.86
Optimal 45-degree angled drilling square grid	$573,000 700 ft	0.96	$754,000 600 ft	0.96	$265,000 1100 ft	0.99	$1,581,000 400 ft	0.98

four types of contact metasomatic deposits of the Cordillera. The parameters of optimized airborne geophysical surveys, ground geophysical surveys, and vertical and angled drilling programs are listed in Tables 7.48–7.50 for the Cu–Fe type; Tables 7.51–7.53 for the Pb–Zn–Cu type; Tables 7.54–7.56 for the Cu–Mo–Au type, and finally Tables 7.57–7.58 for the W–Mo type. All parameters are also assembled for the readers' convenience in one summary tabulation (Table 7.47).

7.7.4.2. Comparison of Airborne Survey Grid Designs. Switching from the parallel to the square grid design results in a modest reduction of the expected loss criterion, averaging 10% for all three types of deposits, as in the instance of the liminal option.

7.7.4.3. Comparison of Drilling Programs. Optimally angled drilling is vastly superior to vertical drilling for all four types of deposits. The reduction in the expected loss criterion varies from 60% for the Cu–Fe type to as much as 86% for the smaller deposits (W–Mo type).

7.7.4.4. Optimization Tables. The reader will find below in pages 203–209 a series of eleven tables covering the optimization of the search for four types of contact metasomatic deposits of the Cordillera by airborne and ground programs.

TABLE 7.48
Optimal Design of Airborne Geophysical Surveys for the Detection of Contact Metasomatic (Cu–Fe) Deposits of the North American Cordillera Belt.

Confidence interval: all results are reported as 95% confidence intervals respectively
Expected target length in feet: 982   1273   1650
Expected shape ratio R = 0.25
Unit Cost: $70/1.ml.

(i) Survey Design: Parallel Grid

Optimal grid spacing in feet:	800	900	1100
Cost in US$ per mile square:	602	551	476
Probability of detection:	0.79	0.82	0.83
Optimal grid orientation:	90 degrees to expected strike		

(ii) Survey Design: Square Grid

Optimal grid spacing in feet:	1275	1610	1950
Cost in US$ per mile square:	720	599	519
Probability of detection:	0.83	0.84	0.88

TABLE 7.49
Optimal Design of Ground Geophysical Surveys for the Detection of Contact Metasomatic (Cu–Fe) Deposits of the North American Cordillera Belt.

Survey Design: Square Grid With Spacing S by S Feet.
Confidence Interval: All results are reported as 95% confidence intervals respectively.
Expected Target Length in Feet: 982   1273   1650
Expected Shape Ratio R = 0.25
Unit Cost: $6/Station

Optimal Grid Spacing in Feet:	500	550	750
Cost in $ Per Mile Square:	2189	1946	1352
Probability of Detection:	0.73	0.92	0.86
Optimal Grid Orientation:	30 degrees + or −10 to expected strike direction of target.		

TABLE 7.50
Optimal Designs for the Detection of Contact Metasomatic (Cu–Fe) Deposits of the North American Cordillera Belt by Drilling to a Vertical Depth of 300 Feet.

Survey design: square grid with spacing S by S feet
Confidence interval: all results are reported as 95% confidence intervals respectively
Expected target length in feet: 982   1273   1650
Unit Cost = $20 per linear ft.

(a) Vertical Drilling:

Optimal grid spacing in feet:	500	550	750
Cost in $thou. per mile square:	803	675	390
Probability of detection:	0.73	0.92	0.86
Optimal grid orientation:	30 degrees to expected strike of target		

(b) Angled drilling at 55 degrees:

Optimal grid spacing in feet:	600	700	900
Cost in $thou. per mile square:	705	536	346
Probability of detection:	0.89	0.92	0.89
Optimal grid orientation:	45 degrees to expected strike of target		

(c) Angled drilling at optimal angle = 50 degrees:

Optimal grid spacing in feet:	600	700	900
Cost in $thou. per mile square:	754	573	370
Probability of detection:	0.93	0.96	0.92
Optimal grid orientation:	45 degrees to expected strike of target		

DETECTION OF CONTACT METASOMATIC DEPOSITS

TABLE 7.51
Optimal Design of Airborne Geophysical Surveys for the Detection of Contact Metasomatic (Pb–Zn–Cu) Deposits of the North American Cordillera Belt.

Confidence interval: all results are reported as 95% confidence intervals respectively
Expected target length in feet: 949   1217   1562
Expected shape ratio R = 0.17
Unit Cost: $70/l.ml.

(i) Survey Design: Parallel Grid

Optimal grid spacing in feet:	700	800	1000
Cost in US$ per mile square:	668	602	510
Probability of detection:	0.79	0.82	0.82
Optimal grid orientation:	90 degrees to expected strike		

(ii) Survey Design: Square Grid

Optimal grid spacing in feet:	1230	1550	1715
Cost in US$ per mile square:	742	617	571
Probability of detection:	0.82	0.83	0.91

TABLE 7.52
Optimal Design of Ground Geophysical Surveys for the Detection of Contact Metasomatic (Pb–Zn–Cu) Deposits of the North American Cordillera Belt.

Survey Design: Square Grid With Spacing S by S Feet.
Confidence Interval: All results are reported as 95% confidence intervals respectively.
Expected Target Length in Feet: 949   121   1562
Expected Shape Ratio R = 0.17
Unit Cost: $6/Station

Optimal Grid Spacing in Feet:	400	500	550
Cost in $ Per Mile Square:	2914	2189	1946
Probability of Detection:	0.72	0.75	0.90
Optimal Grid Orientation:	30 degrees + or -10 to expected strike direction of target.		

TABLE 7.53
Optimal Designs for the Detection of Contact Metasomatic (Pb–Zn–Cu) Deposits of the North American Cordillera Belt by Drilling to a Vertical Depth of 300 Feet.

```
Survey design:  square grid with spacing S by S feet
Confidence interval:  all results are reported as 95% confidence intervals
respectively
Expected target length in feet:        949      1217      1562
Unit Cost = $20 per linear ft.
```

(a) Vertical drilling:

Optimal grid spacing in feet:	400	500	550
Cost in $thou. per mile square:	1212	803	675
Probability of detection:	0.95	0.75	0.90
Optimal grid orientation:	30 degrees to expected strike of target		

(b) Angled drilling at 55 degrees:

Optimal grid spacing in feet:	500	600	700
Cost in $thou. per mile square:	900	705	536
Probability of detection:	0.93	0.93	0.96
Optimal grid orientation:	45 degrees with expected strike of target		

(c) Angled drilling at optimal angle = 50 degrees:

Optimal grid spacing in feet:	500	600	700
Cost in $thou. per mile square:	1048	754	573
Probability of detection:	0.97	0.96	0.98
Optimal grid orientation:	45 degrees to expected strike of target		

DETECTION OF CONTACT METASOMATIC DEPOSITS 207

TABLE 7.54
Optimal Design of Airborne Geophysical Surveys for the Detection of Contact Metasomatic (Cu–Mo–Au) Deposits of the North American Cordillera Belt.

Confidence interval: all results are reported as 95% confidence intervals respectively
Expected target length in feet: 2206   2648   3180
Expected shape ratio R = 0.19
Unit Cost: $70/l.ml.

(i) Survey Design: Parallel Grid

Optimal grid spacing in feet:	1400	1800	2100
Cost in US$ per mile square:	404	345	316
Probability of detection:	0.83	0.82	0.82
Optimal grid orientation:	90 degrees to expected strike		

(ii) Survey Design: Square Grid

Optimal grid spacing in feet:	2270	2560	2910
Cost in US$ per mile square:	465	428	394
Probability of detection:	0.95	0.98	1.00

TABLE 7.55
Optimal Design of Ground Geophysical Surveys for the Detection of Contact Metasomatic (Cu–Mo–Au) Deposits of the North American Cordillera Belt.

Survey Design: Square Grid With Spacing S by S Feet.
Confidence Intervals: All results are reported as 95% confidence intervals respectively.
Expected Target Length in Feet: 2206   2648   3180
Expected Shape Ratio R = 0.19
Unit Cost: $6/Station

Optimal Grid Spacing in Feet:	800	1000	1100
Cost in $ Per Mile Square:	1259	990	898
Probability of Detection:	0.93	0.90	0.96
Optimal Grid Orientation:	30 degrees + or −10 to expected strike direction of target.		

TABLE 7.56
Optimal Designs for the Detection of Contact Metasomatic (Cu–Mo–Au) Deposits of the North American Cordillera Belt by Drilling to a Vertical Depth of 300 Feet.

```
Survey design:  square grid with spacing S by S feet
Confidence interval:  all results are reported as 95% confidence intervals
respectively
Expected target length in feet:         2206        2648        3180
Unit cost = $20 per linear ft.
```

(a) Vertical drilling:

```
Optimal grid spacing in feet:            800        1000        1100
Cost in $thou. per mile square:          348         237         203
Probability of detection:               0.93        0.90        0.96
Optimal grid orientation:        30 degrees to expected strike of target
```

(b) Angled drilling at 55 degrees:

```
Optimal grid spacing in feet:           1000        1100        1300
Cost in $thou. per mile square:          290         247         188
Probability of detection:               0.93        0.98        0.97
Optimal grid orientation:        45 degrees with expected strike of target
```

(c) Angled drilling at optimal angle = 50 degrees:

```
Optimal grid spacing in feet:           1000        1100        1300
Cost in $thou. per mile square:          310         265         201
Probability of detection:               0.96        0.99        0.98
Optimal grid orientation:        45 degrees to expected strike of target
```

TABLE 7.57
Optimal Design of Ground Geophysical Surveys for the Detection of Contact Metasomatic (W–Mo) Deposits of the North American Cordillera Belt.

```
Survey Design:  Square Grid With Spacing S by S Feet.
Confiddence Interval:  All results are reported as 95% confidence intervals
respectively.
Expected Target Length in Feet:           598         724         875
Expected Shape Ratio R = 0.27
Unit Cost:  $6/Station
```

```
Optimal Grid Spacing in Feet:             300         350         400
Cost in $ Per Mile Square                4308        3483        2914
Probability of Detection                 0.79        0.84        0.90
Optimal Grid Orientation:  30 degrees + or -10 to expected strike
direction of target.
```

TABLE 7.58
Optimal Designs for the Detection of Contact Metasomatic (W–Mo) Deposits of the North American Cordillera Belt by Drilling to a Vertical Depth of 300 Feet.

```
Survey design:   square grid with spacing S by S feet
Confidence interval:  all results are reported as 95% confidence intervals
respectively
Expected target length in feet:    598        724        875
Unit Cost = $20 per linear ft.
```

(a) Vertical drilling:

Optimal grid spacing in feet:	300	350	400
Cost in $thou. per mile square:	2072	1554	1212
Probability of detection:	0.79	0.84	0.90
Optimal grid orientation:	30 degrees to expected strike of target		

(b) Angled drilling at an angle of 55 degrees:

Optimal grid spacing in feet:	400	500	500
Cost in $thou. per mile square:	1479	981	981
Probability of detection:	0.77	0.86	0.98
Optimal grid orientation:	45 degrees with expected strike of target		

(c) Angled drilling at the optimal angle = 50 degrees:

Optimal grid spacing in feet:	400	400	500
Cost in $thou. per mile square:	1581	1581	1048
Probability of detection:	0.80	1.00	1.00
Optimal grid orientation:	45 degrees to expected strike of target		

REFERENCES AND SELECTED READINGS

1. BALL, C. W., 1951, The Emerald, Feeney, and Dodger tungsten ore-bodies, Salmo, British Columbia, *Econ. Geol.* **49,** 625–638.
2. BUTLER, B. S., 1913, Geology and ore deposits of the San Francisco and adjacent districts, Utah, *U.S. Geol. Surv. Prof. Pap.* No. 80.
3. CATHRO, R. J., 1969, Tungsten in the Yukon, *West. Miner* 42(4), 23–40.
4. COMPTON, R. R., 1960, Contact metamorphism in the Santa Rosa Range, Nevada, *Bull. Geol. Soc. Am.* **71,** 1383–1416.
5. EINAUDI, M. T., 1982, General features, description and origin of skarns associated with porphyry–copper plutons, in *Advances in the Geology of Porphyry–Copper Deposits of Southwestern North America*, pp. 139–210, University of Arizona Press, Tucson.
6. EVANS, A. M., 1980, *An Introduction to Ore Geology*, Chap. 11, Blackwell, Oxford.

7. GHOSE, N. C., 1966, Behaviour of trace elements during thermal metamorphism and/or granitization of metasediments and basic igneous rocks, *Geol. Rundsch.* **55**, 608–617.
8. GHOSE, N. C., 1970, Geochemistry of thermal metamorphic and granitization processes in aureole rocks around Richughuta, Palamua Dist., Bihar, India, *Geol. Rundsch.* **59**, 686–724.
9. HOLSER, W. T., 1947, Metasomatic processes, *Econ. Geol.* **42**, 384–395.
10. KERR, P. F., 1946, Tungsten mineralization in the United States, *Geol. Soc. Am. Mem.* No. 15.
11. KLICHINOV, V. A., and SEGALOVICH, V. I., 1979, The application of geophysics to exploration for chromite and tungsten, *Geol. Surv. Can. Econ. Geol. Rep.* No. 31, pp. 476–484.
12. KNOPF, A., 1917, Tungsten deposits of northwestern Inyo County, California, *U.S. Geol. Surv. Bull.* No. 640, pp. 229–249.
13. LINDGREN, W., 1933, *Mineral Deposits,* Chap. 28, McGraw-Hill, New York.
14. LITTLE, H. W., 1959, Tungsten deposits of Canada, *Geol. Surv. Can. Econ. Geol. Ser.* No. 17.
15. LOVERING, T. S., 1941, The origin of tungsten ore of Boulder County, Colorado, *Econ. Geol.* **36,** 229–279.
16. LOVERING, T. S., 1949, Rock alteration as a guide to ore—East Tintic District, Utah, *Econ. Geol.* Monograph No. 1.
17. MORRIS, H. T., and LOVERING T. S., 1952, Supergene and hydrothermal dispersion of heavy metals in wall rocks of ore bodies, Tintic District, Utah, *Econ. Geol.* **47**, 685–716.
18. RAMBERG, H., 1952, *The Origin of Metamorphic and Metasomatic Rocks,* The University of Chicago Press, Chicago.
19. RENNIE, C. C., and SMITH, T. S., 1957, Lead–zine and tungsten ore bodies of Canadian Exploration Ltd., Salmo, B.C., Structural Geology of Canadian Ore Deposits, Sixth Commonwealth Congress, Montreal, pp. 116–123.
20. RIDGE, J. D., (Ed.) 1968, *Ore Deposits of the United States,* Vols. 1 and 2, A.I.M.E., New York.
21. SANGSTER, D. F., 1969, The contact metasomatic magnetite deposits of Southwest British Columbia, *Geol. Suv. Can. Bull.* No. 172.
22. SMIRNOV, D. F., 1976, *Geology of Mineral Deposits,* Chap. 7 (translated by H. C. Creighton), MIR Publishers, Moscow.
23. WHITE, L. G., 1963, The Canada Tungsten property, Flat River Area, N.W.T., Canada, *Can. Inst. Min. Metall. Bull.* **56**(614), 390–393.
24. YOUNG, G. A., and UGLOW, W. L., 1926, The iron ores of Canada, Vol. 1, British Columbia and the Yukon, *Geol. Surv. Can. Econ. Geol. Ser.* No. 3.

CHAPTER EIGHT

DETECTION OF Ni–Cu ULTRAMAFIC DEPOSITS OF THE NORTH AMERICAN SHIELD BY OPTIMIZED GEOPHYSICAL SURVEYS AND DRILLING PROGRAMS

8.1. GENERAL GEOLOGICAL BACKGROUND

8.1.1. Main Geological Settings

According to Ross and Travis,[18] there are three main tectonic settings for the sulfide nickel deposits of the World: (1) stable platforms, chiefly of Proterozoic age, (2) remnants of island arcs and mobile belts, mainly of Archean age, referred to as "Greenstone Belts," and (3) Paleozoic and Mesozoic fold belts. The first two types of setting are well represented in the North American Precambrian Shield; the third one is found along the North American Cordillera Belt. Ross and Travis[18] and Naldrett[16] recognize three genetic and lithologic groups of sulfide nickel deposits throughout the world. The first two are associated with layered complexes and undifferentiated intrusives, while the third one has mafic volcanic affiliations. All three are known to occur in North America.

8.1.2. Intrusive Layered Complexes

Most of the world's primary nickel ore produced between the turn of the century and the mid-1950s was gained from large layered mafic–ultramafic intrusive complexes occurring in platforms of Proterozoic age. Mineralization commonly consists of low-grade disseminated sulfide ore which occurs in the basal portion of mafic layers in preference to the ultramafic layers of the complexes. Notable exceptions are known in the North American Shield, including, for examples, the famous Sudbury Basin so well described in many publications, and the more recently discovered gabbro-peridotite layered complex at Lynn Lake, Manitoba, which has yielded some ore of grade higher than the average for the group.

8.1.3. Undifferentiated Mafic Intrusives

A substantial number of sulfide nickel ore deposits generally of moderate tonnage and grade have been discovered in the North American Shield during the 1950s and the 1960s in association with undifferentiated mafic intrusives. The deposits discovered in the provinces of Manitoba and Ontario and in the northwestern portion of the province of Quebec are of Archean age, while the Ungava deposits of New Quebec are of early Proterozoic age. The lithology of the undifferentiated mafic intrusives varies from basic to ultrabasic. In relation to wallrock formations, they assume two main types of geometry: conformable, either continuous (sills) or discontinuous (lenses), and forcibly intrusive (dykes and plugs).[1]

Because of their economic importance as a major new source of sulfide nickel ore, the large deposits of the Mystery–Moak–Thompson Lake Belt discovered in eastern Manitoba in the late 1950s deserve a special mention among the undifferentiated intrusive group. The deposits occur along a "mobile belt" separating the western portion of the Superior Province from the southeastern edge of the Churchill Province. The mobile belt is the seat of a strong regional gravity anomaly which exhibits a complex structure. The anomaly consists of two positive strips bracketing a branching negative strip which coincides with the nickel-bearing peridotite belt.[24]

About one half of the total length of 500 miles of the mobile belt has been prospected to date, and several nickel deposits of commercial and subeconomic importance have been discovered within the 60-mile-long Mystery–Moak–Thompson Lake Belt, along the Nelson River. One type of deposit occurs mainly within the Mystery–Moak section of the belt. The mineralization consists of finely disseminated pyrrhotite and pyrite with minor pentlandite occurring in strongly serpentinized peridotite sills or lenses. The deposits are in the low-grade, large-tonnage category and are considered as subeconomic in the present economic circumstances.

The second type of nickel ore deposits is found in the Thompson Lake section of the belt. The mineralization is of a much higher grade than that of the Mystery–Moak section. The setting is strikingly different from that of other deposits of the undifferentiated deposit group: the ore does not occur within the peridotite sills, but appears to have been remobilized into the sillimanite gneiss wallrock.[24,26]

8.1.4. Extrusive Mafic–Ultramafic Suites

Since the early 1960s, a new type of sulfide nickel ore deposits has begun to make a significant contribution to the World's nickel production. These deposits occur mainly in Archean greenstone belts of the Western Australian and North American shields and are considered to be of volcanogenic affiliation. Coad[3] aptly summarizes the results of much of the research work done on this type of deposits since they were first discovered in Western Australia and subsequently in Canada in the 1960s.

Two types of volcanogenic Ni–Cu mafic–ultramafic deposits are distinguished. The first type occurs in komatiite volcanic suites and comprises two categories: massive sulfide ore occurring in typical extrusive settings and disseminated ore found mainly in intrusive settings. The second type is found in tholeiitic volcanic suites, which are generally far less productive than the komatiite series, with a few notable exceptions in the Scandinavian Shield.

The komatiite suites, originally described in the Komati River region of the Transvaal, are highly magnesian in composition, as expressed by a low FeO/Mgo + FeO ratio, and show a low TiO^2 content, in contrast with the tholeiitic series. Much effort has been devoted to the study of the stratigraphic distribution of sulfide ore within the komatiite flows. It was soon recognized that the mineralization of economic importance tends to concentrate as pockets, pods, and lenses of massive sulfides occurring along corrugations, embayments, and flexures of the lower contact of the extrusive bodies. The massive mineralization commonly grades upward into disseminated sulfides within the middle portion of the flows. In the intrusive type, however, the bulk of the mineralization is of a disseminated nature and occurs throughout the central portion of the bodies or even near the top contact.

8.2. FIELD DETECTION METHODOLOGY

8.2.1. Detection Approaches

All successful sulfide nickel exploration programs, particularly the more recent ones in Western Australia and in the Provinces of Manitoba, Ontario, and Quebec of Canada, have relied on a combination of direct and indirect approaches. The direct approach involves the visual detection of the mineralization and associated clues either on the ground or from airborne platforms or by mechanical means (drilling). The indirect approach, on the other hand, relies on the ground or aerial detection of geophysical signatures of the target mineralization, or on the ground detection of geochemical signatures provided by primary or secondary halos.

8.2.2. Direct Detection

Mafic–ultramafic layered complexes are the only kind of prospective nickel-bearing geological structures large enough to be detectable from airborne platforms, either by photogeology techniques from aircraft or by remote multisensor techniques from satellites. The detection is enhanced by topography, vegetation, and color contrasts between the layered complexes and the surrounding formations. A good example, well known to the first writer, is provided by the Bay of Islands ultrabasic complex of central western Newfoundland.

Direct visual detection from an airborne platform is not helpful in the search for undifferentiated mafic intrusives and even less so if volcanogenic complexes are the targets, because of their relatively small extent and lack of visual contrasts. An integrated approach successfully used in the West Australian and Canadian shields in the search for sulfide nickel deposits of volcanogenic affiliation calls for reconnaissance coverage followed by detailed mapping to locate favorable formations and pinpoint favorable horizons. Guidance is provided by the display of "spinifex" textures, often directly recognizable in the field, which are generally associated with environments favorable for the deposition of nickel ore.[3] Mapping is complemented by small-scale geophysical and geochemical coverages to fill the gaps, and by some reconnaissance drilling to test geological targets.

8.2.3. Indirect Geophysical Detection

The indirect geophysical approach has been widely used in the search for sulfide nickel targets in many areas of the North American and Scandinavian shields where outcrops are scarce and glacial drift cover quite thick. The geophysical environment of most sulfide nickel deposits, particularly those with volcanogenic affiliations, is quite suitable for the application of the indirect geophysical approach. Iron sulfides including pyrite and pyrrhotite are the most widespread mineral associated with nickel–copper mineralizations. Pyrrhotite, for one, is a good electrical conductor of moderate magnetic susceptibility and specific gravity. As a result, nickel–copper deposits themselves are up to 1000 times more conductive, 10 times more magnetic, and 30%–50% more dense than surrounding host rocks. The contrast of physical properties is generally sufficient to provide good geophysical targets.

Airborne magnetic surveys are very useful to delineate the extent and main features of basic–ultrabasic intrusive–extrusive complexes at the initial stage of the search for nickel–copper deposits. But ground follow-up consisting of magnetic surveys and geological mapping is required in order to distinguish the favorable lithologic units from the less prospective ones. The geological interpretation of the surveys is further complicated by widespread occurrences of iron formations which commonly accompany volcanic flows in the greenstone belts of Western Australia and the North American Shield.

At the next stage of the search, detailed electromagnetic surveys both in the frequency and time domains are generally conducted from helicopter platforms in order to test the favorable lithologic units for the presence of sulfides. Finally, induced polarization surveys are used to pinpoint on the ground and delineate sulfide zones of possible economic interest for drill testing. However, the induced polarization has many drawbacks, including slowness and high cost on the one hand, and lack of discrimination on the other hand. Dowsett[5] describes the successful use of the gravimetric method to distinguish sulfides from graphite or serpentinized fractures in areas affected by relatively light and uniform glacial cover such as the Chibougamau region of Northwest Quebec. He notes, however, that marked variations in overburden thickness may generate spurious anomalies of the same order of magnitude as what could be expected from sulfide zones.

8.2.4. Indirect Geochemical Detection

Bedrock sampling for the detection of primary halos is a valid approach in areas of relatively abundant outcrops such as the Abitibi greenstone belts of northeast Ontario and northwest Quebec in the North American Shield. A quick and low-cost field test using the chemical dimethylglioxime is of valuable assistance for the screening of favorable environments for nickel mineralization. But much more sophisticated laboratory techniques are required to detect the variations of abundance of elements such as MgO and sulfur which point to favorable environments for nickel deposition.

The detection of secondary halos by residual soil sampling and geobotanical surveys in the North American Cordillera Belt was described by Miller.[13] Obial[17] describes similar surveys in the tropical environment of the Philippines. These techniques have not proved very useful in the glaciated environments of the subarctic portion of the North American Shield. But their application to the thinly covered northernmost portions of the Shield, such as the Northwest Territories and New Quebec, should not be ruled out.

8.3 STATISTICAL MODELING OF GEOMETRIC PARAMETERS OF Ni–Cu DEPOSITS

8.3.1. Scope of Study

A total of 25 commercial and subeconomic deposits were included in the database to represent the whole statistical population of known and undiscovered

TABLE 8.1
List of Ni–Cu Ultramafic Deposits of the North American Shield Included in the Database.

Manitoba	Quebec
Rankin Inlet	Marbridge
Bird River	Blondeau Twp.
Lynn Lake	Lake Renzy
Bowden Lake	Lorraine
Thompson	
Pipe	
Ontario	**New Quebec**
Alexo	Cross Lake
Sothman	Katiniq
Ajax	Raglan
Langmuir	Expo
Texmont	
Trebor	
Temagami Lake	
Gordon Lake	
Werner Lake	
Kenbridge	
Shebandowan	

Ni–Cu ultramafic deposits of the North American Shield. The deposits are listed by name and region in Table 8.1. An effort was made to obtain the best possible cross section of the three main types of deposits listed in Section 8.1. The very large Sudbury deposits were deliberately omitted from the subsample representing the layered complexes, because their uniqueness would flaw the results of the statistical study. The remainder of the sample is more or less evenly divided between the undifferentiated and the volcanogenic types of deposits.

8.3.2. Statistical Modeling

A total of five parameters are required for the calculation of detection probabilities and the optimization of the search for Ni–Cu deposits of the North American Shield. Three of the parameters are of a dimensional nature, including length and breadth of the horizontal section of deposits within the range of detectors and the shape ratio $R = B/L$. The other two are attitudinal parameters, including the unoriented true dip angle in degrees and the oriented strike direction referred to true North and measured in degrees within the right half-circle.

DETECTION OF Ni–Cu ULTRAMAFIC DEPOSITS

The range of the parameter measurements collated from the 25 deposits of the sample is as follows: 500–6100 feet for length, 20–620 feet for breadth, and 60–85 degrees for dip angle. The parameter measurements were grouped into frequency distributions which are summarized by means of four statistics: arithmetic mean, mode, standard deviation, and coefficient of skewness. The values of the statistics are listed in the lower half of Table 8.2. The degree of asymmetry of the distributions of the dimensional parameters and of the strike direction is moderate, as indicated by the skewness coefficient value of 0.5; but the skewness is much stronger for the distribution of dip angles (coefficient close to unity).

Lognormal models were successfully fitted to the observed distributions of the dimensional parameters, as indicated by the results of χ^2 goodness of fit tests conducted at the 0.05 confidence level (see Chapter 5, Section 5.3.2). A normal model was also satisfactorily fitted at the same confidence level to the observed distribution of unoriented dip angles, and a circular normal one was selected for the oriented strike directions.

TABLE 8.2

Statistical Modeling Summary for Ni–Cu Ultramafic Deposits of the North American Shield (Sample Size: 25 Deposits).

Statistic of fitted model	Length of horiz. sect. in feet	Breadth of horiz. sect. in feet	Shape ratio R=B/L	Strike direct. from T. North in degrees	Dip in degrees
25%ile	750	25	0.02	32	69
95% L.c.l.	815	41	0.03	37	73
Mean	1097	62	0.06	52	77
95% U.c.l.	1475	95	0.09	67	80
75%ile	2000	120	0.14	85	86
Disp. coef.	0.11	0.25	0.62	0.16	0.05

Statistic of observed data					
Arithm. mean	1459	114	0.12	54	77
Mode	750	35	0.03	37	85
Stand. dev.	1283	147	0.18	37	10
Skewness	0.55	0.54	0.51	0.46	0.82

The statistics describing the fitted models are listed in the upper half of Table 8.2. They include the geometric means of the lognormal models, arithmetic means for normal models, and circular means for circular normal models, all with the limits of their 95% confidence intervals. Additional statistics include the 25th and 75th percentiles of the fitted distributions, and the dispersion coefficients. The latter gives an indication of the spread of the data about the mean of the fitted model, which, in turn, reflects the total variability of the data. (See Chapter 5, Section 5.3.2.) The variability is rather small for the length and breadth parameters (range 0.1–0.25), but is substantially greater for the shape parameter R (0.6).

8.4. CONSTRUCTION AND ORGANIZATION OF DETECTION PROBABILITY TABLES

8.4.1. Introduction

For detection purposes, the Ni–Cu ultramafic deposits of the North American Shield are considered as dipping oriented slabs with a horizontal cross section of elliptical shape. As a result, the target dimensions and shape ratios have to be recalculated as shown in Section 5.4 of Chapter 5 by adding the appropriate dip component to the breadth of the horizontal section of each deposit. For example, the mean shape ratio of the actual deposits is inflated from 0.06 (see Table 8.2) to 0.12 for the targets in the context of vertical detection to a depth of 300 feet, and to 0.26 for 55-degree angled detection to the same vertical depth, thus considerably boosting the probability of detection.

The orientation factor was introduced in the construction of the tables because of the resulting improvement in detection success as compared with results obtained under the orientation randomness assumption. A dynamic programming study showed that the probability of detection is maximized in the following manner: (a) in the case of airborne surveys, when the parallel grid is oriented at 90 degrees to the expected strike, (b) for ground surveys (vertical detection) when the grid is laid at an angle of 20 degrees with the expected strike direction; the angle should be 30 degrees when the 55-degree angled detection option is considered. The grid is then laid out at the appropriate angle from the expected strike direction as known from the results of previous surveys or photogeological studies. If little is known of the likely orientation of the target, the mean of the circular normal fitted model is a statistically valid estimate to be considered for the design of the grids (see Table 8.2., upper half).

8.4.2. Description of the Detection Probability Tables

Three kinds of tables are constructed for the detection of Ni–Cu ultramafic deposits of the North American Shield. The first two cover the single detection (at least one intersection of the target) and confirmed detection (at least two target intersections) situations and are constructed in a similar manner. The third kind of table lists grid spacings and corresponding coverage costs per unit of area which are required to obtain prespecified levels of detection probabilities ranging from 0.10 to 0.95.

The heads of the first two kinds of tables display the statistics of the fitted models as obtained from Table 8.2, including the mean target length within its 95% confidence interval and the mean shape ratio. The probabilities are calculated first for the mean length, and then for the upper and lower fiducial limits of the mean, for grid spacings increased in steps of 50–250 feet, depending on the type of field program. As a result, the tables display the probability of detection or confirmed detection within its 95% confidence interval for each grid spacing.

8.4.3. Organization of Tables

For the convenience of the readers, the tables have been grouped into three blocks, each covering a specific kind of field program. The first block deals with probabilities of detection and confirmed detection of Ni–Cu ultramafic deposits by airborne geophysical surveys on parallel and square grids (Tables 8.4–8.7). The second block covers the detection and confirmed detection of Ni–Cu deposits by ground geophysical surveys or vertical drilling to a depth of 300 feet on square grids (Tables 8.9–8.11). The third block pertains to detection and confirmed detection of deposits by 55-degree angled drilling to a vertical depth of 300 feet (Tables 8.13–8.15).

8.5. DESIGNING THREE STRATEGIES FOR THE DETECTION OF Ni–Cu DEPOSITS OF THE NORTH AMERICAN SHIELD

8.5.1. Search Strategies

We wish to consider three options for the detection of Ni–Cu deposits of the North American Shield which are designed to meet the requirements of three common exploration situations. The three options are referred to as the liminal, maximal, and optimal strategies. The first one is tailored to deal with large coverage of areas of little known potential within budget limits by minimizing costs while ensuring at least a 0.500 level of detection probability. The second one provides a high probability of detection, close to certainty (0.980) at a much higher cost, which can be justified for the coverage of small areas of high merit. Finally, the optimal option is a compromise plan based on the maximization of the probability of detection under constraints of cost minimization. Explorationists should favor the third option because it meets the requirements of the majority of exploration situations which call for coverage of areas of moderate size and moderate potential within strict budgetary constraints. The detection characteristics corresponding to the three strategies are summarized and assembled into one single table for the convenience of the readers (Table 8.3.).

8.5.2. Liminal Detection Strategy

The detection characteristics pertaining to the liminal strategy are listed in the leftmost portion of Table 8.3. They include grid spacings and associated coverage costs per unit of area for airborne surveys on two grid designs, ground geophysical surveys, and vertical and angled drilling programs.

The cost-multiplying factor required to obtain confirmed detection of Ni–Cu deposits instead of single detection averages 2.3 for airborne geophysical surveys and for vertical and angled drilling programs, but falls to 1.8 for ground geophysical surveys. Cost factor figures are not available for airborne geophysical surveys on square grids and for optimally angled drilling programs because the probability of confirmed detection could not be satisfactorily calculated.

A comparison of the merits of various grid and program designs is of obvious interest to explorationists for planning purposes. Since the probability of detection is held at a specified level of at least 0.500, the coverage cost per unit of area is a suitable yardstick to use to measure the detection performance of various designs. An inspection of the figures listed in the leftmost portion of Table 8.3 will show that the parallel grid design should be favored for airborne detection because it reduces the coverage cost by 5% as compared to square grid coverage cost. The preference for angled drilling over vertical drilling to the same depth is justified by a 35% reduction of coverage cost.

8.5.3. Maximal Strategy

Grid spacings and corresponding coverage costs for various field programs and grid designs pertaining to the maximal option are listed in the rightmost portion of Table 8.3, in the same manner as for the liminal strategy. The cost multiplying factor required to obtain confirmation of detection averages 1.4 for both airborne and ground geophysical surveys. It escalates to 1.8 for vertical drilling programs and reaches 2.4 for angled drilling programs. A comparison

TABLE 8.3
Summary of Detection Characteristics of Ni–Cu Ultramafic Deposits of the North American Shield for Three Types of Strategies.

Type of survey and grid geometry	Liminal detection		Optimal detection		Maximal detection	
	Coverage cost: US$/mile sq. Grid size: feet		Coverage cost: US$/mile sq. Grid size: ft	Detection probability	Coverage cost: US$/mile sq. Grid size: feet	
	Detection	Confirmed detection			Detection	Confirmed detection
Airborne geophysical parallel grid	$380 1500 ft	$905 525 ft	$448 1200 ft	0.75	$1990 200 ft	$2570 150 ft
Airborne geophysical square grid	$398 3400 ft	—	$494 2090 ft	0.76	$805 1300 ft	—
Ground geophysical square grid	$2500 450 ft	$4650 275 ft	$2190 500 ft	0.72	$7800 200 ft	$12,000 150 ft
Vertical drilling square grid	$974,000 450 ft	$2,280,000 275 ft	$803,000 500 ft	0.72	$4,507,000 200 ft	$7,860,000 150 ft
55-degree angled drilling square grid	$632,000 650	$1,425,000 415 ft	$621,000 700 ft	0.81	$1,420,000 430 ft	$3,520,000 280 ft
Optimal 45-deg angled drilling square grid	—	—	$491,000 800 ft	0.80	—	—

of the figures on both sides of Table 8.3 shows the relative advantage of the maximal option over the liminal strategy when specifying confirmation of detection by geophysical surveys and vertical drilling; but there is little difference between the two options when angled drilling is considered.

An inspection of the rightmost portion of Table 8.3 shows that the square grid design should be preferred to the parallel design for airborne geophysical surveys, because selection of the former leads to a 60% saving in coverage costs. When comparing the merits of the two types of drilling program for field planning purposes, angled drilling against the dip direction should be favored over vertical drilling to the same depth because of the resulting 68% saving in coverage costs. If the dip direction cannot be ascertained from previous surveys or photogeological studies, vertical drilling may be the preferred option in order to avoid the risk of drilling down-dip.

8.5.4. Optimal Strategy

Design optimization based on the efficiency criterion was carried out for each kind of field program by dynamic programming, as described in Section 5.5.4 of Chapter 5. The optimal parameters include (a) optimal grid orientation with respect to the expected target strike direction within its confidence limit, (b) optimal grid spacing and corresponding coverage cost per unit of area, (c) optimal drill inclination, if the angled drilling option is chosen, and (d) the optimal detection probability. The parameters of optimized airborne geophysical surveys on parallel and square grid designs are listed in Table 8.8, while Tables 8.12 and 8.16 cover the optimization of ground geophysical surveys and drilling programs, respectively. All results are summarized and listed in the central portion of Table 8.3 for the reader's convenience.

Inspection of Table 8.3 shows that the optimal probabilities of detection for all types of field programs and grid designs range between the 0.72 and 0.81 levels, averaging 0.75. The advantage of the optimal option over the maximal strategy is emphasized by the 66% coverage cost reduction in favor of the former, as compared to a drop of only 23% in detection probability level.

As mentioned in the previous chapters 6 and 7, we cannot use the coverage cost per unit of area as a criterion for the comparison of detection performances of various field program or grid designs because the optimal probability levels vary. We shall use instead the expected loss criterion, which is described in Section 6.5.4 of Chapter 6. As a result, we find that there is no incentive in selecting the square grid design for airborne surveys because of the 6% expected loss increase incurred. On the other hand, the optimally angled drilling option should be preferred to the vertical drilling option because of the resulting 56% decrease in expected loss for the coverage of a unit of area.

8.6. DETECTION PROBABILITY AND OPTIMIZATION TABLES

The reader will find in the following section a series of thirteen tables covering the probabilities of detection and optimal designs of the search for Ni–Cu ultramafic deposits of the North American Shield by airborne and ground programs (pages 223–232).

TABLE 8.4
Probabilities of Detection of Ni–Cu Ultramafic Deposits of the North American Shield by Airborne Geophysical Surveys (Continuous Readings) on Parallel Grids.

```
Survey design:  parallel lines with spacing S feet
Elliptical targets with expected major axis = L feet
in the confidence interval: l.c.l. ( geom. mean ( u.c.l.
and minor axis = B feet
R is defined as the ratio B/L with geometric mean = 0.12
Flight lines at 90 degrees with strike line.
```

Grid spacing S feet	Probability of detection		
	l.c.l. L= 815	geom. mean 1097	u.c.l. 1475 feet
200	0.947	0.961	1.000
400	0.893	0.922	0.943
600	0.836	0.882	0.915
800	0.775	0.841	0.886
1000	0.705	0.798	0.856
1200	0.656	0.751	0.825
1400	0.562	0.697	0.793
1600	0.492	0.675	0.759
1800	0.437	0.600	0.722
2000	0.393	0.540	0.740
2200	0.358	0.491	0.673
2400	0.328	0.450	0.617
2600	0.303	0.415	0.569
2800	0.281	0.386	0.529
3000	0.262	0.360	0.494
3200	0.246	0.337	0.463
3400	0.231	0.317	0.435
3600	0.219	0.300	0.411
3800	0.207	0.284	0.390
4000	0.197	0.270	0.370
4200	0.187	0.257	0.353
4400	0.179	0.245	0.337
4600	0.171	0.235	0.322
4800	0.164	0.225	0.308
5000	0.157	0.216	0.296
5200	0.151	0.208	0.285

TABLE 8.5
Determination of Grid Size for Specified Probability Levels in the Detection of Ni–Cu Ultramafic Deposits of the North American Shield by Airborne Geophysical Surveys on Parallel Grids.

Survey design: parallel lines with spacing S feet
Unit Cost = US$70 per line-mile

Specified detection probability level	Required grid spacing S in feet	Corresponding cost in $ per mile square
0.10	10020	$ 176.89
0.15	6880	$ 193.72
0.20	5260	$ 210.27
0.25	4240	$ 227.17
0.30	3560	$ 243.82
0.35	3060	$ 260.78
0.40	2700	$ 276.89
0.45	2400	$ 294.00
0.50	2160	$ 311.11
0.55	1980	$ 326.67
0.60	1820	$ 343.08
0.65	1680	$ 360.00
0.70	1360	$ 411.76
0.75	1180	$ 453.22
0.80	960	$ 525.00
0.85	720	$ 653.33
0.90	480	$ 910.00
0.95	220	$1820.00

TABLE 8.6
Probabilities of Confirmed Detection of Ni–Cu Ultramafic Deposits of the North American Shield by Airborne Geophysical Surveys (Continuous Readings) on Parallel Grids.

```
Survey design:  parallel lines with spacing S feet,
Randomly orientated elliptical targets with expected major axis = L feet
in the confidence interval:  l.c.l. (geom. mean ( u.c.l.
and minor axis = B feet
R is defined as the ratio B/L with geometric mean = 0.12
```

	Probability of confirmed detection		
Grid spacing S feet	l.c.l. L= 815	geom. mean 1097	u.c.l. 1475 feet
300	0.858	0.905	0.938
400	0.828	0.870	0.900
500	0.559	0.802	0.870
600	0.391	0.712	0.840
700	0.279	0.539	0.800
800	0.205	0.414	0.747
900	0.156	0.323	0.607
1000	0.127	0.255	0.498
1100	0.000	0.204	0.412
1200	0.000	0.166	0.343
1300	0.000	0.140	0.288
1400	0.000	0.124	0.243
1500	0.000	0.000	0.206
1600	0.000	0.000	0.177
1700	0.000	0.000	0.154
1800	0.000	0.000	0.137
1900	0.000	0.000	0.125
2000	0.000	0.000	0.000

TABLE 8.7
Probabilities of Detection of Ni–Cu Ultramafic Deposits of the North American Shield by Airborne Geophysical Surveys (Continuous Readings) on Square Grids.

Survey design: square grid with spacing S by S feet
Randomly orientated elliptical targets with expected major axis = L feet
in the confidence interval: l.c.l. (geom. mean (u.c.l.
and minor axis = B feet
R is defined as the ratio B/L with geometric mean = 0.12

Grid Spacing S feet	Probability of detection		
	l.c.l. L= 815	geom. mean 1097	u.c.l. 1475 feet
1000	1.000	1.000	1.000
1200	0.972	1.000	1.000
1400	0.921	1.000	1.000
1600	0.863	0.979	1.000
1800	0.807	0.945	1.000
2000	0.754	0.904	0.994
2200	0.707	0.862	0.979
2400	0.664	0.820	0.955
2600	0.626	0.781	0.926
2800	0.591	0.744	0.895
3000	0.560	0.710	0.865
3200	0.532	0.678	0.834
3400	0.506	0.649	0.805
3600	0.482	0.621	0.777
3800	0.461	0.596	0.750
4000	0.441	0.573	0.724
4200	0.423	0.551	0.700
4400	0.407	0.531	0.677
4600	0.391	0.512	0.655
4800	0.377	0.494	0.635
5000	0.363	0.477	0.616
5200	0.351	0.462	0.597

TABLE 8.8
Optimal Design of Airborne Geophysical Surveys (Continuous Readings) for the Detection of Ni–Cu Ultramafic Deposits of the North American Shield.

Confidence interval: all results are reported as 95% confidence intervals respectively:
Expected target length in feet: 815   1097   1475
Expected shape ratio $R = 0.12$
Unit cost: $70/l.ml.

(i) Survey Design:
    Parallel Grid

Optimal grid spacing in feet:	800	1200	1400
Cost in US$ per mile square:	602	448	404
Probability of detection:	0.78	0.75	0.79
Optimal grid orientation:	90 degrees to expected strike		

(ii) Survey Design:
    Square Grid

Optimal grid spacing in feet:	1755	2090	2385
Cost in US$ per mile square:	561	494	450
Probability of detection:	0.67	0.75	0.86

TABLE 8.9
Probabilities of Detection of Ni–Cu Ultramafic Deposits of the North American Shield by Ground Geophysical Surveys (Discrete Readings) or Vertical Drilling to a Depth of 300 Feet.

```
Survey design: square grid with spacings S by S feet.
The expected shape of the ore deposits is elliptical
with major axis = L in the 95% confidence interval:
L.c.l. ( geometric mean ( U.c.l., and minor axis = B feet,
R is defined as the ratio B/L with geometric mean = 0.12
grid orientation 20 degrees + or - 10 from the expected
strike line.
```

	Probability of detection		
Spacing S in feet	L.c.l. $L = 815$	G. mean 1097	U.c.l. 1475ft
150	1.000	1.000	1.000
200	0.962	1.000	1.000
300	0.913	0.943	1.000
400	0.627	0.935	0.948
500	0.419	0.723	0.978
600	0.292	0.539	0.859
700	0.215	0.403	0.701
800	0.165	0.310	0.566
900	0.130	0.245	0.457
1000	0.105	0.198	0.372
1200	0.073	0.138	0.259
1600	0.041	0.077	0.146
2000	0.026	0.050	0.093

TABLE 8.10
Determination of Grid Size for Specified Probability Levels in the Detection of Ni–Cu Ultramafic Deposits of the North American Shield by Vertical Drilling to a Depth of 300 Feet.

Survey Design: Square Grid With Spacings S by S Feet
Unit Cost = $6000 Per Diamond Drill-Hole

Probability Level Of Detection	Grid Size In Feet	Drilling Cost Per Mile Square In U.S.$. Thousands
0.05	1850	89.58
0.15	1150	188.25
0.25	900	283.73
0.35	750	388.81
0.45	700	438.91
0.55	600	577.42
0.65	550	675.43
0.75	500	803.19
0.85	450	974.35
0.95	300	2077.99

TABLE 8.11
Probabilities of Confirmed Detection of Ni–Cu Ultramafic Deposits of the North American Shield by Ground Geophysical Surveys or Vertical Drilling to a Depth of 300 Feet.

Survey design: square grid with spacings S by S feet.
The expected shape of the ore deposits is elliptical
with major axis = L in the 95% confidence interval:
L.c.l. (geometric mean (U.c.l., and minor axis = B feet,
R is defined as the ratio B/L with geometric mean = 0.12
grid orientation 0 degrees + or - 10 from the expected
strike line.

Spacing S in feet	Probability of confirmed detection		
	L.c.l. L = 815	G. mean 1097	U.c.l. 1475ft
100	1.000	1.000	1.000
150	0.987	1.000	1.000
200	0.863	0.951	1.000
250	0.627	0.829	0.944
300	0.329	0.608	0.846
350	0.201	0.348	0.701
400	0.119	0.230	0.426
450	0.069	0.150	0.305
500	0.039	0.198	0.368
550	0.022	0.060	0.153
600	0.012	0.037	0.106

TABLE 8.12
Optimal Design of Ground Geophysical Surveys on Square Grids for the Detection of Ni–Cu Ultramafic Deposits of the North American Shield.

```
Survey Design:  Square Grid With Spacing S by S Feet.
Confidence Interval:  All results are reported as 95% confidence intervals
respectively.
Expected Target Length in Feet:         815    1097    1475
Expected Shape Ratio R = 0.12
Unit Cost:  $6/Station
```

```
Optimal Grid Spacing in Feet:           300     500     600
Cost in $ Per Mile Square:             4308    2189    1752
Probability of Detection:              0.91    0.72    0.86
Optimal Grid Orientation:  20 degrees + or -10 to expected strike
direction of target.
```

TABLE 8.13
Probabilities of Detection of Ni–Cu Ultramafic Deposits of the North American Shield by 55-Degree Angled Drilling to a Vertical Depth of 300 Feet.

```
Survey design: square grid with spacings S by S feet.
The expected shape of the ore deposits is elliptical
with major axis = L in the 95% confidence interval:
L.c.l. ( geometric mean ( U.c.l., and minor axis = B feet,
R is defined as the ratio B/L with geometric mean = 0.26
grid orientation 30 degrees + or - 10 from the expected
strike line.
```

	Probability of detection		
Spacing S in feet	L.c.l. L = 815	G. mean 1097	U.c.l. 1475ft
350	1.000	1.000	1.000
400	0.996	1.000	1.000
500	0.838	1.000	1.000
600	0.626	0.968	1.000
700	0.466	0.815	1.000
800	0.356	0.658	0.981
900	0.282	0.529	0.888
1000	0.228	0.429	0.767
1100	0.189	0.355	0.656
1200	0.158	0.298	0.559
1300	0.135	0.254	0.470
1400	0.116	0.219	0.412
1500	0.101	0.191	0.359
1600	0.089	0.168	0.316
1700	0.079	0.149	0.280
1800	0.070	0.133	0.249
1900	0.068	0.119	0.224
2000	0.057	0.107	0.202

TABLE 8.14

Determination of Grid Size for Specified Probability Levels in the Detection of Ni–Cu Ultramafic Deposits of the North American Shield by 55-Degree Angled Drilling to a Vertical Depth of 300 Feet.

Survey Design: Square Grid With Spacings S by S Feet
Unit Cost = $7325 Per Diamond Drill-Hole

Probability Level Of Detection	Grid Size In Feet	Drilling Cost Per Mile Square In U.S.$. Thousands
0.05	2800	61.42
0.15	1800	113.90
0.25	1400	167.46
0.35	1100	247.26
0.45	1000	289.81
0.55	900	346.39
0.65	800	424.21
0.75	750	474.68
0.85	700	535.83
0.95	650	611.00

TABLE 8.15

Probabilities of Confirmed Detection of Ni–Cu Ultramafic Deposits of the North American Shield by 55-Degree Angled Drilling to a Vertical Depth of 300 Feet.

Survey design: square grid with spacings S by S feet. The expected shape of the ore deposits is elliptical with major axis = L in the 95% confidence interval: L.c.l. (geometric mean (U.c.l., and minor axis = B feet, R is defined as the ratio B/L with geometric mean = 0.26 grid orientation 0 degrees + or - 10 from the expected strike line.

Spacing S in feet	Probability of confirmed detection		
	L.c.l. L = 815	G. mean 1097	U.c.l. 1475ft
200	1.000	1.000	1.000
300	0.941	1.000	1.000
400	0.543	1.000	1.000
500	0.279	0.735	1.000
600	0.125	0.428	0.896
700	0.052	0.258	0.699
800	0.019	0.144	0.462
900	0.005	0.077	0.325
1000	0.001	0.039	0.219
1200	0.000	0.008	0.091
1500	0.000	0.000	0.020

TABLE 8.16
Optimal Designs for the Detection of Ni–Cu Ultramafic Deposits of the North American Shield by Drilling to a Vertical Depth of 300 Feet.

```
Survey design:   square grid with spacing S by S feet
Confidence interval:   all results are reported as 95% confidence intervals
respectively
Expected target length in feet:          815       1097      1475
Unit Cost = $20 per linear ft.
```

```
(a)   Vertical Drilling:

Optimal grid spacing in feet:            300        500       600
Cost in $thou. per mile square:         2078        803       577
Probability of detection:               0.91       0.72      0.86
Optimal grid orientation:           20 degrees to expected strike of target
```

```
(b)   Angled drilling at 55 degrees:

Optimal grid spacing in feet:            500        700      1000
Cost in $thou. per mile square:          817        621       491
Probability of detection:               0.84       0.81      0.77
Optimal grid orientation:           30 degrees to expected strike of target
```

```
(c)   Angled drilling at optimal angle = 45 degrees:

Optimal grid spacing in feet:            600        800      1000
Cost in $thou. per mile square:          817        491       336
Probability of detection:               0.88       0.80      0.79
Optimal grid orientation:           30 degrees to expected strike of target
```

REFERENCES AND SELECTED READINGS

1. CAMERON, E. M., SIDDELEY, G., and DURHAM, C. G., 1971, Distribution of ore elements in rocks for evaluating ore potential: Nickel, copper and sulfur in ultra-mafic rocks of the Canadian Shield, *Can. Inst. Min. Metall. Spec. Vol.* **11**, 298–313.
2. CHAYER, T. P., 1960, Some critical differences between Alpine-type and stratiform peridotite-gabbro complexes, 21st International Geological Congress, Pt. XIII, pp. 247–259.

3. COAD, P. R., 1979, Nickel sulfide deposits associated with ultramafic rocks of the Abitibi Belt and economic potential of mafic-ultramafic intrusions, Ontario Geological Survey, Study No. 20, Toronto, Canada.
4. CORNWALL, H. R., 1966, Nickel deposits of North America, *U.S. Geol. Surv. Bull.* No. 1223.
5. DOWSETT, J. S., 1979, Geophysical exploration methods for nickel, *Geol. Surv. Can. Econ. Geol.* Rep. No. 31, 310–321.
6. ECKSTRAND, O. R., 1972, Ultramafic flows and nickel sulfide deposits in the Abitibi Orogenic Belt, *Geol. Surv. Can. Pap.* 72.-1, part A, 75–81.
7. IRVINE, T. N., and SMITH, C. H., 1967, The ultrabasic rocks of the Muskox intrusion, N.W.T., Canada, in *Ultramafic and Related Rocks,* pp. 38–49, Wiley, New York.
8. JACKSON, E. D., 1961, Primary textures and mineral associations in the ultramafic zone of the Stillwater complex, Montana, *U.S. Geol. Surv. Prof. Pap.* No. 358.
9. KILBURN, L. C., WILSON, H. D. B., GRAHAM, A. R., and OGURA, Y., 1969, Nickel sulfide ores related to ultrabasic intrusions in Canada, Symposium on magmatic ore deposits, Economic Geology Publishing Co, pp. 276–293.
10. LIEBENBERG, L., 1970, The sulfides in the layered sequence of the Bushveld igneous complex, Geological Society of South Africa, Special Publication No. 1, pp. 108–207.
11. LUSK, J., 1976, A possible volcanic-exhalative origin for lenticular nickel sulfide deposits of volcanic association, with special references to those of Western Australia; *Can. J. Sci.* **13,** 451–458.
12. MACKENZIE, B. W., 1968, Nickel: Canada and the World, Mining Rept No. 16, Department of Energy, Mines and Resources, Ottawa.
13. MILLER, C. P., 1959, A comparison of plant and soil prospecting for nickel, A.I.M.E. Preprint No. 59-L-40.
14. NALDRETT, A. J., 1966, The role of sulfurization in the genesis of iron–nickel sulfide deposits of the Porcupine District, *Can. Inst. Min. Metall. Bull.* **59,** 489–497.
15. NALDRETT, A. J., and GASPARRINI, E. L., 1971, Archean nickel sulfide deposits in Canada: Their classification, geological setting and genesis, with some suggestions as to exploration, *Geol. Soc. Aust. Spec. Pub.* No. 3, 201–226.
16. NALDRETT, A. J., 1973, Nickel sulfide deposits—Their classification and genesis with special emphasis on deposits of volcanogenic association, *Can. Inst. Min. Metall. Bull.* **66**(739), 45–63.
17. OBIAL, R., APELO, M., and SANTOS, G., 1972, Geochemical prospecting for nickel sulfides—An orientation survey, *J. Geol. Soc. Philipp.* **26,** 1–36.
18. ROSS, J. R., and TRAVIS, G. A., 1981, Nickel sulfide deposits of Western Australia in global perspective, *Econ. Geol.* **76,** 1291–1329.
19. SHLANKA, R., 1969, Copper, nickel, lead and zinc deposits of Ontario, Ontario Department of Mines, M.R.C. 12, Toronto.
20. TAYLOR, R. B., 1964, Geology of the Duluth gabbro complex near Duluth, Minnesota, *Minn. Geol. Surv. Bull.* No. 44.
21. WAGER, L. R., and BROWN, G. M., 1968, *Layered Igneous Rocks,* Oliver & Boyd, London.
22. WAGER, L. R., VINCENT, E. A., and SMALES, A. A., 1957, Sulfides in the Skaergaard intrusion, East Greenland, *Econ. Geol.* **52,** 855–903.
23. WALKER, F., 1940, Differentiation of the Palisades diabase, New Jersey, *Bull. Geol. Soc. Am.* **51,** 1059–1106.
24. WILSON, H. D. B., and BRISBIN, W. C., 1961, Regional structure of the Thompson–Moak Lake Nickel Belt, *Can. Inst. Min. Metall. Bull.* **54**(596), 815–821.
25. WYLIE, P. J. (Ed.) 1967, *A Review of the Geology of the Ultra-mafic and Related Rocks,* Wiley, New York.
26. ZURBRIGG, H. F., 1963, Thompson mine geology, *Can. Inst. Min. Metall. Bull.* **56,** 451–460.

CHAPTER NINE

OPTIMIZED AIRBORNE AND GROUND SEARCH FOR VOLCANOGENIC MASSIVE SULFIDE DEPOSITS OF THE NORTH AMERICAN SHIELD AND CORDILLERA BELT

9.1. GENERAL STATEMENT

Although volcanogenic massive sulfide deposits have been mined since antiquity and through medieval times, the deposit type has been recognized and defined as such only quite recently. Prior to the mid 1950s, the volcanogenic sulfide deposits were classified as hydrothermal, genetically related to often hypothetical granitic intrusives in favorably fractured environments. There has been much confusion between the two terms "stratabound" and "stratiform," which have been used interchangeably during the past two decades, until Sangster's attempt to clarify the matter.[36] He defined as "stratabound," deposits such as the volcanogenic sulfides which occur as discontinuous bodies, generally conformable within specific horizons of a series, but locally cutting across stratification, as opposed to "stratiform" deposits, such as the Mississippi Valley type described in Chapter 10, which are quite continuous and always concordant with encasing strata.

Hutchinson[19] recognizes three categories of volcanogenic sulfide deposits which appear to have gradually evolved through the geological time scale. The first category, considered as a rather primitive and poorly differentiated type, yields mainly zinc–copper ore with minor gold and silver and is characteristically of early to mid-Archean age, with a resurgence in early Proterozoic and again in Paleozoic times. The second category, considered as more differentiated than the first one, consists of complex lead–zinc–copper–silver mineralization. It first appeared in the early Proterozoic period and reached its peak in the Paleozoic era, while continuing strongly through Mesozoic and Cenozoic times. The third category, referred to as "ophiolitic," consists mainly of pyrite and chalcopyrite mineralization genetically related to basic volcanism; it is a much later type which appeared only in Paleozoic times and extended through Mesozoic times.

The geographic spread of the volcanogenic sulfide deposits is as extensive as their chronologic span. The regions of greatest abundance of volcanogenic sulfide deposits are the Precambrian Shields of North America, Scandinavia, and Western Australia. The second largest concentration lies in the Paleozoic Belts of the Appalachian of North America, Scandinavia, and eastern Australia (Tasman Geosyncline). The third concentration is found in the Mesozoic to Cenozoic foldbelts, including the Cordillera of the Americas, the South Pacific Island Arcs, and the Mediterranean region of Europe. The present study covers only two regions of occurrence, namely, the North American Shield and Cordillera Belt.

9.2. GENERAL GEOLOGICAL BACKGROUND: VOLCANOGENIC SULFIDE DEPOSITS OF THE NORTH AMERICAN SHIELD

9.2.1. Lithologic Environment

The deposits consist of one or several lenticular bodies of massive iron and base metal sulfides grossly conformable with encasing series of acid volcanic flows and pyroclastic of submarine volcanogenic origin. The deposits are stratigraphically controlled and are believed to be genetically related to two types of volcanic events; breaks in volcanic geochemistry marked by changes in composition from acidic to more basic, and changes from volcanism to sedimentation, commonly marked by the presence of cherty iron-rich horizons. Although the Archean volcanism produced many cycles of tholeiitic and calc-alkaline volcanic suites, Fox[10] points out that massive sulfide deposition occurred only in the calc-alkaline cycles, where the sulfur content of the intermediate to acid flows is abnormally higher than that of the basic members of the cycles. Similarly, Riddler[33,34] draws attention to the predominance of sulfide over carbonate or oxide facies in the exhalites intervening between volcanic–sedimentary cycles, as a clue to the proximity of volcanogenic sulfide deposits.

Generally, volcanogenic massive sulfide deposits terminate abruptly along the hangingwall, with only a minimal transgression of disseminated pyrite and chloritic alteration within a few feet above the sulfide contact.[11] In relatively little disturbed environments, the two longest dimensions of the deposits, strike length and dip length, parallel stratigraphic features of the volcanic host formation, such as flow contacts and bedding planes. On the footwall side, however, massive sulfide ore grades into "stringer ore" and then into disseminated pyrite within a funnel-shaped alteration pipe consisting of chlorite–sericite and epidote. The pipe generally stands at right angles to the longest dimension of the deposits and tapers off within a distance of 300–400 feet below the massive sulfide bodies. Geochemically, the footwall zone of North American Archean volcanogenic sulfide deposits shows a marked increase in Fe, Mg, and S and a decrease in Si, Na, and K, as compared to the hanging wall formations (see Chapter 12, Ref. 31). In many cases, the alteration pipes of Archean volcanogenic sulfide deposits have been truncated or deformed by post-ore tectonics, or even obliterated by post-ore metamorphism.[36]

Because of the occurrence of repeated cycles of folding and metamorphism which have affected Archean volcanogenic sulfide deposits since their formation, the principal lithologic and geochemical features controlling ore deposition have been disturbed and can be surmised only by analogy from observations made on more recent deposits of Mesozoic and Cenozoic ages. Post-ore alteration affecting ore controlling features belongs to three main types: regional metamorphism, contact metamorphism, and supergene alteration. Sangster[36] reports that under the influence of regional metamorphism the chloritic material of the syn-ore alteration pipes becomes unstable and is altered into an assemblage of hornblend, biotite, and anthophyllite; if sufficient calcium is present, epidote, actinolite, and tremolite appear. Examples of the influence of post-ore contact metamorphism on syn-ore alteration have been described in the Noranda district of North western Quebec. Large subcircular cordierite aggregates form at the expense of the chlorite and give the footwall rock a peculiar spotted appearance (dalmatianite). Supergene alteration, which is quite extensive in tropical and subtropical climates, is not well developed in the North American region and results only in the formation of gossans which assist direct detection from the air and on the ground.

9.2.2. Metallogeny

Two kinds of ore material commonly make up the volcanogenic massive sulfide deposits: massive sulfides comprising from 50% to 60% of the rock by volume, and "stringer ore" with 20% to 30% sulfides only. Sangster[37] reports that 50% of the 70 deposits of his database feature both types of ore material, 45% include massive ore only, while the remaining 5% exhibit stringer ore only, possibly due to the truncation of the massive sulfide portion by post-ore tectonics.

A few volcanogenic sulfide deposits are nearly monomineralic, with pyrite making up nearly 100% of the sulfides by volume (Spain, Philippines). But most deposits are polymineralic: the ore material consists of an admixture of pyrite, pyrrhotite, chalcopyrite, sphalerite with or without galena, minor gold and silver values. The proportion of iron sulfide may be as high as 68% in volume (Geco, northwest Ontario) or as low as 36% (Coronation, Manitoba); pyrrhotite is always present in Archean deposits, but generally subordinate to pyrite. The amount of chalcopyrite varies from as little as 10% of the sulfides by volume (Geco) to as much 60% (Coronation). Pyrrhotite and chalcopyrite are usually predominant in the stringer ore of the footwall zone of the deposits. Sphalerite is nearly always of the iron-rich, high-temperature marmatite variety. Pyrite and sphalerite, with or without galena, prevail in the massive sulfide portion of the deposits.

9.2.3. Mode of Occurrence

There are some 120 known volcanogenic massive sulfide deposits in the Canadian portion of the Precambrian Shield, to which should be added two deposits recently discovered in the U.S. portion of the shield (State of Wisconsin) and 10 deposits occurring in Arizona, within a large Precambrian window included in the Cordillera Belt. In his 1980 study, Sangster[37] shows that the Canadian volcanogenic sulfide deposits are aggregated into eight large clusters varying in size from 90 to 600 square miles. The clusters are made up of as few as four deposits (Manitouwage cluster) or as many as 20 deposits (Noranda cluster). The average number is 12, and the average size of the clusters is equivalent to a circle 20 miles in diameter. A second order of clustering, often directional in nature, is again evident with each regional group.

Among the eight regional clusters, a total of four including the Timmins, Noranda, Joutel-Normetal, and Mattagami districts occur in the eastern portion of the Superior Province of the Canadian Shield. There are two districts in western Ontario including the Manitouwage and the newly discovered Sturgeon Lake districts, and one each in Southeastern Manitoba and in the Arctic region of the Northwest Territories.

9.3. FIELD DETECTION METHODOLOGY

9.3.1. Detection Approaches

Volcanogenic massive sulfide deposits have been one of the most sought for exploration targets during the past three decades, particularly in the Precambrian Shields of North America, Scandinavia, and Western Australia. The recent development of the volcanogenic model for these base metal sulfide deposits has proved very fruitful and has been a main contributing factor to the surge of interest in volcanogenic sulfide exploration. A second factor was the steady improvement of airborne and ground geophysical detection techniques, particularly in Canada and the Scandinavian countries, at a time when the more obvious deposits have all been discovered after several decades of thorough prospecting.

The various field detection methods used in the search for volcanogenic sulfide deposits can be grouped into two main categories, including a direct approach based on visual detection on the ground or from airborne platforms and assisted by mechanical probes (drilling). The second approach, an indirect one, relies on the detection of the signatures of the volcanogenic sulfide deposits by geophysical or geochemical methods.

9.3.2. Direct Detection

Volcanogenic sulfide deposits are much too small targets for direct detection by remote sensing from satellite platforms. The most effective direct approach on the ground consists of thorough prospecting backed by detailed mapping and complemented by some information drilling. According to Boldy,[5] the above approach is responsible for the discovery of 47 deposits, or 43% of the total known to date, mostly during the "pregeophysics" period (1920–1950). In the Noranda district of Northwest Quebec the score is seven deposits or 32% of the total known to date.

Specialized mapping at a semiregional scale backed by conceptual geological models is presently the most successful direct approach, as the days of easy prospecting are over. For instance, mapping based on the exhalite concept[33,34] seeks to define the proximal environment of volcanogenic sulfide deposits on the basis of exhalite facies change from carbonate and oxide to sulfide. Fox[9] describes how detailed mapping of facies changes in pyroclastic formations may be used to locate former volcanic vents as possible sources of base metal sulfides.

The geological insight gained from the guidance by the volcanogenic model in the greenstone belts of northeastern Ontario and northwestern Quebec has been successfully applied to new districts such as Western Ontario and the Arctic region of the Northern Territories, resulting in the discovery of substantial tonnages of sulfide ores. In the well-prospected districts of eastern Canada, the trend of the future should be to rely on information drilling followed by systematic drilling to search for subjacent deposits beyond the average 300 foot detection range of present-day geophysical instrumentation, as demonstrated by the Lake Dufault success story of recent years.

9.3.3. Indirect Geophysical Detection

The detection ability of indirect geophysical methods largely depends on the magnitude of the contrast of physical properties between ore deposits and wallrock. As shown by Paterson,[28] the geophysical environment of volcanogenic sulfide deposits is generally suitable for indirect detection, featuring a high conductivity contrast (500–10,000 times that of wallrock), a moderately high specific gravity contrast (2–2.5 times), accompanied by a moderate magnetic susceptibility contrast (3–100 times), if sufficient pyrrhotite or magnetite (Mattagami deposit) is present. Furthermore, the presence of pyrite disseminations along the margin and footwall of deposits results in polarizable sulfide targets substantially larger than the deposits themselves, which enhances the effectiveness of airborne methods such as INPUT or ground methods such as induced polarization surveys.

Despite the favorable characteristics of volcanogenic sulfide deposits for geophysical detection purposes, these deposits still are elusive exploration targets. According to Patterson's summing up of the results of several years of airborne and ground search for volcanogenic sulfide deposits in the North American Shield,[28] only one ground confirmed geophysical anomaly in 2000 leads to a productive ore deposit. Hindrances to a successful detection of such deposits are many: a relatively unfavorable geometry (small size and steep dip) is one of them. Others include the confounding of sulfide signatures with that of graphite or water-logged fractures, and the masking of sulfide responses by the effect of extensive surficial conductive layers of water-logged glacial clay and tills so prevalent in the Shield.

However, as a result of much recent research work, the discriminating power of geophysical instrumentation has considerably improved. Recent developments include the use of multifrequency, multicoil, and multichannel single instruments, and of hybrid technologies such as pulse electromagnetic and the magnetic induced polarization.[39] Explorationists increasingly rely on combined surveys to eliminate the recording of spurious anomalies masking sulfide signatures. Two of the more successful combinations are (a) electromagnetics + magnetics + gravimetry and (b) induced polarization + magnetics + gravimetry. As a result, according to Boldy,[5] 54% of the volcanogenic deposits discovered since 1950 were detected by airborne and ground geophysical surveys.

9.3.4. Indirect Geochemical Detection

Much interest has recently developed in the detection of primary halos of volcanogenic sulfide deposits by means of bedrock sampling as a method of targeting "blind" sulfide deposits. Holmes,[15] Barragar,[4] and Fox[10] showed that major economic elements such as Cu, Zn, Pb cannot be used as proximal indicators for sulfide deposits because they do not transgress the hanging wall markers (hematite tuff, magnetite jasper, etc.). However, elements such as S, As, F, and some rare earths are good proximal indicators for sulfide ore deposits, along with ratios such as Mg + Fe/Ca + Na, which reflect the geochemical changes occurring in the vicinity of sulfide deposits. A major drawback is the general scarcity of outcrops which prevails over large areas of the Shield.

The detection of secondary halos is hampered in the shield by the prevalence of transported glacial alluvium. The interpretation of soil anomalies is further complicated in the northernmost portions of the Shield by the influence of the permafrost. As a result, the success ratio of the indirect geochemical approach is rather low.

9.4. STATISTICAL MODELING OF GEOMETRIC PARAMETERS OF SHIELD VOLCANOGENIC SULFIDE DEPOSITS

9.4.1. Scope of Study

The sample selected for inclusion in the database comprises a total of 102 commercial and subeconomic volcanogenic sulfide deposits, all located in the Canadian portion of the North American Shield. The deposits are listed by name and province of occurrence in Table 9.1. The geographic breakdown of the sample is as follows: greenstone belts of northeastern Ontario and northwestern Quebec: 57 deposits (56% of total); greenstone belts of western Ontario: 10 deposits (9% of total); southeastern Manitoba: 31 deposits (30% of total); and Northwest Territories: 4 deposits, or one half of the total discovered to date in that region. Because of its large size and wide geographic spread, the sample should offer a statistically satisfactory representation of the population of volcanogenic sulfide deposits of the North American Shield.

9.4.2. Statistical Modeling

Measurements of the five geometric parameters required for the calculation of detection probabilities were obtained from each of the 102 volcanogenic sulfide deposits included in the sample. The geometric parameters are of two kinds. The first three are of a dimensional nature; they include the length, breadth, and shape ratio ($R = B/L$) of the horizontal section of the portion of the deposits which lies within the average 300-foot range of commonly used detectors. The statistical range of the measured parameters is as follows: from to 225 to 4850 feet for length and 20 to 600 feet for breadth. The last two parameters are

TABLE 9.1(a)
List of Volcanogenic Massive Sulfide Deposits of the North American Shield (N.W. Quebec Section) Included in the Database.

Noranda (Horne)	Barvue
Quemont	Vendome
McDonald	New Formaque
Amulet "A"	Belfort
Iso	Barvallee
Joliet	Normetal
Mobrun	Poirier
Corbett	Joutel
Norbec	Northern Exploration
Millenbach	Brouillan
Aldermac	Lessard
East Waite	Lemoine
West Waite	Scott
New Insco	Jay Copper
Amulet "C"	Lynx Yellowknife
Delbridge	Dumagami
Vauze	Pershcourt
Amulet "F"	Louvicourt
Waite-Dufault	Lake Berrigan
East Sullivan	Mattagami
Manitou-Barvue	Orchan
Louvem	New Hosco
Abitibi Copper	Norita
Dunraine	Phelps Dodge
Quebec Manitou	Bell Allard
	Garon Lake
	Radiore

TABLE 9.1(b)
List of Volcanogenic Massive Sulfide Deposits of the North American Shield (Ontario, Manitoba, and N.W.T. Sections) Included in the Database.

Ontario

Ecstall	Willecho
Kamkotia	Big Nama
Jameland	Mattabi
Jamieson	N.B.U.
Munro	Lyon Lake
Geco	Creek
Willroy	Coldstream
	South Bay

Manitoba

Flin-flon	Chisel
Pine Bay	Stall
Coronation	Osborne
Centennial	Freeport Reed
Schist Lake	Anderson
West Schist	H.B. Reed
Cuprus	Wim
White Lake	Dickstone
Birch Lake	Little Stall
North Star	Rail
Mandy	Ghost Lake
	Lost Lake
	Copperman

Northwest Territories

Hackett	Ruttan
Isok	Fox Lake
High Lake	
Taki Lake	

attitudinal in nature; they include the unoriented true dips and the strike orientation referred to true north and measured in degrees within the right half-circle. The range of the measured dip angles is from 48 to 85 degrees.

The measurements of the geometric parameters are grouped into classes and assembled into frequency distributions. The statistics summarizing the distributions of observed data are listed in the lower half of Table 9.2. Included are arithmetic means, modes, standard deviations, and coefficient of skewness. The latter is constructed as the ratio (mean-mode)/standard deviation to indicate the degree of asymmetry of the frequency distributions. Skewness is slight (0.1)

TABLE 9.2
Statistical Modeling Summary for Volcanogenic Massive Sulfide Deposits of the North American Shield (Sample Size: 102 Deposits).

Statistic of fitted model	Length of horiz. sect. in feet	Breadth of horiz. sect. in feet	Shape ratio R=B/L	Strike direct. from T. North in degrees	Dip in degrees
25%ile	530	25	0.02	62	50
95%L.c.l.	810	48	0.05	84	66
Mean	947	61	0.06	93	69
95% U.c.l.	1106	76	0.08	102	72
75%ile	1650	150	0.12	130	83
Disp. Coef.	0.12	0.31	0.34	0.21	0.13

Statistic of observed data					
Arith.mean	1286	118	0.14	92	69
Mode	700	50	0.06	97	80
Stand.dev.	1175	152	0.20	45	16
Skewness	0.51	0.45	0.41	0.11	0.73

for the strike orientation parameter, moderate (average 0.45) for length, breadth, and shape ratio, but is quite strong (0.75) for the dip angle parameter.

Lognormal models were successfully fitted to the observed frequency distributions of the dimensional parameters, as confirmed by the results of the χ^2 goodness of fit testing at the 0.05 confidence level. The statistical models fitted to the dip angle and strike direction distributions are normal and circular normal, respectively. The statistics of the fitted models are assembled in the upper half of Table 9.2. They include the geometric means and their 95% confidence limits for the lognormal models, as well as the arithmetic and circular means and their confidence limits for the attitudinal parameters. Additional listings include the 25th and 75th percentiles and dispersion coefficients. The latter are constructed as the ratios of standard deviations over means in order to give a dimensionless indication of the variability of the data. The coefficient is small (0.1) for the length parameter, and moderately large (0.33) for the breadth and shape ratio parameters.

9.5. CONSTRUCTION AND ORGANIZATION OF DETECTION PROBABILITY TABLES FOR SHIELD VOLCANOGENIC SULFIDE DEPOSITS

9.5.1. Introduction

For detection purposes, the volcanogenic sulfide deposits are considered as oriented dipping slabs with horizontal cross sections of elliptical shape. The target dimensions are obtained by introducing a dip component calculated in the manner described in Sections 2.2.3 of Chapter 2 and 5.4.2 of Chapter 5 for both vertical and angled detection situations. As a result, the mean shape ratio is inflated from 0.06 for horizontal sections of deposits to 0.19 and 0.32 for target sections, when vertical and 55-degree angled detection are considered, thus considerably boosting the probability of successful detection of volcanogenic sulfide deposits.

The latter is further increased when the control grid orientation with respect to the expected strike direction of targets is optimized according to the results of a dynamic programming study conducted as described in Section 5.5.4 of Chapter 5. The optimal grid orientations for the detection of volcanogenic sulfide deposits by various types of field programs are as follows: (a) parallel grids for airborne geophysical surveys are to be laid at 90 degrees ± 10 to the expected target strike, (b) ground square grids for vertical and angled detection are to be oriented at angles of 30 and 45 degrees ± 10 with target strike, respectively. The expected target strike direction can be estimated from previous surveys or photogeological studies. If no prior information is available, the mean of the fitted circular normal model is a statistically valid estimate of the strike direction (see Table 9.2).

9.5.2. Description and Organization of Tables

Two similar sets of tables were constructed to provide the probabilities of single detection (at least one target intersection by the grids) and of confirmed detection (at least two intersections) of volcanogenic sulfide deposits of the North American Shield. The grid spacings are varied from 200 to 6000 feet in steps of 50 to 200 feet, as appropriate. For each spacing, the probability of detection is calculated and listed within a 95% confidence interval, based on the mean target length and its fiducial interval.

A third set of tables is constructed to provide the grid spacing and associated coverage cost per unit of area required to ensure a prespecified level of detection probability, which should prove most useful for planning purposes. The probability tables are organized into three blocks, one for each type of field program. Tables 9.4.–9.7 cover detection by airborne geophysical surveys on parallel and square grids. Vertical ground detection is covered by Tables 9.9–9.11, and 55-degree angled detection by Tables 9.13–9.15.

9.6. DESIGN OF THREE STRATEGIES FOR THE DETECTION OF SHIELD VOLCANOGENIC SULFIDE DEPOSITS

9.6.1. Detection Strategies

Three detection strategies may be considered for the detection of volcanogenic sulfide deposits in the North American Shield, each applied to a specific exploration situation. The first one, referred to as the liminal option, ensures a minimum level of detection probability of 0.500 at a cost low enough to carry out large-scale coverages of area of little known potential. The second one, the maximal option, provides near certain detection (0.980) at a much higher cost than the liminal option, thus restricting its application to small areas of high potential. Finally, the optimal option is an attractive compromise between the two options listed above, which should prove suitable for the majority of exploration situations.

9.6.2. Liminal Option

The detection characteristics pertaining to the liminal option are listed in the leftmost portion of Table 9.3, covering both single and confirmed detection for three types of field programs. The coverage cost multiplier required to ensure confirmed detection rather than single detection is 2.3 for airborne geophysical surveys. It falls to 1.95 for ground geophysical surveys and rises again to 2.5 for both vertical and angled drilling programs.

The relative detection performances of various field programs or grid designs may be measured by means of the coverage cost per unit of area as a yardstick.

TABLE 9.3
Summary of Detection Characteristics of Volcanogenic Massive Sulfide Deposits of the North American Shield for Three Types of Detection Strategies.

Type of survey and grid geometry	Liminal detection		Optimal detection		Maximal detection	
	Coverage cost: US$/mile sq. Grid size: feet		Coverage cost: US$/mile sq. Grid size: ft	Detection probability	Coverage cost: US$/mile sq. Grid size: feet	
	Detection	Confirmed detection			Detection	Confirmed detection
Airborne geophysical parallel grid	$448 1320 ft	$1025 475 ft	$633 750 ft	0.79	$1990 200 ft	$3825 100 ft
Airborne geophysical square grid	$510 2050 ft	—	$533 1870 ft	0.62	$975 880 ft	—
Ground geophysical square grid	$2620 430 ft	$4950 265 ft	$3483 350 ft	0.95	$4410 280 ft	$7900 180 ft
Vertical drilling square grid	$1,120,000 430 ft	$2,450,000 265 ft	$1,554,000 350 ft	0.85	$2,080,000 280 ft	$4,508,000 180 ft
55-degree angled drilling square grid	$572,200 675 ft	$1,350,300 440 ft	$705,000 600 ft	0.82	$1,010,000 480 ft	$1,685,000 370 ft
Optimal 40-deg angled drilling square grid	—	—	$683,000 700 ft	0.83	—	—

We find that the parallel grid design is preferable to the square design for airborne geophysical searches because the coverage cost for the former is 13% cheaper. Similarly, the 55-degree angled drilling option should be preferred to the vertical option because the coverage cost by the former is 50% cheaper.

9.6.3. Maximal Option

The rightmost portion of Table 9.3 lists the detection parameters pertaining to the maximal option. The cost multiplier required to ensure confirmed detection by airborne geophysical surveys is 1.92, and for ground geophysical surveys as well as vertical and angled drilling programs the average is 1.80. It should be noted that the maximal multiplying factors are lower than their liminal counterparts; thus, when opting for the maximal strategy, confirmation of detection is, relatively speaking, a more attractive requirement than in the liminal context.

If we wish to compare the relative merits of the parallel and square grid designs for airborne surveys, we find that selection of the square design is the more attractive option because it leads to a 54% coverage cost reduction. Angled drilling is also much more attractive an option than vertical detection because the coverage cost is 51% cheaper.

9.6.4. Optimal Option

Based on the results of a three-stage dynamic programming study (see Section 5.5.4 of Chapter 5), three grid design parameters were computed to ensure the optimal detection of volcanogenic sulfide deposits of the North American Shield. They include (a) the optimal grid orientation with respect to the expected strike direction, (b) optimal grid spacing and corresponding coverage cost per unit of area, as well as the associated detection probability, and (c) the optimal drilling angle when the likely dip direction may be estimated. The optimal parameters are listed within their 95% confidence intervals in Table 9.8 for airborne geophysical surveys on two grid designs, in Table 9.12 for ground geophysical surveys, and in Table 9.16 for vertical and angled drilling programs. For the reader's convenience, the optimal detection characteristics were assembled and are listed in the central portion of Table 9.3.

If we wish to compare the detection performances of various field programs or grid designs, we have to rely on the "expected loss" criterion as described in Section 6.5.4 of Chapter 6, instead of the coverage cost criterion used above, because the optimal probabilities of detection vary with the type of survey. We find that the expected loss associated with the parallel grid design for airborne

SEARCH FOR VOLCANOGENIC MASSIVE SULFIDE DEPOSITS 249

surveys is 53% lower than in the square grid case; therefore, the former should be our choice. The optimally angled (40 degrees) drilling program is a clear-cut choice over the vertical program, because of the 50% reduction in expected loss which can be expected.

9.6.5. Multipurpose Optimal Grids

As indicated in Sections 8.14 of Chapter 8 and 9.2.1 of the present chapter, both Ni–Cu ultramafic deposits of volcanogenic affiliations and volcanogenic massive sulfide deposits are known to occur in "island arc" remnants referred to commonly as "greenstone belts" of the North American Archean Shield. Admittedly, the two types are not spatially correlated within the Archean eugeosynclinal edifices of the island arcs. However, a joint coverage for both types of deposits in little-prospected greenstone belts of unknown potential should be considered in the context of the optimal option.

A careful comparison of Tables 8.8 and 9.8 (airborne surveys), Tables 8.12 and 9.12 (ground geophysical surveys), and finally Tables 8.16 and 9.16 (drilling programs) shows overlaps between the paired 95% confidence intervals for the optimal grid spacings. The regions common to the paired confidence intervals define narrow ranges of spacings which are jointly optimal for the search for both types of deposits within the same greenstone belts.

The overlaps of confidence intervals are listed as optimal grid spacing ranges in Table 9.17, along with the corresponding coverage cost and associated optimal detection probabilities for each type of field program or grid design. For example, one can see from Table 9.17 that parallel grids with a 1300-foot spacing, and an 1800-foot square grid size are optimal for the joint detection of the two types of deposits by airborne surveys. Likewise, a 400-foot square grid size is jointly optimal for ground geophysical surveys. Finally, a 700-foot spacing is jointly optimal for the search of both deposits by drilling optimally angled between 40 and 45 degrees.

9.7. DETECTION PROBABILITY AND OPTIMIZATION TABLES FOR SHIELD VOLCANOGENIC SULFIDE DEPOSITS

The following sequence of fourteen tables deals with the probabilities of detection and optimal designs of airborne and ground programs in the search for volcanogenic massive sulfide deposits of the Shield (pages 250–261).

TABLE 9.4
Probabilities of Detection of Volcanogenic Massive Sulfide Deposits of the North American Shield by Airborne Geophysical Surveys on Parallel Grids.

```
Survey design:  parallel lines with spacing S feet
Elliptical targets with expected major axis = L feet
in the confidence interval:  l.c.l. ( geom. mean ( u.c.l.
and minor axis = B feet
R is defined as the ratio B/L with geometric mean = 0.19
Orientation of flight lines:  90 degrees with strike-line.
```

	Probability of detection		
Grid spacing S feet	l.c.l. L= 810	geom. mean 947	u.c.l. 1106 feet
200	0.938	0.948	1.000
400	0.874	0.895	0.915
600	0.805	0.839	0.871
800	0.724	0.778	0.824
1000	0.616	0.732	0.773
1200	0.513	0.610	0.714
1400	0.440	0.523	0.643
1600	0.385	0.458	0.563
1800	0.342	0.407	0.500
2000	0.308	0.366	0.450
2200	0.280	0.333	0.409
2400	0.257	0.305	0.375
2600	0.237	0.282	0.346
2800	0.220	0.262	0.322
3000	0.205	0.244	0.300
3200	0.192	0.229	0.281
3400	0.181	0.215	0.265
3600	0.171	0.203	0.250
3800	0.162	0.193	0.237
4000	0.154	0.183	0.225
4200	0.147	0.174	0.214
4400	0.140	0.166	0.205
4600	0.134	0.159	0.196
4800	0.128	0.153	0.188
5000	0.123	0.146	0.180
5200	0.118	0.141	0.173

TABLE 9.5
Determination of Grid Size for Specified Levels of Probability in the Detection of Volcanogenic Massive Sulfide Deposits of the North American Shield by Airborne Geophysical Surveys on Parallel Grids.

```
Survey design:  parallel lines with spacing S feet
Unit cost = US$70 per line-mile
```

Specified detection probability level	Required grid spacing S in feet	Corresponding cost in $ per mile square
0.10	6580	$ 196.17
0.15	4520	$ 221.77
0.20	3460	$ 246.82
0.25	2780	$ 272.95
0.30	2340	$ 297.95
0.35	2020	$ 322.97
0.40	1780	$ 347.64
0.45	1580	$ 373.92
0.50	1420	$ 400.28
0.55	1300	$ 424.31
0.60	1200	$ 448.00
0.65	1100	$ 476.00
0.70	1020	$ 502.35
0.75	860	$ 569.77
0.80	700	$ 668.00
0.85	540	$ 824.44
0.90	360	$1166.67
0.95	200	$1988.00

TABLE 9.6
Probabilities of Confirmed Detection of Volcanogenic Massive Sulfide Deposits of the North American Shield by Airborne Geophysical Surveys on Parallel Grids.

Survey design: parallel lines with spacing S feet,
Randomly orientated elliptical targets with expected major axis = L feet
in the confidence interval: l.c.l. (geom. mean (u.c.l.
and minor axis = B feet
R is defined as the ratio B/L with geometric mean = 0.19

Grid spacing S feet	Probability of confirmed detection		
	l.c.l. L= 810	geom. mean 947	u.c.l. 1106 feet
200	0.908	0.918	0.958
300	0.806	0.821	0.836
400	0.604	0.635	0.716
500	0.401	0.537	0.603
600	0.293	0.395	0.571
700	0.228	0.302	0.432
800	0.194	0.242	0.343
900	0.000	0.205	0.280
1000	0.000	0.000	0.236
1100	0.000	0.000	0.207
1200	0.000	0.000	0.191

TABLE 9.7
Probabilities of Detection of Volcanogenic Massive Sulfide Deposits of the North American Shield by Airborne Geophysical Surveys on Square Grids.

Survey design: square grid with spacings S by S feet
Randomly orientated elliptical targets with expected major axis = L feet
in the confidence interval: l.c.l. (geom. mean (u.c.l.
and minor axis = B feet
R is defined as the ratio B/L with geometric mean = 0.19

	Probability of detection		
Grid spacing S feet	l.c.l. L= 810	geom. mean 947	u.c.l. 1106 feet
800	1.000	1.000	1.000
1000	0.866	0.960	1.000
1200	0.765	0.861	1.000
1400	0.682	0.776	0.890
1600	0.615	0.703	0.816
1800	0.558	0.642	0.751
2000	0.511	0.590	0.695
2200	0.471	0.546	0.645
2400	0.437	0.507	0.602
2600	0.407	0.474	0.564
2800	0.381	0.444	0.530
3000	0.358	0.418	0.500
3200	0.338	0.395	0.473
3400	0.320	0.374	0.449
3600	0.303	0.355	0.427
3800	0.288	0.338	0.407
4000	0.275	0.323	0.389
4200	0.263	0.309	0.372
4400	0.252	0.296	0.357
4600	0.241	0.284	0.343
4800	0.232	0.273	0.330
5000	0.223	0.263	0.318
5200	0.215	0.253	0.307

TABLE 9.8
Optimal Design of Airborne Geophysical Surveys for the Detection of Volcanogenic Massive Sulfide Deposits of the North American Shield.

```
Confidence interval:  all results are reported as 95% confidence intervals
respectively
Expected target length in feet:      810        947       1106
Expected shape ratio R = 0.19
Unit Cost:  $70/l.ml.
```

```
  (i)   Survey Design:
        Parallel Grid

Optimal grid spacing in feet:        700        750       1050
Cost in US$ per mile square:         668        633        476
Probability of detection:            0.77       0.79       0.75
Optimal grid orientation:         90 degrees to expected strike
```

```
  (ii)  Survey Design:
        Square Grid

Optimal grid spacing in feet:       1645       1870       1990
Cost in US$ per mile square:         589        536        512
Probability of detection:            0.60       0.62       0.70
```

TABLE 9.9
Probabilities of Detection of Volcanogenic Massive Sulfide Deposits of the North American Shield by Ground Geophysical Surveys or Vertical Drilling to a depth of 300 Feet.

Survey design: square grid with spacings S by S feet. The expected shape of the ore deposits is elliptical with major axis = L in the 95% confidence interval: L.c.l. (geometric mean (U.c.l., and minor axis = B feet, R is defined as the ratio B/L with geometric mean = 0.19 grid orientation 30 degrees + or - 10 from the expected strike line.

	Probability of detection		
Spacing S in feet	L.c.l. L = 810	G. mean 947	U.c.l. 1106ft
200	1.000	1.000	1.000
250	0.998	1.000	1.000
300	0.939	0.997	1.000
350	0.793	0.951	1.000
400	0.635	0.837	0.981
450	0.510	0.700	0.921
500	0.416	0.579	0.820
600	0.289	0.408	0.606
700	0.212	0.300	0.453
800	0.163	0.230	0.348
900	0.128	0.182	0.275
1000	0.104	0.147	0.222
1100	0.086	0.122	0.184
1200	0.072	0.102	0.154
1300	0.062	0.087	0.132
1400	0.053	0.075	0.114
1500	0.046	0.065	0.099

TABLE 9.10
Determination of Grid Size for Specified Levels of Probability in the Detection of Volcanogenic Massive Sulfide Deposits of the North American Shield by Vertical Drilling to a Depth of 300 Feet.

Survey Design: Square Grid With Spacings S by S Feet
Unit Cost = $6000 Per Diamond Drill-Hole

Probability Level Of Detection	Grid Size In Feet	Drilling Cost Per Mile Square In U.S.$. Thousands
0.05	1600	111.46
0.15	1000	237.38
0.25	800	347.47
0.35	650	500.48
0.45	600	577.42
0.55	550	675.43
0.65	500	803.19
0.75	450	974.35
0.85	400	1211.54
0.95	350	1554.43

TABLE 9.11

Probabilities of Confirmed Detection of Volcanogenic Massive Sulfide Deposits of the North American Shield by Ground Geophysical Surveys or Vertical Drilling to a Depth of 300 Feet.

Survey design: square grid with spacings S by S feet.
The expected shape of the ore deposits is elliptical
with major axis = L in the 95% confidence interval:
L.c.l. (geometric mean (U.c.l., minor axis = B feet,
R is defined as the ratio B/L with geometric mean = 0.19
grid orientation 0 degrees + or - 10 from the expected
strike line.

	Probability of confirmed detection		
Spacing S in feet	L.c.l. L = 810	G. mean 947	U.c.l. 1106ft
100	1.000	1.000	1.000
150	0.987	1.000	1.000
200	0.863	0.951	1.000
250	0.627	0.829	0.944
300	0.329	0.608	0.846
350	0.201	0.348	0.701
400	0.119	0.230	0.426
450	0.069	0.150	0.305
500	0.039	0.095	0.218
550	0.022	0.060	0.153
600	0.012	0.037	0.106
700	0.003	0.014	0.050
800	0.000	0.004	0.023
900	0.000	0.001	0.010
1000	0.000	0.000	0.004

TABLE 9.12

Optimal Design of Ground Geophysical Surveys for the Detection of Volcanogenic Massive Sulfide Deposits of the North American Shield.

Survey Design: Square Grid With Spacing S by S Feet.
Confidence Interval: All results are reported as 95% confidence intervals respectively.
Expected Target Length in Feet: 810   947   1106
Expected Shape Ratio R = 0.19
Unit Cost: $6/Station

Optimal Grid Spacing in Feet:	300	350	500
Cost in $ Per Mile Square:	4308	3483	2189
Probability of Detection:	0.94	0.95	0.82

Optimal grid Orientation: 30 degrees + or -10 to expected strike
direction of target.

TABLE 9.13
Probabilities of Detection of Volcanogenic Massive Sulfide Deposits of the North American Shield by 55-Degree Angled Drilling to a Vertical Depth of 300 Feet.

Survey design: square grid with spacings S by S feet. The expected shape of the ore deposits is elliptical with major axis = L in the 95% confidence interval: L.c.l. (geometric mean (U.c.l., and minor axis = B feet, R is defined as the ratio B/L with geometric mean = 0.32 grid orientation 45 degrees + or - 10 from the expected strike line.

Spacing S in feet	Probability of detection		
	L.c.l. L = 810	G. mean 947	U.c.l. 1106ft
300	1.000	1.000	1.000
350	0.997	1.000	1.000
400	0.930	1.000	1.000
450	0.810	0.971	1.000
500	0.688	0.883	0.999
550	0.578	0.779	0.974
600	0.487	0.678	0.907
700	0.358	0.506	0.739
800	0.274	0.387	0.584
900	0.216	0.306	0.463
1000	0.175	0.248	0.375
1100	0.145	0.205	0.310
1200	0.122	0.172	0.260
1300	0.104	0.147	0.222
1400	0.089	0.126	0.191
1500	0.078	0.110	0.167
1600	0.068	0.097	0.146
1700	0.061	0.086	0.130
1800	0.054	0.076	0.116
1900	0.049	0.069	0.104
2000	0.044	0.062	0.094
2500	0.028	0.040	0.060

TABLE 9.14

Determination of Grid Size for Specified Levels of Probability in the Detection of Volcanogenic Massive Sulfide Deposits of the North American Shield by 55-Degree Angled Drilling to a Vertical Depth of 300 Feet.

Survey Design: Square Grid With Spacings S by S Feet
Unit Cost = $7325 Per Diamond Drill-Hole

Probability Level Of Detection	Grid Size In Feet	Drilling Cost Per Mile Square In U.S.$. Thousands
0.15	1300	188.27
0.25	1000	289.81
0.35	900	346.39
0.45	800	424.21
0.55	700	535.83
0.65	650	611.00
0.75	600	704.45
0.85	550	824.03
0.95	500	979.89

TABLE 9.15

Probabilities of Confirmed Detection of Volcanogenic Massive Sulfide Deposits of the North American Shield by 55-Degree Angled Drilling to a Vertical Depth of 300 Feet.

Survey design: square grid wth spacings S by S feet.
The expected shape of the ore deposits is elliptical
with major axis = L in the 95% confidence interval:
L.c.l. (geometric mean (U.c.l., and minor axis = B feet,
R is defined as the ratio B/L with geometric mean = 0.32
grid orientation 0 degrees + or - 10 from the expected
strike line.

Spacing S in feet	Probability of confirmed detection		
	L.c.l. L = 810	G. mean 947	U.c.l. 1106ft
350	1.000	1.000	1.000
400	0.854	1.000	1.000
450	0.420	0.974	1.000
500	0.269	0.580	1.000
550	0.169	0.375	0.980
600	0.101	0.258	0.735
700	0.025	0.114	0.325
800	0.001	0.040	0.174
900	0.000	0.007	0.085
1000	0.000	0.000	0.034

TABLE 9.16
Optimal Designs for the Detection of Volcanogenic Massive Sulfide Deposits of the North American Shield by Drilling to a Vertical Depth of 300 Feet.

```
Survey Design:  square grid with spacing S by S feet
Confidence interval:  all results are reported as 95% confidence intervals
respectively
Expected target length in feet:      810         947        1106
Unit Cost = $20 per linear ft.
```

(a) Vertical drilling:

```
Optimal grid spacing in feet:        300         350         500
Cost in $thou. per mile square:      2078        1554        803
Probability of detection:            0.67        0.82        0.81
Optimal grid orientation:       30 degrees to expected strike of target
```

(b) Angled drilling at 55 degrees:

```
Optimal grid spacing in feet:        600         600         700
Cost in $thou. per mile square:      705         705         536
Probability of detection:            0.88        0.83        0.83
Optimal grid orientation:       45 degrees to expected strike of target
```

(c) Angled drilling at optimal angle = 40 degrees:

```
Optimal grid spacing in feet:        600         700         800
Cost in $thou. per mile square:      898         683         541
Probability of detection:            0.94        0.95        0.82
Optimal grid orientation:       45 degrees to expected strike of target
```

TABLE 9.17
Parameters of Multipurpose Optimal Grids for the detection of Ni–Cu Ultramafic and Volcanogenic Massive Sulfide Deposits of the North American Shield.

Type of survey & grid geometry	Range of grid size	Range of coverage cost ($/ml.sq.)	Range of probability of detection
A.B. geophysical parallel grid	1290 to 1350 ft	$415 to $430	0.56 to 0.72
A.B. geophysical square grid	1760 to 1990 ft	$510 to $540	0.62 to 0.93
Grd. geophysical square grid	300 to 500 ft	$2200 to $4310	0.77 to 0.93
Vertical drilling square grid	300 to 500 ft	$803,000 to $2,078,000	0.77 to 0.93
Opt. angled (40 to 45 deg.) drilling square grid	600 to 800 ft	$424,000 to $705,000	0.75 to 0.98

9.8. GEOLOGICAL SYNOPSIS FOR VOLCANOGENIC SULFIDE DEPOSITS OF THE NORTH AMERICAN CORDILLERA BELT

Volcanogenic massive sulfide deposits ranging in age from Paleozoic to Mesozoic occur in two main regions of the North American Cordillera Belt. The first one lies in the Canadian section of the belt, in the west central and southern portion of the Province of British Columbia and in Vancouver Island. The second region extends through the northern and northcentral portions of the state of California, in the U.S. section of the belt. A third region lies in the state of Arizona, but is not covered here because the deposits occur in a large Precambrian window included in the Cordillera Belt (see Table 9.18). The most common geological environment for the Cordillera volcanogenic sulfide deposits consists of roof pendants of metavolcanics and sediments of Paleozoic to Mesozoic age engulfed in the coastal batholits of British Columbia or the Sierra Nevada of California.

There are some 25 known commercial or subeconomic occurrences of volcanogenic sulfide ores in British Columbia. Most of them are found in a Permo–Triassic volcanic environment. The stratigraphic control for the sulfide deposits appears to be the contact between series of lapilli-tuff, lithic tuffs, or dacitic tuffs, often altered into quartz-sericite schists, on the one hand, and acid pyroclastic piles, on the other hand. One deposit of possible, if doubtful, volcanogenic affiliation, known as "Sam Goosely," in north central British Columbia, is thought to have been subjected to remobilization and redeposition in sedimentary piles. The large Granduc deposit, presently encased in formations mylonitized beyond recognition, is also thought to have originally formed in a volcanogenic environment.

Two types of volcanogenic sulfide deposits are found in the Californian section of the Cordillera Belt.[1,21] The earlier type consists of disseminated to massive copper–zinc mineralization occurring in middle Devonian to Permian volcanic series of the West Shasta District. A younger and more complex type consisting of massive Zn–Pb–Cu–Ag mineralization is found in volcanic series of Triassic to Jurassic age which stretch southward from the East Shasta District and along the western edge of the Sierra Nevada over a length of 300 miles.

There are some 20 known commercial or subeconomic volcanogenic sulfide deposits in the whole Shasta District, most of them of modest size but very high grade; a few of them are of the large tonnage and low grade type, mainly in the West Shasta District. Deposits of the early type show the typical chlorite–sericite footwall alteration described in Section 9.2.1 as pertaining to Archean volcanogenic deposits of the Shield. Some of the younger deposits of the East Shasta District are notable for the development of barite and anhydrite in the gangue, which could establish a genetic link with the much more recent "kuroko" type of volcanogenic deposits of Japan, Fiji, and the Philippines.

9.9. STATISTICAL MODELING OF GEOMETRIC PARAMETERS OF CORDILLERA VOLCANOGENIC SULFIDE DEPOSITS

A total of 21 commercial and subeconomic deposits were selected as a representative sample of the whole population of known and undiscovered volcanogenic sulfide deposits of the North American Cordillera Belt. The British Columbia section of the belt is represented by 13 deposits and the Californian section by 8. The deposits are listed by name in Table 9.18.

The geometry of the deposits is described by several dimensional and attitudinal parameters. The former include length, breadth, and shape ratio of the

TABLE 9.18
List of Volcanogenic Massive Sulfide Deposits of the North American Cordillera Belt Included in the Database.

British Columbia (Canada)

Tulsequah	Lynx
Anyox	H.W.
Chuchua	Myra
Sam Goosly	Price
Goldstream	
Granduc	
Kutcho	
Seneca	

California (U.S.A.)	Arizona (U.S.A.) (*)
Shasta-King	Iron King
Afterthought	United Verde
Iron Mountain	Copper Queen
Bully Hill + Rising Star	Old Dick
Mammoth	
Balaklala	
Copper Hill	
Keystone	

(*) Precambrian window; not included in sample.

horizontal sections of deposits, and the latter, unoriented dip angle and strike orientation with respect to true north measured in degrees within the right half-circle. The statistical range of the length measurements for the 21 deposits of the sample is from 300 to 7000 feet; breadth measurements vary from 25 to 220 feet, and dip angles from 28 to 85 degrees.

The parameter measurements are grouped into classes from which frequency distributions are constructed. The statistics summarizing the distributions of observed data are listed in the lower half of Table 9.19, including arithmetic means, modes, standard deviations, and coefficients of skewness. The latter give an indication of the degree of asymmetry of the distributions of observed data. The skewness is slight (0.13) for the strike orientation, moderate for lengths, breadths, and dips, but quite strong for the shape ratios (0.6).

Lognormal models were successfully fitted to the observed distributions of the dimensional parameters, while normal and circular normal models were used to fit the distributions of dips and strike directions. The models are summarized by statistics which are listed in the upper half of Table 9.19, including geometric means for the log-normal models, and arithmetic and circular means for the attitudinal models, with their 95% confidence limits. Additional statistics are the 25th and 75th percentiles and the dispersion coefficients. The latter give an indication of the variability of the data. Dispersion is slight for the lengths (0.12), and moderate for the breadths and shaped ratios (0.25).

9.10. CONSTRUCTION AND ORGANIZATION OF DETECTION PROBABILITY TABLES FOR CORDILLERA VOLCANOGENIC SULFIDE DEPOSITS

Volcanogenic sulfide deposits of the Cordillera are treated in the same manner as the shield deposits for detection purposes (see Section 9.5.1). Two sets of tables are constructed to provide the probabilities of detection or confirmed detection of the volcanogenic sulfide deposits by means of three types of field programs, based on grid spacings varying from 200 to 6000 feet in steps of 50 to 200 feet. A different set of tables is constructed for planning purposes, providing grid spacings and corresponding coverage costs per unit of area for specified levels of detection probability. Tables 9.21–9.24 cover airborne geophysical surveys on parallel and square grids. Ground geophysical detection probabilities are provided by Tables 9.26–9.28. Finally, Tables 9.30–9.32 deal with detection by angled drilling.

9.11. DESIGNING THREE STRATEGIES FOR THE DETECTION OF CORDILLERA VOLCANOGENIC SULFIDE DEPOSITS

9.11.1. Liminal and Maximal Strategies

Two strategies, referred to as liminal and maximal, were described in Section 9.6.2 and are applied to the detection of volcanogenic sulfide deposits of the Cordillera. The detection characteristics pertaining to the two strategies are listed in the leftmost and rightmost sections of Table 9.20, respectively.

Cost multipliers required to ensure confirmation of single detection are 2.2 for airborne and ground geophysical surveys and 2.5 for vertical and angled drilling, for the liminal option. The maximal cost multipliers are all smaller than their liminal counterparts, making the confirmation requirement more attractive than in the liminal context, relatively speaking.

If we wish to compare the merits of two types of grid designs for airborne

SEARCH FOR VOLCANOGENIC MASSIVE SULFIDE DEPOSITS 265

TABLE 9.19
Statistical Modeling Summary for Volcanogenic Massive Sulfide Deposits of the North American Cordillera Belt (Sample Size: 22 Deposits).

Statistic of fitted model	Length of horiz.sect. in feet	Breadth of horiz.sect. in feet	Shape ratio R=B/L	Strike direct. from T. North in degrees	Dip in degrees
25%ile	1200	35	0.02	65	50
95%L.c.l.	1201	40	0.02	72	53
Mean	1842	61	0.03	88	62
95%U.c.l.	2836	92	0.05	103	70
75%ile	3350	120	0.06	110	73
Disp.coef.	0.13	0.23	0.26	0.18	0.12
Statistic of observed data					
Arithm.mean	2508	85	0.05	85	62
Mode	2000	35	0.02	93	68
Stand.dev.	1875	71	0.05	65	19
Skewness	0.31	0.28	0.62	0.13	0.33

geophysical surveys using the coverage cost per unit area as a yardstick, we find that, as is the case for the Shield volcanogenic deposits, the square grid design should be rejected in the liminal context because it results in a 12% cost increase. In the maximal context, however, the square grid design is more attractive than the parallel one, as it results in a 56% reduction of coverage cost per unit area.

An evaluation of the respective merits of angled and vertical drilling programs based on the yardstick mentioned above leads to the selection of angled drilling over vertical detection in both the liminal and maximal contexts. Under the former option, a 50% cost reduction is obtained by choosing angled drilling; under the latter option, the saving in coverage cost is as much as 58%.

9.11.2. Optimal Strategy

The optimization of control grids and field program designs for the detection of volcanogenic sulfide deposits of the Cordillera was conducted in the same manner as mentioned in Section 9.6.4 for the Shield deposits. The resulting

TABLE 9.20
Summary of Detection Characteristics of Volcanogenic Massive Sulfide Deposits of the North American Cordillera Belt for Three Types of Detection Strategies.

Type of survey and grid geometry	Liminal detection				Maximal detection	
	Coverage cost: US$/mile sq. Grid size: feet		Coverage cost US$/mile sq. Grid size: ft	Detection probability	Coverage cost: US$/mile sq. Grid size: feet	
	Detection	Confirmed detection			Detection	Confirmed detection
Airborne geophysical parallel grid	$325 2000 ft	$676 700 ft	$424 1300 ft	0.80	$1525 300 ft	$2905 150 ft
Airborne geophysical square grid	$360 3380 ft	–	$493 2100 ft	0.89	$670 1400 ft	–
Ground geophysical square grid	$1435 710 ft	$3125 440 ft	$1752 600 ft	0.90	$2520 450 ft	$4480 275 ft
Vertical drilling square grid	$438,900 710 ft	$1,092,200 440 ft	$577,000 600 ft	0.90	$975,000 450 ft	$2,280,000 275 ft
55-degree angled drilling square grid	$214,380 1200 ft	$564,2000 680 ft	$424,000 800 ft	0.97	$424,000 800 ft	$725,000 650 ft
Optimal 40-deg angled drilling square grid	–	–	$454,000 800 ft	0.99	–	–

SEARCH FOR VOLCANOGENIC MASSIVE SULFIDE DEPOSITS

optimal grid parameters are displayed in Tables 9.25 for airborne geophysical surveys, 9.29 for ground geophysical surveys, and 9.33 for vertical and angled drilling programs, and are assembled in the central portion of Table 9.20 for the readers' convenience.

Comparisons of detection performances of various grid designs and field programs can be based on the expected loss criterion (see Section 6.5.4 of Chapter 6). The choice of the square grid is recommended because it leads to a 35% reduction of the expected loss which could be incurred if the parallel design was selected. As far as drilling programs are concerned, the choice between angled and vertical detection should be in favor of optimally angled drilling (40 degrees), because the expected loss is 91% lower than in the case of the vertical option.

9.11.3. Multipurpose Optimal Grid Designs

As mentioned in the previous section 9.3.5, it is advantageous to design grids which may be jointly optimal for the detection of several types of ore targets likely to occur within specific geological environments. An example is the simultaneous search for volcanogenic massive sulfides and contact metasomatic deposits of the Cu–Fe and Pb–Zn–Cu types in the North American Cordillera Belt.

If we carefully compare the following three sets of tables in Chapters 7 and 9: Tables 7.9, 7.24, and 9.24 for airborne geophysical surveys, Tables 7.13, 7.28, and 9.29 for ground geophysical surveys, and Tables 7.17, 7.32, and 9.33 for drilling programs, we find that the 95% confidence intervals of grid spacings overlap to a small extent, thus defining narrow ranges of jointly optimal grid spacings for the detection of the three types of ore targets mentioned above.

From Table 9.34, which displays all jointly optimal grid parameters, we can see that optimal joint detection is provided by a 1100-foot grid spacing for parallel airborne grids, increased to a 1700-foot spacing for square grids. For ground geophysical surveys and vertical drilling programs the spacing is 525 feet, while optimally angled drilling (45 degrees) requires a 650-foot spacing.

9.12. DETECTION PROBABILITY AND OPTIMIZATION TABLES FOR CORDILLERA VOLCANOGENIC SULFIDE DEPOSITS

The following fourteen tables cover the probabilities of detection and the optimization of the search for volcanogenic sulfide deposits of the Cordilleran Belt by airborne and ground programs (pages 268–279).

TABLE 9.21
Probabilities of Detection of Volcanogenic Massive Sulfide Deposits of the North American Cordillera Belt by Airborne Geophysical Surveys on Parallel Grids.

```
Survey design:   parallel lines with spacing S feet
Elliptical targets with expected major axis = L feet
in the confidence interval:   l.c.l. ( geom. mean ( u.c.l.
and minor axis = B feet
R is defined as the ratio B/L with geometric mean = 0.19
Orientation of flight lines:   90 degrees with strike-line
```

	Probability of detection		
Grid spacing S feet	l.c.l. L=1201	geom. mean 1842	u.c.l. 2826 feet
200	1.000	1.000	1.000
400	0.923	0.942	0.950
600	0.883	0.913	0.925
800	0.842	0.884	0.899
1000	0.799	0.853	0.873
1200	0.751	0.821	0.846
1400	0.718	0.788	0.818
1600	0.628	0.751	0.789
1800	0.558	0.710	0.758
2000	0.502	0.671	0.723
2200	0.457	0.610	0.701
2400	0.419	0.559	0.643
2600	0.387	0.516	0.593
2800	0.359	0.480	0.551
3000	0.335	0.448	0.514
3200	0.314	0.420	0.482
3400	0.296	0.395	0.454
3600	0.279	0.373	0.429
3800	0.264	0.353	0.406
4000	0.251	0.336	0.386
4200	0.239	0.320	0.367
4400	0.228	0.305	0.351
4600	0.218	0.292	0.335
4800	0.209	0.280	0.321
5000	0.201	0.269	0.309
5200	0.193	0.258	0.297

TABLE 9.22
Determination of Grid Size for Specified Levels of Probability in the Detection of Volcanogenic Massive Sulfide Deposits of the North American Cordillera Belt by Airborne Geophysical Surveys on Parallel Grids.

Survey design: parallel lines with spacing S feet
Unit Cost = US$70 per line-mile

Specified detection probability level	Required grid spacing S in feet	Corresponding cost in $ per mile square
0.10	12360	$ 169.90
0.15	8500	$ 183.48
0.20	6480	$ 197.04
0.25	5240	$ 210.53
0.30	4400	$ 224.00
0.35	3780	$ 237.78
0.40	3320	$ 251.33
0.45	2960	$ 264.86
0.50	2680	$ 277.91
0.55	2440	$ 291.48
0.60	2240	$ 305.00
0.65	2060	$ 319.42
0.70	1800	$ 345.33
0.75	1560	$ 376.92
0.80	1280	$ 428.75
0.85	960	$ 525.00
0.90	640	$ 717.50
0.95	280	$1460.00

TABLE 9.23
Probabilities of Confirmed Detection of Volcanogenic Massive Sulfide Deposits of the North American Cordillera Belt by Airborne Geophysical Surveys on Parallel Grids.

```
Survey design:  parallel lines with spacing S feet,
Randomly orientated elliptical targets with expected major axis = L feet
in the confidence interval:  l.c.l. ( geom. mean ( u.c.l.
and minor axis = B feet
R is defined as the ratio B/L with geometric mean = 0.19
```

	Probability of confirmed detection		
Grid spacing S feet	l.c.l. L=1201	geom. mean 1842	u.c.l. 2826 feet
800	0.410	0.675	0.841
900	0.330	0.555	0.699
1000	0.271	0.462	0.587
1100	0.228	0.389	0.499
1200	0.197	0.331	0.427
1300	0.178	0.285	0.369
1400	0.000	0.249	0.322
1500	0.000	0.220	0.283
1600	0.000	0.198	0.251
1700	0.000	0.182	0.225
1800	0.000	0.172	0.205
1900	0.000	0.000	0.189
2000	0.000	0.000	0.177

TABLE 9.24
Probabilities of Detection of Volcanogenic Massive Sulfide Deposits of the North American Cordillera Belt by Airborne Geophysical Surveys on Square Grids.

Survey design: square grid with spacings S by S feet
Randomly orientated elliptical targets with expected major axis = L feet
in the confidence interval: l.c.l. (geom. mean (u.c.l.
and minor axis = B feet
R is defined as the ratio B/L with geometric mean = 0.19

	Probability of detection		
Grid spacing S feet	l.c.l. L=1201	geom. mean 1842	u.c.l. 2826 feet
1200	1.000	1.000	1.000
1400	0.955	1.000	1.000
1600	0.882	1.000	1.000
1800	0.816	1.000	1.000
2000	0.757	0.919	1.000
2200	0.706	0.866	0.943
2400	0.660	0.817	0.895
2600	0.619	0.773	0.850
2800	0.583	0.732	0.809
3000	0.551	0.695	0.770
3200	0.522	0.661	0.735
3400	0.496	0.630	0.702
3600	0.472	0.602	0.672
3800	0.450	0.576	0.644
4000	0.430	0.552	0.618
4200	0.412	0.530	0.594
4400	0.396	0.500	0.572
4600	0.380	0.490	0.552
4800	0.366	0.473	0.532
5000	0.353	0.456	0.514
5200	0.340	0.441	0.497

TABLE 9.25
Optimal Design of Airborne Geophysical Surveys for the Detection of Volcanogenic Massive Sulfide Deposits of the North American Cordillera Belt.

```
Confidence interval:  all results are reported as 95% confidence intervals
respectively
Expected target length in feet:      1201      1842      2826
Expected shape ratio R = 0.19
Unit Cost:  $70/l.ml.
```

(i) Survey Design: Parallel Grid			
Optimal grid spacing in feet:	1050	1300	1500
Cost in US$ per mile square:	492	424	386
Probability of detection:	0.79	0.80	0.80
Optimal grid orientation:	90 degrees to expected strike		

(ii) Survey Design: Square Grid			
Optimal grid spacing in feet:	1660	2100	2315
Cost in US$ per mile square:	586	493	459
Probability of detection:	0.86	0.89	0.92

TABLE 9.26
Probabilities of Detection of Volcanogenic Massive Sulfide Deposits of the North American Cordillera Belt by Ground Geophysical Surveys or Vertical Drilling to a Depth of 300 Feet.

Survey design: square grid with spacings S by S feet. The expected shape of the ore deposits is elliptical with major axis = L in the 95% confidence interval: L.c.l. (geometric mean (U.c.l., and minor axis = $\hat{B}$ feet, R is defined as the ratio B/L with geometric mean = 0.19 grid orientation 30 degrees + or - 10 from the expected strike line.

	Probability of detection		
Spacing S in feet	L.c.l. L = 1201	G. mean 1842	U.c.l. 2826ft
350	1.000	1.000	1.000
400	0.992	1.000	1.000
450	0.957	1.000	1.000
500	0.879	0.999	1.000
600	0.682	0.958	0.995
700	0.514	0.834	0.949
800	0.397	0.684	0.839
900	0.314	0.553	0.709
1000	0.255	0.452	0.588
1100	0.210	0.375	0.492
1200	0.177	0.316	0.416
1300	0.151	0.269	0.355
1400	0.130	0.232	0.306
1500	0.113	0.202	0.267

TABLE 9.27
Determination of Grid Size for Specified Levels of Probability in the Detection of Volcanogenic Massive Sulfide Deposits of the North American Cordillera Belt by Vertical Drilling to a Depth of 300 Feet.

Survey Design: Square Grid With Spacings S by S Feet
Unit Cost = $6000 Per Diamond Drill-Hole

Probability Level Of Detection	Grid Size In Feet	Drilling Cost Per Mile Square In U.S.$. Thousands
0.15	1700	101.64
0.25	1350	145.30
0.35	1150	188.25
0.45	1000	237.38
0.55	900	283.73
0.65	850	312.92
0.75	750	388.81
0.85	700	438.91
0.95	600	577.42

TABLE 9.28
Probabilities of Confirmed Detection of Volcanogenic Massive Sulfide Deposits of the North American Cordillera Belt by Ground Geophysical Surveys or Vertical Drilling to a Depth of 300 Feet.

Survey design: square grid with spacings S by S feet.
The expected shape of the ore depsoits is elliptical
with major axis = L in the 95% confidence interval:
L.c.l. (geometric mean (U.c.l., and minor axis = B feet,
R is defined as the ratio B/L with geometric mean = 0.19
grid orientation 0 degrees + or - 10 from the expected
strike line.

	Probability of confirmed detection		
Spacing S in feet	L.c.l. L = 1201	G. mean 1842	U.c.l. 2826ft
200	1.000	1.000	1.000
250	0.985	1.000	1.000
300	0.918	1.000	1.000
350	0.824	0.972	1.000
400	0.686	0.919	0.973
450	0.432	0.851	0.929
500	0.321	0.767	0.873
550	0.239	0.638	0.805
600	0.176	0.435	0.718
700	0.092	0.279	0.417
800	0.047	0.177	0.284
900	0.024	0.110	0.192
1000	0.011	0.067	0.127

TABLE 9.29
Optimal Design of Ground Geophysical Surveys for the Detection of Volcanogenic Massive Sulfide Deposits of the North American Cordillera Belt.

```
Survey Design:   Square Grid With Spacing S by S Feet.
Confidence Interval:  All results are reported as 95% confidence intervals
respectively.
Expected Target Length in Feet:           1201      1842      2826
Expected Shape Ratio R = 0.19
Unit Cost:  $6/Station
```

```
Optimal Grid Spacing in Feet:              450       600       650
Cost in $ Per Mile Square:                 2501      1752      1594
Probability of Detection:                  0.51      0.90      0.98
Optimal Grid Orientation:  30 degrees + or -10 to expected strike
direction of target.
```

TABLE 9.30
Probabilities of Detection of Volcanogenic Massive Sulfide Deposits of the North American Cordillera Belt by 55-Degree Angled Drilling to a Vertical Depth of 300 Feet.

Survey design: square grid with spacings S by S feet.
The expected shape of the ore deposits is elliptical
with major axis = L in the 95% confidence interval:
L.c.l. (geometric mean (U.c.l., and minor axis = B feet,
R is defined as the ratio B/L with geometric mean = 0.32
grid orientation 45 degrees + or - 10 from the expected
strike line.

	Probability of detection		
Spacing S in feet	L.c.l. L = 1201	G. mean 1842	U.c.l. 2826ft
550	1.000	1.000	1.000
600	0.991	1.000	1.000
650	0.944	1.000	1.000
700	0.876	1.000	1.000
800	0.727	0.991	1.000
1000	0.479	0.804	0.942
1200	0.333	0.593	0.754
1400	0.245	0.436	0.576
1600	0.187	0.334	0.441
1800	0.148	0.264	0.349
2000	0.120	0.214	0.282
2200	0.099	0.177	0.233
2400	0.083	0.149	0.196
2600	0.071	0.127	0.167
2800	0.061	0.109	0.144
3000	0.053	0.095	0.126

TABLE 9.31

Determination of Grid Size for Specified Levels of Probabilities in the Detection of Volcanogenic Massive Sulfide Deposits of the North American Belt by 55-Degree Angled Drilling to a Vertical Depth of 300 Feet.

Survey Design: Square Grid With Spacings S by S Feet
Unit Cost = $7325 Per Diamond Drill-Hole

Probability Level Of Detection	grid Size In Feet	Drilling Cost Per Mile Square In U.S.$. Thousands
0.15	2400	75.48
0.25	1900	105.16
0.35	1600	136.07
0.45	1400	167.46
0.55	1300	188.40
0.65	1200	214.39
0.75	1050	267.10
0.85	1000	289.81
0.95	900	346.39

TABLE 9.32

Probabilities of Confirmed Detection of Volcanogenic Massive Sulfide Deposits of the North American Cordillera Belt by 55-Degree Angled Drilling to a Vertical Depth of 300 Feet.

Survey design: square grid with spacings S by S feet.
The expected shape of the ore deposits is elliptical
with major axis = L in the 95% confidence interval:
L.c.l. (geometric mean (U.c.l., and minor axis = B feet,
R is defined as the ratio B/L with geometric mean = 0.32
grid orientation 0 degrees + or - 10 from the expected
strike line.

Spacing S in feet	Probability of confirmed detection		
	L.c.l. L - 1201	G. mean 1842	U.c.l. 2826ft
450	1.000	1.000	1.000
500	0.843	1.000	1.000
550	0.667	1.000	1.000
600	0.496	1.000	1.000
700	0.295	0.773	1.000
800	0.165	0.499	0.784
900	0.087	0.343	0.530
1000	0.042	0.224	0.386
1100	0.018	0.144	0.269
1200	0.006	0.089	0.185
1300	0.001	0.052	0.124

TABLE 9.33
Optimal Designs for the Detection of Volcanogenic Massive Sulfide Deposits of the North American Cordillera Belt by Drilling to a Vertical Depth of 300 Feet.

```
Survey Design:   square grid with spacing S by S feet
Confidence interval:  all results are reported as 95% confidence intervals
respectively
Expected target length in feet:    1201      1842      2826
Unit Cost = $20 per linear ft.
```

(a) Vertical drilling:

Optimal grid spacing in feet:	500	550	650
Cost in $thou. per mile square:	804	675	501
Probability of detection:	0.70	0.82	0.86
Optimal grid orientation:	30 degrees to expected strike of target		

(b) Angled drilling at 55 degrees:

Optimal grid spacing in feet:	550	600	700
Cost in $thou. per mile square:	843	705	536
Probability of detection:	0.97	0.96	0.97
Optimal grid orientation:	45 degrees to expected strike of target		

(c) Angled drilling at optimal angle = 45 degrees:

Optimal grid spacing in feet:	600	700	800
Cost in $thou. per mile square:	817	621	491
Probability of detection:	0.95	0.99	0.99
Optimal grid orientation:	45 degrees to expected strike of target		

TABLE 9.34
Parameters of Multipurpose Optimal Grids for the Detection of Contact-Metasomatic and Volcanogenic Massive Sulfide Deposits of the North American Cordillera Belt.

Type of survey & grid geometry	Range of grid size	Range of coverage cost ($/ml.sq.)	Range of probability of detection
A.B. geophysical parallel grid	1050 to 1100 ft	$460 to $480	0.66 to 0.82
A.B. geophysical square grid	1660 to 1720 ft	$615 to $635	0.86 to 0.88
Grd. geophysical square grid	500 to 550 ft	$1950 to $2200	0.73 to 0.90
Vertical drilling square grid	500 to 550 ft	$675,000 to $803,000	0.73 to 0.90
Optimally angled (50 degr.) drilling square grid	600 to 700 ft	$564,000 to $705,000	0.97 to 0.99

REFERENCES AND SELECTED READINGS

1. ALBERS, J. P., and ROBERTSON, J. E., 1961, Geology and ore deposits of the East Shasta copper–zinc district, Shasta County, California, *U.S. Geol. Surv. Prof. Pap.* 338.
2. ANDERSON, C. A., 1969, Massive sulfide deposits and volcanism; *Econ. Geol.* **64**(2), 129–146.
3. ANNIS, R. C., CRANSTONE, D. A., and VALLEE, M., 1981, A survey of known mineral deposits in Canada that are not being mined in 1980, Mineral Bulletin No. 181, Department of Energy, Mines and Resources, Ottawa.
4. BARAGAR, W. R. A., 1968, Major element geochemistry of the Noranda volcanic belt, *Can. J. Earth Sci.* **5**, 773–790.
5. BOLDY, J., 1977, (Un)certain exploration facts and figures, *Can. Inst. Min. Metall. Bull.* **70**(781), 86–95.
6. CAMERON, E. M., 1975, Geochemical methods of exploration for massive sulfide mineralization in the Canadian Shield, *Geochemical Exploration 1974*, pp. 21–49, Elsevier, Amsterdam.

7. DESCARREAUX, J., 1973, A petrochemical study of the Abitibi volcanic belt and its bearing on the occurrences of massive sulfide ores, *Can. Inst. Min. Metall. Bull.* **66,** 61–69.
8. EASDON, M. M., 1970, A compilation of graphitic occurrences in the Archean part of Northwest Quebec, Ministère des Richesses Naturelles, Quebec, Open File Report G.M. 25662.
9. FOX, J. S., 1977, Rapid pyroclastic mapping in base metal exploration, *Can. Inst. Min. Metall. Bull.* **70,** 173–178.
10. FOX, J. S., 1979, Host rock geochemistry and massive sulfide volcanogenic ore, *Can. Inst. Min. Metall. Bull.* **72**(804), 127–134.
11. GILMOUR, P., 1965, The origin of massive sulfide mineralization in the Noranda district, Northwestern Quebec, *Geol. Assoc. Canada Proc.* **16,** 1239–1249.
12. GOODWIN, A. M., 1967, Volcanic studies in the Timmins–Kirkland Lake Noranda region of Ontario and Quebec, *Geol. Surv. Can. Pap.* 67-1, Pt. A, 138–142.
13. GRAF, J. L., 1977, Rare earth elements as hydrothermal tracers during the formation of massive sulfide deposits in volcanic rocks, *Econ. Geol.* **72,** 527–545.
14. HODGSON, C. J., and LYDON, J. W., 1977, Geological setting of volcanogenic massive sulfide deposits and active hydrothermal systems: Some implications for exploration, *Can. Inst. Min. Metall. Bull.* **70,** 95–106.
15. HOLMES, R., and TOOMS, J. S., 1973, Dispersion from a submarine/exhalative body, in *Geochemical Exploration 1972,* pp. 193–202, Institute of Mining and Metallurgy, London.
16. HOPWOOD, T. P., 1975, "Quartz-eye"-bearing porphyroidal rocks and volcanogenic massive sulfide deposits, *Econ. Geol.* **71,** 589–612.
17. HUTCHINSON, R. W., 1965, Genesis of Canadian massive sulfides reconsidered by comparison to Cyprus deposits, *Can. Inst. Min. Metall. Trans.* **68,** 286–300.
18. HUTCHINSON, R. W., RIDDLER, R. H., and SUFFEL, G. G., 1971, Metallogenic relationships in the Abitibi Belt, Canada: A model for Archean metallogeny, *Can. Inst. Min. Metall. Bull.* **64,** 48–57.
19. HUTCHINSON, R. W., 1973, Volcanogenic sulfide deposits and their metallogenic significance, *Econ. Geol.* **68,** 1223–1246.
20. JENKS, W. F., 1971, Tectonic transport of massive sulfide deposits in submarine volcanic and sedimentary host rocks, *Econ. Geol.* **66,** 1215–1224.
21. KINKEL, A. R. Jr., 1956, Geology and base metal deposits of West Shasta copper–zinc district, Shasta County, California, *U.S. Geol. Surv. Prof. Pap.* 285.
22. LARSON, L., and WEBBER, G. R., 1977, Chemical and petrographic variations in rhyolitic zones in the Noranda area, *Can. Inst. Min. Metall. Bull.* **70**(784), 80–90.
23. LATULIPPE, M., 1966, The relationship of mineralization to Precambrian stratigraphy in the Mattagami and Val d'Or districts of Quebec, *Geol. Assoc. Can., Spec. Pap.* **3,** pp. 21–42.
24. MACGECHAN, P. J., 1981, Exploration significance of the emplacement and genesis of massive sulfide in the Main Zone at the Norita Mine, Mattagami, Quebec, *Can. Inst. Min. Metall. Bull.,* **84,** 59–75.
25. MILLER, L. J., 1960, Massive sulfide deposits in eugeosynclinal belts, *Econ. Geol.* **55,** 1327–1332.
26. MILLER, R. J. M., 1973, The morphology of some Canadian massive sulfide deposits, A.I.M.E., Annual meeting, Chicago.
27. OHMOTO, H., 1978, Submarine caldera: A key to the formation of volcanogenic massive sulfide deposits? *Min. Geol.* **28,** 219–231.
28. PATERSON, N. R., 1971, Airborne electromagnetic methods applied to the search for sulfide deposits, *Can. Inst. Min. Metall. Bull.* **64**(705), 29–38.

29. PEARSON, D. E., 1977, Volcanic suites of southwestern British Columbia, Trip No 6, Geological Association of Canada Guidebook, pp. 20–25.
30. PELTON, W. H., 1977, New I.P. method may discriminate between graphite and massive sulfides, *North. Miner* **63**, 27, September 15.
31. PIRIE, J. D., and NICHOL, I., 1981, Geochemical dispersion in wallrocks associated with the Norbec deposit, Noranda, Quebec, *J. Geochem. Expl.* **15**, 159–180.
32. PODOLSKY, G., and SLANKIS, J., 1979, Izok Lake deposit, N.W.T., Canada: A geophysical case history, *Geol. Surv. Can. Rep.* **31**, 641–652.
33. RIDDLER, R. H., 1971, Analysis of Archean volcanic basins in the Canadian Shield using the "exhalite" concept, *Can. Inst. Min. Metall. Bull.* **64**(714), 20 (abstract).
34. RIDDLER, R. H., 1973, The "Exhalite" concept: A new tool for exploration, *North. Miner* November 29, 59–61.
35. SANGSTER, D. F., 1972, Precambrian volcanogenic massive sulfide deposits in Canada: A review, *Geol. Surv. Can. Pap.* No. 72-22.
36. SANGSTER, D. F., and SCOTT, S. D., 1976, Precambrian strata-bound massive Cu–Pb–Zn sulfide ores of North America, in *Handbook of Strata-bound and Stratiform Ore Deposits*, pp. 129–222, Elsevier, Amsterdam.
37. SANGSTER, D. F., 1980, Quantitative characteristics of volcanogenic massive sulfide deposits, *Can. Inst. Min. Metall. Bull.* **73**(814), 74–81.
38. SAWKINS, F. J., 1972, Sulfide ore deposits in relation to plate tectonics, *J. Geol.* **80**, 377–397.
39. SEIGEL, H. O., 1974, The magnetic induced polarization (M.I.P.) method, *Geophysics* **39**, 321–339.
40. SOLOMON, M., 1976, "Volcanic" massive sulfide deposits and their host rocks: A review and an explanation, in *Handbook of Strata-bound and Stratiform Ore Deposits*, pp. 21–54, Elsevier, Amsterdam.
41. SPENCE, C. D., and DE ROSEN-SPENCE, A. F., 1975, The place of sulfide mineralization in the volcanic sequence at Noranda, Quebec, *Econ. Geol.* **70**, 90–101.
42. STANTON, R. L., 1960, General features of the comformable pyritic orebodies; Part 1: Field associations, Part 2: Mineralogy, *Can. Inst. Min. Metall. Bull.* **53**, 24–29, 66–74.
43. WALKER, G. P. L., 1971, Grain-size characteristics of pyroclastic deposits, *J. Geol.* **79**, 696–714.
44. WOLFE, W. J., 1973, Geochemical exploration in Archean metavolcanic and meta-sedimentary belts of the Pukaskwa Region, Ontario, *Can. Inst. Min. Metall. Bull.* **66**(735), 75–83.
45. WOLFE, W. J., 1975, Zinc abundance in early Precambrian volcanic rocks; Its relationship to exploitable levels of zinc in sulfide deposits of volcanogenic–exhalative origin, in *Geochemical Exploration 1974*, pp. 261–278, Elsevier, Amsterdam.

CHAPTER TEN

DESIGNING OPTIMIZED FIELD PROGRAMS FOR THE DETECTION OF MISSISSIPPI VALLEY-TYPE Pb–Zn DEPOSITS OF THE NORTH AMERICAN ARCTIC PALEOZOIC PLATFORM

10.1. GENERAL GEOLOGICAL BACKGROUND

10.1.1. General Setting

The Mississippi Valley-type Pb–Zn deposits are one of the principal sedimentary sources of lead–zinc ores of the world. The deposits are generally tabular in shape, subhorizontal in attitude, and, although occurring at rather shallow depths, difficult to detect because they have little or no surface expression. They are truly "stratiform" deposits, i.e., occurring as continuous and extensive blankets concordant with the encasing sediments, as opposed to "stratabound" deposits such as the volcanogenic massive sulfide ore bodies described above in Chapter 9.

Lead–zinc deposits of the "Mississippi Valley" type were first discovered in several districts lying within the watershed of the Mississippi River, in the central portion of the U.S.A., hence the name. Since the original 18th century discovery, similar deposits have been found in other areas of the North American Paleozoic Platform and in other regions of the world.

The Paleozoic Platform, made up mainly of little-disturbed sediments, stretches around the North American Precambrian Shield, between the Appalachian Belt to the east and the Cordillera Belt to the west. There are seven main districts of occurrence of Mississippi Valley-type deposits in the North American Paleozoic Platform. Four of these districts—viz., East Tennessee, Rosiclar, Southeast Missouri-Tri-State, and Upper Mississippi Valley—are not covered by the present study. The three remaining districts, all situated in the Arctic portion of the Paleozoic Platform, include the Great Slave Lake, Selwyn Mountain, and Arctic Islands areas, which are dealt with in this chapter.

The age of the sedimentary formations encasing the Mississippi Valley Pb–Zn deposits of North America is Paleozoic and varies from Upper Cambrian in the Southeast Missouri and in the Selwyn Mountains districts, to Ordovician in the Rosiclar, Upper Mississippi Valley, East Tennessee, and Arctic Islands districts. More recent settings are known in the Great Slave Lake area (Devonian) and in the Tri-State area (Lower to middle Carboniferous). A special mention should be made of the Black Angel deposit of the west coast of Greenland, in the Arctic region, which occurs in late Precambrain metamorphosed carbonate formations.

The Mississippi Valley-type deposits generally occur within carbonate series composed of cherty or reefal limestones at various stages of dolomitization with pelitic intercalations (Lower Paleozoic) or clastic horizons (middle and Upper Paleozoic). Callahan[2] recognizes two types of structural control which have a regional significance: (a) underlying unconformities (Southeast Missouri, Selwyn Mountains, and Great Slave Lake) and (b) overlying unconformities (East Tennessee, Upper Mississippi Valley, and Tri-State). Local depositional controls include biologically induced unconformities (reefs), sedimentary unconformities (pinch outs, draping around buried ridges, and mudslides), and, finally, chemically induced unconformities (solution-thinning generating collapse breccias).

The mineralogy of the Mississippi Valley Pb–Zn deposits is rather simple. The ore material is an admixture of coarsely crystallized marcasite, rather than pyrite, iron-poor sphalerite, and galena generally low in silver content. In a few districts such as Southeast Missouri and the eastern margin of the Upper Mississippi Valley zinc–lead area, finely disseminated chalcopyrite accompanies the iron and zinc–lead sulfides. The gangue is composed of coarsely crystallized and vuggy calcite and dolomite with minor barite and fluorite. Locally, as in Arkansas, barite is the chief economic mineral; in other areas, such as Rosiclar in Illinois, fluorite is the main ore material. Mississippi Valley-type ore is amenable to low-cost concentration methods. Mining is by low-cost excavation methods including open-cut (Great Slave Lake area) and underground room and pillar trackless mining.

10.1.2. Geological Synopsis for the North American Arctic Deposits

The Arctic section of the North American Paleozoic Platform stretches over a length of 1200 miles and a width of 500 miles east of the Cordillera Belt, extending west and north of the Precambrian Shield through the Arctic Islands to Greenland. The discovery and development of Mississippi Valley-type Pb–Zn deposits in the North American Arctic are quite recent events. Investigations of the Great Slave deposits in the Pine Point area started in earnest only in the early 1950s. Further discoveries were made in the Arctic Islands in the 1960s, but the Selwyn Mountain mineralization was disclosed only in the early 1970s.

In the Pine Point area of the Great Slave region, as many as 40 deposits totalling some 100 million tons of lead–zinc ore have been discovered to date within a roughly rectangular area extending in an east–west direction over a length of 20 miles and a width of 3 miles. The stratigraphic setting is that of a 500-foot-thick middle Devonian carbonate formation with many vertical and horizontal facies variations which belongs to a reefal complex of regional extent. The carbonates are overlain by pelitic series (Buffalo River Shales) and underlain by thick evaporites. The ore deposits are either pipelike, prismatic, or cylindrical in shape and subvertical in attitude, or tabular and subhorizontal. The local structural control of ore deposition appears to be a series of three parallel southwesterly trending "hinges."[4, 5, 7]

Farther west and north of the Great Slave region, the newly discovered Selwyn Mountain lead–zinc district stretches along the eastern edge of the Cordillera Belt into the Northwest Territories and the Yukon.[9, 10, 17] A number of promising lead–zinc discoveries have been made and investigated in the Bonnet Plume area of the Selwyn Mountains. So far the deposits showing the best potential are those of Goz Creek and Cypress. The ore material consists of pyrite, light-colored sphalerite, and galena with inclusions of boulangerite which boosts the silver content of the mineralization. The sulfides occur in breccia zones that extend along structural hinges in Lower Cambrian dolomite.

Because of the remoteness of the region and the rigor of the climate, more than 10 years elapsed between the initial discoveries of promising lead–zinc mineralization in the Arctic Islands, in the early 1960s, and the full appraisal and development of the large Nanisivik and Polaris deposits. The Nanisivik mineralization forms a tabular body of massive pyrite carrying varying amounts of sphalerite and galena with attractive silver values occurring in Ordovician dolomites. The Polaris ore consists of early marcasite followed by sphalerite and galena in a gangue of vuggy and friable calcite with 5% ice content. On the Greenland west coast, farther east, the large Black Angel deposit was investigated in the late 1960s and is presently in production. The stratiform mineralization is up to 60 feet thick and occurs in tremolite-talc marble of late Precambrian age. The ore consists of massive sphalerite with about 20% pyrite and subordinate galena low in silver.

10.2. FIELD DETECTION METHODOLOGY

10.2.1. Detection Approaches

Both direct and indirect approaches have been successfully used in the search for Mississippi Valley-type Pb–Zn deposits. The direct approach relies on the visual detection of the mineralization or associated clues on surface (prospecting and mapping) or below surface by mechanical means (drilling). The indirect approach seeks to detect geophysical signatures of the ore deposits themselves or the primary or secondary halos associated with them. Commonly the two approaches are used simultaneously or sequentially in order to reduce the financial risk of exploration.

10.2.2. Direct Geological Detection

The direct geological detection of "blind" stratiform ore deposits of the Mississippi Valley type is a great challenge for prospectors and geologists alike. The task is particulary difficult inasmuch as the deposits have little or no surface expression in areas of prevailing flat and featureless topography. Geological detection is made easier in regions where the topography is more deeply incised and more likely to expose the mineralization, such as in the Selwyn Range and the Arctic Islands. Photogeologic studies from aircraft platforms may be of assistance in detecting old karstic features in the subsurface which could lead to ore.

10.2.3. Direct Detection by Pattern Drilling

Blind, tabular, stratiform lead–zinc deposits of the Mississippi Valley type are prime candidates for detection by systematic vertical drilling, much as are the "Colorado" type of uranium deposits of the Western U.S.A. Generally non-coring drilling methods including percussion and old-style "churn" drilling are preferred to the much more expensive coring methods. The pattern-drilling approach was developed and perfected in the older lead–zinc districts of the central portion of the North American Platform.

The initial stage of the programs calls for drilling for stratigraphic and mineralogic information on widely spaced lines laid at right angles to the geological grain of the region. If the regional structural trend is not apparent, the information drilling is carried on square patterns. Subsequently, the net is tightened progressively over smaller areas selected on the strength of the information-drilling results. At the final stage, closely spaced drilling on square grids is used to delineate and sample the deposits detected in the previous stages.

10.2.4. Indirect Geophysical Detection

Callahan's 1967 assessment of the merits of indirect geophysical techniques for the detection of Mississippi Valley-type Pb–Zn deposits was quite pessimistic.[3] He based his verdict on the failure of tests of various techniques including electromagnetic, self-potential, induced polarization, and seismic surveys to produce diagnostic responses to known deposits in several central U.S. zinc–lead districts. Gravity was not tested, and the only successful technique used was the "mise a la masse" from existing drill holes.

However, since the late 1960s, much research work has been conducted on the geophysical environment of Mississippi Valley Pb–Zn deposits and, as a result, the detection technology has been considerably improved. It is now appreciated that such deposits do offer good geophysical targets because of the presence of sizable masses of high specific gravity sulfides with moderate to good conductivity and high polarizability, occurring at a shallow depth. No case history of detection of Mississippi Valley-type deposits by airborne geophysical techniques such as INPUT could be found in the available literature. But we feel that, on the basis of the amount of geophysical research work presently done, some new techniques will be perfected in the near future. Thus we are presenting probability tables for the airborne detection of Mississippi Valley-type deposits in Section 10.6 of this chapter.

Induced polarization appears to be the most promising ground geophysical method for the detection of the Mississippi Valley-type deposits, as confirmed by successful surveys combining induced polarization (I.P.) and gravimetric work in the Pine Point area of the Great Slave Lake region.[8, 14, 15] Drawbacks of the induced polarization technique are (a) slowness and high cost and (b) strong interference by telluric noise in Arctic locations. Gravimetric work which has proved useful for I.P. confirmation and tonnage calculations is hampered by sudden variations in overburden thickness and by the presene of buried karstic features. Modern, fully computerized seismic instrumentation could prove useful in delineating large buried collapse structures of possible significance as ore traps.

10.2.5. Indirect Geochemical Approach

Advantage may be taken of the different degrees of mobility of the Zn and Pb cations both in primary and secondary dispersion environments to design a sequential approach for the geochemical detection of Mississippi Valley-type deposits. Broad zinc anomalies, which are relatively easy to detect, define target areas within which lead halos detection will pinpoint smaller target areas for coverage by more expensive combined geophysical surveys and test drilling. Bedrock sampling for the detection of primary halos and residual soil and groundwater sampling have been used with varying degrees of success in the search for lead–zinc deposits of the Mississippi Valley type in the Paleozoic Platform of North America.

Bedrock sampling has proved useful in the Arctic region, including the

Selwyn Mountains,[10, 17] the Arctic Islands, and the Great Slave region.[4] In well prospected districts such as the Upper Mississippi Valley zinc–lead area, drill core or cuttings may be used to advantage as sampling medium.[1] The sampling of residual soil to detect secondary halos has proved quite successful in nonglaciated areas of moderate relief (Upper Mississippi Valley) but would be of little assistance when dealing with glacial alluvium in the Arctic region. Groundwater sampling for zinc and copper has been used successfully by the first writer in the Upper Mississippi Valley area (see Chapter 12, Ref. 35), and should prove useful in other regions of moderately well dissected topography with temperate climate and rainfall above 32 inches per year.

10.3. STATISTICAL MODELING OF THE GEOMETRIC PARAMETERS OF NORTH AMERICAN ARCTIC DEPOSITS

10.3.1. Scope of the Study

A total of 14 commercial deposits were included in the database to represent the population of Mississippi Valley-type deposits of the Arctic region. The sample includes ten deposits from the Pine Point area, one from the Selwyn Mountains, and three from the Arctic Islands, as shown in Table 10.1. The sample size is near the lower limit of adequacy for satisfactory statistical processing, but this is a risk which has to be accepted when dealing with newly discovered districts.

TABLE 10.1

List of Mississippi Valley-type Pb–Zn Deposits of the North American Arctic Paleozoic Platform Included in the Database.

Northwest Territories & Arctic Islands	
Pyramid No. 1	Polaris
Pyramid No. 2	Nanisivik
Coronet	Goz Creek
Newconex	
X-25	Greenland
R-190	Black Angel
M-40	
N-815	
A-70	
K-57	

10.3.2. Statistical Modeling

Measurements of three geometric parameters including length and breadth of the horizontal sections of deposits and shape coefficients were collected from the 14 deposits included in the sample. The observed data range from 550 to 10,500 feet for lengths, and 225 to 1860 feet for breadths. As for the attitude parameters of the deposits, the dips are either horizontal or vertical, and, owing to the inadequacy of available orientation data, no strike orientation measurements are included in the database.

The observed data were grouped into classes, and frequency distributions were constructed for each parameter. The statistics summarizing these distributions including arithmetic means, modes, standard deviation, and coefficients of skewness are listed in the lower half of Table 10.2. The coefficient of skewness which is constructed as the ratio (mean-mode)/standard deviation indicates in a dimensionless manner the extent of the departure of observed frequency distributions from the symmetry of the standard normal model. The asymmetry is slight (coefficient near 0.1) for the length and for the shape ratio, but is quite strong for the breadth parameter (coefficient near 0.6).

TABLE 10.2

Summary of Statistical Modeling of Geometric Parameters of Mississippi Valley-Type Pb–Zn Deposits of the North American Arctic Paleozoic Platform (Sample Size: 14 Deposits).

Statistic of fitted model	Length of horiz. section in feet	Breadth of horiz. section in feet	Shape ratio $R = B/L$
25%ile	835	325	0.20
95%L.c.l.	927	392	0.27
G.mean	1337	539	0.40
95% U.c.l.	1927	742	0.60
75%ile	1800	1020	0.60
Disp. coef.	0.09	0.09	0.74

Statistic of observed data			
Arithm. mean	1879	662	0.49
Mode	1700	360	0.51
Stand.dev.	2388	468	0.23
Skewness	0.07	0.64	0.10

The three distributions of observed dimensional data were fitted with lognormal models. The fit was successfully tested at the 0.05 confidence level by means of the χ square test. The statistics summarizing the fitted models are listed in the upper half of Table 10.2. The statistics include (a) geometric means and the upper and lower limits of their 95% fiducial intervals, (b) 25th and 75th percentiles, and (c) dispersion coefficients. The latter is constructed as the ratio standard deviation/mean and is used to indicate in a dimensionless manner the spread of the data about the mean of the fitted models as a reflection of the total variability of the data. The variability of the length and breadth parameters is rather small (coefficient near 0.1), but that of the shape coefficient R is much larger (coefficient near 0.8).

10.4. CONSTRUCTION AND ORGANIZATION OF DETECTION PROBABILITY TABLES

10.4.1. Introduction

For detection purposes, the Mississippi Valley-type deposits are considered as either vertical pipelike or tabular subhorizontal bodies with randomly orientated horizontal sections of elliptical shape, lying entirely within the average 300-foot vertical range of most commonly used detectors. Therefore, no dip component is added to the horizontal dimensions of the actual deposits in order to obtain the target dimensions required for the calculation of detection probabilities (see Section 5.4.2. of Chapter 5).

Only two parameters are required for the construction of the probability tables, including the mean shape coefficient, and the mean target length within its 95% fiducial limits. The latter enable the calculation of 95% confidence limits for the detection probabilities.

10.4.2. Construction of the Probability Tables

Two similar sets of tables were constructed for the display of detection probabilities. One set covers single detection situations (at lest one target intersection by the detector) and the second one deals with confirmed detection (at least two target intersections). The grid spacings are varied within a range extending from 200 to 6000 feet in steps of 50–250 feet, depending on the nature of field programs. For each spacing, the probability of single or confirmed detection is calculated and displayed in the tables within 95% confidence limits.

A third set of tables was constructed to deal with a converse situation, i.e., calculating the grid spacing and associated coverage cost per unit of area required to ensure a prespecified detection probability level. These tables should prove very useful to explorationists for planning purposes. The range of probability levels extends from 0.10 to 0.95, using steps of 0.05.

10.4.3. Organization of Tables

The tables are grouped into two blocks, each covering a specific type of exploration situation. The first block covers the detection and confirmed detection of Mississippi Valley-type deposits by airborne geophysical surveys on parallel and square grids (Tables 10.4–10.7). The second block deals with ground vertical detection by geophysical surveys or drilling to a depth of 300 feet (Tables 10.9–10.12).

10.5. DESIGNING THREE STRATEGIES FOR THE DETECTION OF ARCTIC MISSISSIPPI VALLEY-TYPE Pb–Zn DEPOSITS

10.5.1. Detection Strategies

Three types of strategies are considered for the detection of Mississippi Valley-type deposits of the North American Arctic to suit three types of exploration situations. The liminal strategy, on the one hand, minimizes the coverage cost while ensuring an acceptable level of detection probability of at least 0.500, which is attractive when planning the coverage of large areas with little known potential. The maximal strategy, on the other hand, provides near-certain detection (0.980 detection probabilility level) at three to four times the liminal coverage cost, which restricts its use to the coverage of small areas with high potential. Finally, the optimal strategy offers a satisfactory compromise between the two cases listed above, resulting in the highest achievable detection probability level under strict cost constraints, which fulfils the requirements of the majority of exploration programs. The detection characteristics pertaining to the three options are summarized and assembled into Table 10.3 for the reader's convenience.

10.5.2. Liminal Strategy

The detection parameters pertaining to the liminal strategy are listed in the leftmost portion of Table 10.3. They include the grid spacings and corresponding coverage costs associated with the 0.500 detection probability level for various grid and field program designs. The latter cover airborne geophysical surveys on parallel and square grids, ground geophysical surveys, and vertical drilling on square grids.

TABLE 10.3
Summary of Detection Characteristics of Mississippi Valley-Type Pb–Zn Deposits of the North American Arctic Paleozoic Platform for Three Types of Detection Strategies.

Type of survey and grid geometry	Liminal detection		Optimal detection		Maximal detection	
	Coverage cost: US$/mile sq. Grid size: feet		Coverage cost US$/mile sq. Grid size: ft	Detection probability	Coverage cost: US$/mile sq. Grid size: feet	
	Detection	Confirmed detection			Detection	Confirmed detection
Airborne geophysical parallel grid	$404 1380 ft	$756 575 ft	$416 1340 ft	0.78	$1525 300 ft	$1925 230 ft
Airborne geophysical square grid	$438 2500 ft	—	$504 2030 ft	0.77	$885 100 ft	—
Ground geophysical square grid	$1465 780 ft	$2720 460 ft	$1353 750 ft	0.89	$2750 420 ft	$3285 365 ft
Vertical drilling square grid	$440,000 780 ft	$1,090,000 460 ft	$389,000 750 ft	0.89	$1,092,000 420 ft	$1,520,000 365 ft

DETECTION OF MISSISSIPPI VALLEY-TYPE Pb–Zn DEPOSITS 293

The cost multiplying factor required to confirm single detection averages 1.8 for both types of geophysical surveys but reaches 2.2 for vertical drilling. The cost of coverage per unit of area may be used as a convenient yardstick of detection performance for various grid designs which are considered at the planning stage. As shown on Table 10.3, there is no incentive to select the square grid instead of the parallel design for airborne geophysical surveys, because of the resulting 10% coverage cost increase.

10.5.3. Maximal Strategy

The detection parameters corresponding to the maximal option are listed in the rightmost portion of Table 10.3 in the same manner as in the case of the liminal option. The cost multiplier required to obtain confirmation of detection for airborne and ground geophysical surveys is 1.2. The cost factor is slightly higher (1.5) for vertical drilling programs. It should be noted that both results are substantially lower than for the liminal option. Thus, for the less cost-conscious explorationist leaning toward the maximal strategy, it would be appropriate to go a step further and plan for confirmed detection as well.

When comparing the merits of the two airborne survey grid designs using coverage cost per unit area as yardstick, the results are quite different from those of the liminal option. The selection of the square grid design becomes an attractive proposition as it leads to a coverage cost reduction of nearly 42% as compared to the parallel grid cost.

10.5.4. Optimal Strategy

Optimization procedures based on the efficiency criterion were carried out for each type of field program, in a manner fully described in Sections 4.4.2 of Chapter 4 and Section 5.5.4 of Chapter 5. The parameters describing optimal field program designs are (a) optimal grid spacings and corresponding coverage costs per unit area and (b) associated probabilities of optimal detection. The optimal parameters pertaining to airborne geophysical surveys on two grid designs are listed in Table 10.8. Those regarding ground geophysical surveys are shown in Table 10.12, and those relating to vertical drilling programs in Table 10.13. Finally, all parameters are assembled and listed in the central portion of Table 10.3 for the reader's convenience.

In order to compare the detection performance of the two airborne grid designs, we must use the "expected loss" criterion as described in Section 6.5.4. of Chapter 6, instead of the usual coverage cost criterion because the optimal detection probability levels are different. As a result, we find that the parallel grid design should be favored over the square design because the choice of the latter would lead to a 26% increase in expected loss per mile square.

10.6. DETECTION PROBABILITY AND OPTIMIZATION TABLES

The reader will find in the following section a series of ten tables covering the probabilities of detection and the optimization of the search for Mississippi Valley-type Pb–Zn deposits of the Arctic by airborne and ground programs (pages 294–302).

TABLE 10.4
Probabilities of Detection of Mississippi Valley-Type Pb–Zn Deposits of the North American Arctic Paleozoic Platform by Airborne Geophysical Surveys of Parallel Grids.

Survey design: parallel lines with spacing S feet,
Randomly orientated elliptical targets with expected major axis = L feet
in the confidence interval: l.c.l. (geom. mean (u.c.l.
and minor axis = B feet
R is defined as the ratio B/L with geometric mean = 0.40

Grid spacing S feet	Probability of detection		
	l.c.l. L= 927	geom. mean 1337	u.c.l. 1927 feet
200	1.000	1.000	1.000
400	0.916	1.000	1.000
600	0.871	0.913	1.000
800	0.821	0.882	0.920
1000	0.725	0.849	0.898
1200	0.604	0.812	0.877
1400	0.518	0.747	0.854
1600	0.453	0.653	0.829
1800	0.403	0.581	0.802
2000	0.362	0.523	0.753
2200	0.329	0.475	0.685
2400	0.302	0.436	0.628
2600	0.279	0.402	0.580
2800	0.259	0.373	0.538
3000	0.242	0.348	0.502
3200	0.227	0.327	0.471
3400	0.213	0.307	0.443
3600	0.201	0.290	0.419
3800	0.191	0.275	0.397
4000	0.181	0.261	0.377
4200	0.173	0.249	0.359
4400	0.165	0.238	0.342
4600	0.158	0.227	0.328
4800	0.151	0.218	0.314
5000	0.145	0.209	0.301
5200	0.139	0.201	0.290

TABLE 10.5
Determination of Grid Size for Specified Probability Levels in the Detection of Mississippi Valley-Type Pb–Zn Deposits of the North American Arctic Paleozoic Platform by Airborne Geophysical Surveys on Parallel Grids.

```
Survey design:  parallel lines with spacing S feet
Unit Cost = US$70 / line mile
```

Specified detection probability level	Required grid spacing S in feet	Corresponding cost in $ per mile square
0.10	8800	$ 182.00
0.15	6200	$ 199.61
0.20	4800	$ 217.00
0.25	3900	$ 234.77
0.30	3300	$ 252.00
0.35	2900	$ 267.45
0.40	2500	$ 287.84
0.45	2300	$ 300.70
0.50	2100	$ 316.00
0.55	1900	$ 334.53
0.60	1700	$ 357.41
0.65	1600	$ 371.00
0.70	1500	$ 386.40
0.75	1400	$ 404.00
0.80	1200	$ 448.00
0.85	900	$ 550.67
0.90	600	$ 756.00
0.95	300	$1372.00

TABLE 10.6
Probabilities of Confirmed Detection of Mississippi Valley-Type Pb–Zn Deposits of the North American Arctic Paleozoic Platform by Airborne Geophysical Surveys on Parallel Grids.

Survey design: parallel lines with spacing S feet,
Randomly orientated elliptical targets with expected major axis = L feet
in the confidence interval: l.c.l. (geom. mean (u.c.l.
and minor axis = B feet
R is defined as the ratio B/L with geometric mean = 0.40

Grid spacing S feet	Probability of confirmed detection		
	l.c.l. L= 927	geom. mean 1337	u.c.l. 1927 feet
200	1.000	1.000	1.000
400	0.769	1.000	1.000
600	0.519	0.738	1.000
800	0.421	0.556	0.801
1000	0.000	0.462	0.638
1200	0.000	0.413	0.537
1400	0.000	0.000	0.472
1600	0.000	0.000	0.431
1800	0.000	0.000	0.407
2000	0.000	0.000	0.000

TABLE 10.7
Probabilities of Detection of Mississippi Valley-Type Pb–Zn Deposits of the North American Arctic Paleozoic Platform by Airborne Geophysical Surveys on Square Grids.

```
Survey design:  square grid with spacings S by S feet
Randomly oriented elliptical targets with expected major axis = L feet
in the confidence interval:  l.c.l. ( geom. mean ( u.c.l
and minor axis = B feet
R is defined as the ratio B/L with geometric mean = 0.40
```

Grid spacing S feet	Probability of detection		
	l.c.l. L= 927	geom. mean 1337	u.c.l. 1927 feet
800	1.000	1.000	1.000
1000	0.942	1.000	1.000
1200	0.855	1.000	1.000
1400	0.776	0.955	1.000
1600	0.708	0.894	1.000
1800	0.649	0.836	1.000
2000	0.598	0.781	0.958
2200	0.554	0.732	0.916
2400	0.516	0.688	0.875
2600	0.482	0.648	0.834
2800	0.453	0.612	0.796
3000	0.427	0.580	0.761
3200	0.403	0.550	0.727
3400	0.382	0.524	0.697
3600	0.364	0.499	0.668
3800	0.346	0.477	0.641
4000	0.331	0.457	0.616
4200	0.316	0.438	0.593
4400	0.303	0.421	0.572
4600	0.291	0.405	0.551
4800	0.280	0.390	0.533
5000	0.270	0.376	0.515
5200	0.260	0.363	0.498

TABLE 10.8
Optimal Design of Airborne Geophysical Surveys for the Detection of Mississippi Valley-Type Pb–Zn Deposits of the North American Arctic Paleozoic Platform.

```
Confidence interval:  all results are reported as 95% confidence intervals
respectively
Expected target length in feet:      927       1337      1927
Expected shape ratio R = 0.40
Unit Cost:  $70/l.ml.
```

(i) Survey design: Parallel Grid			
Optimal grid spacing in feet:	929	1342	1968
Cost in US$ per mile square:	538	415	328
Probability of detection:	0.78	0.78	0.77
(ii) Survey design: Square Grid			
Optimal grid spacing in feet:	1561	2032	2853
Cost in US$ per mile square:	614	504	399
Probability of detection:	0.72	0.77	0.79

TABLE 10.9
Probabilities of Detection of Mississippi Valley-Type Pb–Zn Deposits of the North American Arctic Paleozoic Platform by Ground Geophysical Surveys of Vertical Drilling to a Depth of 300 Feet.

Survey design: square grid with spacings S by S feet. The expected shape of the ore deposits is elliptical with major axis = L in the 95% confidence interval: L.c.l. (geometric mean (U.c.l., and minor axis = B feet, R is defined as the ratio B/L with geometric mean = 0.40.

Spacing S in feet	Probability of detection		
	L.c.l. L = 927	G. mean 1337	U.c.l. 1927ft
300	1.000	1.000	1.000
400	0.979	1.000	1.000
500	0.853	1.000	1.000
600	0.680	0.966	1.000
700	0.529	0.878	1.000
800	0.417	0.758	0.990
900	0.333	0.640	0.950
1000	0.270	0.538	0.885
1200	0.187	0.388	0.718
1400	0.138	0.287	0.565
1600	0.105	0.219	0.448
2000	0.067	0.140	0.292
2400	0.047	0.097	0.203
3000	0.030	0.062	0.130
4000	0.017	0.035	0.073

TABLE 10.10
Determination of Grid Size for Specified Probability Levels in the Detection of Mississippi Valley-Type Pb–Zn Deposits of the North American Arctic Paleozoic Platform by Vertical Drilling to a Depth of 300 Feet.

```
          Survey Design:  Square Grid With Spacings S by S Feet
                  Unit Cost = $6000 Per Diamond Drill-Hole
                            = $2000 Per Percussion Drill-Hole
```

Probability Level Of Detection	Grid Size In Feet	Drilling Cost Per Mile Square In U.S.$. Thousands	
		Percussion	Diamond
0.05	3100	14.72	44.17
0.15	1900	28.71	86.14
0.25	1500	41.04	123.12
0.35	1250	54.79	164.37
0.45	1100	67.51	202.54
0.55	1000	79.13	237.38
0.65	900	94.58	283.73
0.75	800	115.82	347.47
0.85	750	129.60	388.81
0.95	650	166.81	500.48

TABLE 10.11
Probabilities of Confirmed Detection of Mississippi Valley-Type Pb–Zn Deposits of the North American Arctic Paleozoic Platform by Ground Geophysical Surveys of Vertical Drilling to a Depth of 300 Feet.

Survey design: square grid with spacings S by S feet. The expectes shape of the ore deposits is elliptical with major axis = L in the 95% confidence interval: L.c.l. (geometric mean (U.c.l., and minor axis = B feet, R is defined as the ratio B/L with geometric mean = 0.40.

Spacing S in feet	Probability of confirmed detection		
	L.c.l. $L = 927$	G. mean 1337	U.c.l. 1927ft
300	1.000	1.000	1.000
350	0.996	1.000	1.000
400	0.865	1.000	1.000
450	0.445	1.000	1.000
500	0.226	0.999	1.000
550	0.127	0.927	1.000
600	0.070	0.735	1.000
650	0.039	0.438	1.000
700	0.022	0.268	1.000
750	0.011	0.180	0.985
800	0.005	0.120	0.914
1000	0.000	0.023	0.282
1200	0.000	0.009	0.162
1500	0.000	0.000	0.016

TABLE 10.12
Optimal Design of Ground Geophysical Surveys for the Detection of Mississippi Valley-Type Pb–Zn Deposits of the North American Arctic Paleozoic Platform.

```
Survey Design:   Square Grid With Spacing S by S Feet.
Confidence Interval:   All results are reported as 95% confidence intervals
respectively.
Expected Target Length in Feet:        927      1337     1927
Expected Shape Ratio R = 0.40
Unit Cost:  $6/Station
```

Optimal Grid Spacing in Feet:	550	750	1050
Cost in $ Per Mile Square:	1946	1353	941
Probability of Detection:	0.83	0.89	0.92

TABLE 10.13
Optimal Design of Vertical Drilling Programs for the Detection of Mississippi Valley-Type Pb–Zn Deposits of the North American Arctic Paleozoic Platform.

```
Survey design:   square grid with spacing S by S feet
Confidence interval:   all results are reported as 95% confidence intervals
respectively
Expected target length in feet:        927      1337     1927
```

(i) Percussion drilling: Unit Cost = $2000/hole			
Optimal grid spacing in feet:	550	750	1050
Cost in $thou. per mile square:	225	130	73
Probability of detection:	0.83	0.89	0.92
(ii) Diamond drilling: Unit Cost = $6000/hole			
Optimal grid spacing in feet:	550	750	1050
Cost in $thou. per mile square:	675	389	219
Probability of detection:	0.83	0.89	0.92

REFERENCES AND SELECTED READINGS

1. BARNES, H. L., and LAVERY, N. G., 1977, Use of primary dispersion for exploration of Mississippi Valley-type deposits, *J. Geochem. Expl.* **8,** 105–115.
2. CALLAHAN, W. H., 1964, Paleophysiographic premises for prospecting for strata-bound base metal mineral deposits in carbonate rocks, *Nev. Bur. Mines Rep.* No. 13, Part C, 5–50.
3. CALLAHAN, W. H., and McMURRY, H. V., 1967, Geophysical exploration of Mississippi Valley-Appalachian type strata-bound zinc–lead deposits, *Geol. Surv. Can. Rep.* No. 26, 350–360.
4. CAMERON, E. M., 1969, Regional geochemical study of Slave Point carbonates, Western Canada, *Can. J. Earth Sci.* **6,** 247–268.
5. CAMPBELL, N. 1957, Stratigraphy and structure of Pine Point area, Northwest Territories, Canada, in *Structural Geology of Canadian Ore Deposits,* Vol. 2, Canadian Institute of Mining and Metallurgy, Montreal, pp. 161–174.
6. JACKSON, S. A., and BEALS, F. W., 1967, An aspect of sedimentary basin evolution: The concentration of Mississippi Valley-type ore during the late stages of diagenesis, *Can. Pet. Geol. Bull.* **15,** 383–433.
7. JACKSON, S. A., and FOLINSBEE, R. E., 1969, The Pine Point lead–zinc deposits, N.W.T., Canada: Introduction and paleoecology of the Presqu'ile Reef, *Econ. Geol.* **64,** 711–717.
8. LAJOIE, J. J., and KLEIN, J., 1979, Geophysical exploration at the Pine Point Mines Ltd. zinc–lead property, N.W.T., Canada, *Geol. Surv. Can. Econ. Geol. Rep.* No. 31, 653–664.
9. MACQUEEN, R. W., 1976, Sedimentary zinc–lead deposits, Rocky Mountain Belt, Canadian Cordillera, *Geosci. Can.* 3(2) 71–81.
10. MURPHY, J. D., and SINCLAIR, W. D., 1974, Lead–zinc in the Selwyn and McKenzie Mountains, *Can. Min. J.* **95**(4), 40–44.
11. NEWHOUSE, W. H., 1933, The temperature of formation of the Mississippi Valley lead–zinc deposits, *Econ. Geol.* **28,** 744–750.
12. OHLE, E. L., 1959, Some considerations in determining the origin of ore deposits of the Mississippi Valley type, *Econ. Geol.* **54,** 769–789.
13. ROEDDER, E., 1967, Environment of deposition of stratiform (Mississippi Valley-type) ore deposits from studies of fluid inclusions, *Econ. Geol. Monograph* 3, 349–361.
14. SEIGEL, H. O., 1952, Ore body size determination in electrical prospecting, *Geophysics* **17**(4), 907–914.
15. SEIGEL, H. O., HILL, H. L., and BAIRD, J. G., 1968, Discovery case history of the Pyramid ore bodies, Pine Point, N.W.T. Canada, *Geophysics* **33,** 645–656.
16. SKINNER, B. J., 1967, Precipitation of Mississippi Valley-type ores: A possible mechanism, *Econ. Geol. Monograph* 3, 363–369.
17. SMITH, C. L., 1974, A newly discovered Mississippi Valley-type lead province, Yukon-N.W.T., Canada, A.I.M.E. Preprint No. 74-1-306.

CHAPTER ELEVEN

DETECTION OF VEIN-GOLD DEPOSITS OF THE NORTH AMERICAN SHIELD BY OPTIMIZED GROUND PROGRAMS

11.1. GENERAL GEOLOGICAL BACKGROUND

11.1.1. Economic Geology Summary

Gold plays a very prominent role among the "minor" metals recovered from varied geological sources, because of its importance on the world financial scene and increasing industrial and ornamental uses. Two types of deposits are the main sources of gold in the world: vein-type deposits and alluvial deposits. The vein deposits are as widespread geographically as they are chronologically. They are found in the Archean Shields of North America, Southern Africa, Western Australia, India, and Brazil; along the Paleozoic foldbelts of the Appalachian of North America, Tasman Geosyncline of eastern Australia, and the Urals of the U.S.S.R.; and finally along the Mesozoic and Cenozoic foldbelts of the Americas, Southwest Pacific Island Arcs, and East Asia.

The alluvial deposits consist of mechanically disseminated gold in alluvial turbidites of recent and modern ages (Siberia, Yukon, Alaska, California, etc.) or in fossil placers (South Africa), possibly subjected to remobilization or metasomatic processes (Nevada, U.S.A). Gold is also won in substantial amounts from various types of deposits including porphyry–copper–molybdenum and often spatially related contact metasomatic (Cu–Mo–Au) deposits, skarn deposits (Cu–Fe–Au) or (W–Mo–Au), volcanogenic massive sulfides, and vein deposits (Cu–Au), (W–Au), (Sb–Au).

The most prolific period of the history of gold mining occurred on a worldwide basis between 1845 and 1910. The ensuing decline lasted until 1936, when a price rise from 20 to 36 U.S. dollars/oz. generated much activity up to the early postwar years. A steady decline was again reversed in the late 1970s because of a surge of the price of gold to the peak value of 800 U.S. dollars/oz., settling to 400 U.S. dollars/oz. in the 1980s. According to Elevatorski,[5] the present annual production of over 40 million ounces is geographically distributed as follows: South Africa: 54%, U.S.S.R.: 25%, Canada: 5%, U.S.A.: 2.5%, Australia: 2.5%, the remainder (11%) being distributed unevenly between 20 countries throughout the world.

11.1.2. Geological Synopsis for the Archean Vein Gold Deposits of the North American Shield

11.1.2.1. Economic and Geographic Considerations. The North American share of world gold production comes mainly from vein gold deposits occurring in the Canadian portion of the North American Precambrian Shield, and from the Precambrian horst of the Black Hills in the U.S.A.[14] The Canadian share of world gold production declined from about 5 million ounces in the 1930s (13% of world total) to about 1.6 million ounces, or about 5% of world production, some 50 years later. However, at the present time, there is a great surge of exploration activity taking in older districts as well as some new ones, such as the Hemlo area near Lake Superior in central Ontario.

A total of 150 known commercial vein gold deposits occur in Canada, including 120 in the Precambrian Shield. About 55% of the present Canadian production is won in the Province of Ontario, about 25% is mined in the Province of Quebec, 15% from the Northwest Territories, and 5% from the Cordilleran Belt. The Northwest Territory deposits are mainly concentrated in the greenstone belts of the Yellowknife region, north of Great Slave Lake, but there are scattered deposits in the remote and hostile Arctic region (Contwoyto Lake). Very few productive gold deposits occur in the intervening Precambrian regions of the Provinces of Saskatchewan and Manitoba.

There are 33 known gold districts in the Precambrian portion of the Province of Ontario. A total of 22 districts lie in the area extending between the 45th and 52nd parallels, which has been most intensively prospected because of its good accessibility; the remaining 11 districts are scattered north of the 52nd Parallel, a region far less accessible. The largest gold district is the Porcupine Gold Fields, near the town of Timmins, in northeastern Ontario. In its heydays of the 1940s, the Porcupine district boasted of 36 operating mines annually producing 1 million ounces of gold, the leading mines being the Hollinger, McIntyre, and Dome. The gold mineralized belt extends farther east through the Kirkland Lake district, into Northwest Quebec, where about 45 vein gold deposits are known.

11.1.2.2. Geological Considerations. The most common geological environment for the vein gold deposits of the North American Shield consists of greenstone belts of Archean age. These belts are made up of intricately folded and sheared series of metasediments and metavolcanic formations showing a well-developed "green schist" facies of regional metamorphism. The belts are arcuate in shape; they are encased in granite–gneiss and migmatites and are cut across by swarms of quartz (felspar) porphyry dykes and stocks.

The veins are generally tabular in shape: they are extensive along strike and down dip but are very limited in their third dimension. They occur along shears, fault zones, geological contacts, and in zones of crumpling and dragfolding, within wide envelopes of hydrothermal alteration consisting mainly of silicification and carbonatization.

Two main classes of vein gold deposits are generally recognized: (1) cavity-filling and (2) replacement, while in many deposits the two types occur together. The cavity-filling type has well-defined walls and the gold mineralization is restricted to the vein material. In the replacement type, however, the mineralization replaces the wallrock and vein material as well, showing ill-defined boundaries. The Kirkland Lake "Break" of northeastern Ontario and the Homestake deposit of the Black Hills are good examples of the replacement type of vein gold deposit. The replacement veins are generally much longer (several thousand feet) and wider (up to 100 feet) than the cavity-filling veins, which are only 2000–3000 feet in length and 15–30 feet in width at the most.

The Porcupine Gold Fields of northeastern Ontario, being the most productive and the most intensely investigated gold district of the Canadian Shield, will serve as an example for the study of the structural control of gold deposition in the North American Shield. The Porcupine structural setting is that of an overturned, steeply dipping, tightly folded synclinal structure made up of intricately dragfolded Keewatin basic flows uncomfortably overlain by Temiskaming greywacke and associated clastic sediments. The Porcupine Trough is cut across by swarms of quartz (felspar) porphyry dykes and pipelike stocks of Algoman age. The principal locus of the ore is the Pearl Lake porphyry stock, which is most intensely sheared and hydrothermally altered, almost beyond recognition. The structural controls of the deposits are as follows: (a) openings developed along zones of weakness related to the stratigraphic succession, such as flow contacts, bedding planes, and also along intrusive contacts; (b) openings developed by differential movement affecting rocks of differing competencies; basic flows, conglomerate and greywacke beds and porphyry intrusive behave as competent formations in contrast with tuffs, slate, soapstone, chlorite–carbonate schist, and pillowed lava flows.

The mineralogy of the deposits is rather simple. The gangue material consists mainly of quartz with subordinate amounts of carbonates, mainly ankerite and, locally, tourmaline (Porcupine), accompanied by minor chlorite, sericite, and graphite, as well as sporadic sulfides. Among the sulfides, pyrite is the most common and abundant, particularly in the replacement-type deposits; it is followed in order of decreasing importance by arsenopyrite, pyrrhotite, and base metal sulfides. The latter may be quite abundant (Western Ontario), requiring expensive smelting of the "refractory" ore. Likewise, arsenopyrite is troublesome as it requires a costly roasting prior to the cyanadation process. In most deposits of the North American Shield, a large proportion of the gold is associated with the pyrite and is largely "free milling," occurring as small grains of native gold or electrum along crystal boundaries, cleavages, or fracture planes. The rest occurs as tellurides.

11.2. FIELD DETECTION METHODOLOGY

11.2.1. Detection Approaches

The organization of field exploration for vein gold deposits in the Archean Shield of North America was recently reviewed by Hutson.[7] Two main kinds of detection approaches are usually combined: (1) a direct approach based on the visual detection and recognition of deposits or closely related features or on the mechanical detection of the deposits (drilling), and (2) an indirect approach relying on the detection of geophysical or geochemical signatures of the deposits or closely related features. The general sequence of the two approaches is as follows:

1. Direct approach: a. *Airborne surveys:* photogeological mapping for structural information purposes;
   b. *Ground surveys:* surface, prospecting, reconnaissance, and systematic mapping for lithological information; subsurface; trenching and drilling;
2. Indirect approach: a. *Airborne surveys:* aeromagnetic mapping for structural and lithological information;
   b. *Ground surveys:* geophysical and geochemical.

As indicated by Hutson, the sequential methodology combining the two approaches appears to be the most cost-effective search procedure for small and elusive targets such as the vein-gold deposits of the North American Shield.

11.2.2. Direct Detection

Any attempt at visual detection of vein-gold deposits or their immediate geological environment from satellite platforms has to be ruled out because of the small size of the targets. But photogeological surveys from aircraft platforms, are quite useful for regional structural studies. Combined photogeological and aeromagnetic surveys have greatly assisted the delineation of high priority areas in the newly discovered Hemlo Gold District of Ontario.

Visual detection from the ground has proved highly effective over the past 50 years. Up to 90% of the gold deposits discovered to date in the North American Shield were detected by a combination of prospecting, systematic mapping, and some information drilling. In the present circumstances of high gold price, the systematic drilling of small areas of high potential of less than a square mile in extent should be considered in the near future as a valid and effective approach, since the most obvious deposits have already been discovered.

11.2.3. Indirect Geophysical Detection

The past three decades have witnessed many major discoveries of base metal deposits with considerable assistance from airborne and ground geophysical methods; not so for gold deposits. During the same period, only a handful of gold discoveries were made, with geophysical methods playing only a minor role. An obvious reason is that, whereas gold possesses high specific gravity and electrical conductivity, its presence in a widely disseminated state in the vein environment cannot be readily detected by most known geophysical techniques relying on gravity or magnetic or electromagnetic contrast between vein and wallrock. Therefore, two strategies have to be considered: (1) search for structural clues using magnetics, electromagnetics (VLF), or resistivity surveys, and (2) search for associated minerals which are more abundant and widespread than gold itself and susceptible of easier geophysical detection, such as the iron sulfides.

Examples of the successful application of the first strategy were described by Kelly in the Porcupine Gold District.[8] The program combined magnetic and resistivity surveys to take advantage of the high resistivity of the quartz material in the cavity-filling type of vein gold deposits. A combination of magnetic and electromagnetic (VLF) surveys is presently being used to advantage in the Hemlo district mentioned above. The application of computerized induced polarization equipment, both in the time and frequency (phase angle) domains, is an example of the second strategy.[11] Seigel warns, however, that only low amplitude and localized responses can be expected from such small targets as vein gold deposits usually offer; thus much more care and attention to detail are required than is customary in the search for base metal targets.

11.2.4. Indirect Geochemical Detection

The search for primary halos by bedrock sampling is quite difficult because of the small extent of the halos and the general scarcity of outcrops. Some assistance could be gained from drill core sampling in well prospected areas. Geochemical detection of secondary halos in residual soils is not applicable in most regions of the North American Shield because of the extent and thickness of transported glacial alluvium.

11.3. STATISTICAL MODELING OF GEOMETRIC PARAMETERS OF VEIN-GOLD DEPOSITS

11.3.1. Scope of Study

A total of 50 vein gold deposits were selected and included in the database to represent the statistical population of all known and undiscovered vein gold deposits of the North American Shield. The deposits are listed by name and province of occurrence in Table 11.1. The geographic breakdown of the sample is as follows: the largest subsample consisting of 31 deposits covers the northeast, northcentral, and western portions of the Province of Ontario; the second largest subsample including 16 deposits represents the northwest portion of the Province of Quebec, Canada's second largest producer; the third subsample, comprising three deposits, covers the Northwest Territory portion of the shield. The size of the sample is quite satisfactory for statistical processing purposes.

11.3.2. Statistical Modeling

For detection purposes, the geometry of vein-gold deposits is quantitatively described by means of three parameters of a dimensional nature and two attitudinal parameters. The dimensional parameters include the length, breadth, and shape ratio of the horizontal section of the portion of deposits which lies within an average 300-foot detection range. The attitudinal parameters include the unoriented true dip of the deposits and the strike direction referred to True North and measured in degrees within the right half-circle.

The range of the measurements of geometric parameters from the 50 deposits included in the sample is as follows: from 355 feet to 6500 feet for length, 10 (cavity-filling type) to 200 feet (replacement type) for breadth, and 45–85 degrees for dip. The measurements were grouped into classes and frequency distributions were constructed. The statistics summarizing these distributions are listed in the lower half of Table 11.2. They include arithmetic means, modes, standard deviations, and coefficients of skewness. The latter gives a useful indication of the degree of departure from normal symmetry shown by the distributions of observed data. The skewness of the distributions is rather slight for the length and shape

TABLE 11.1
List of Vein-Gold Deposits of the North American Shield Included in the Database.

Ontario

Central Patricia	Hallnor
Cochenour Willans	Dome
Gold Eagle	Preston East Dome
Madsen Red Lake	Paymaster
McKenzie Red Lake	Buffalo Ankerite
Hasaga	Aunor
Little Long Lac	Delnite
Jellicoe	Coniaurum
Tombill	Moneta Porcupine
North Empire	Hollinger
Hard Rock	McIntyre
Renabie	Kerr Addison
Jerome	Sylvanite
Leitch	Wright Hargeaves
Pamour	Bidgood
Broulan	

N.W. Quebec

Omega	Arntfield
Lamaque	Camflo
East Malartic	Canadian Malartic
Central Cadillac	Granada
O'Brien	Agnico
Belleterre	

Northwest Territories

MacWatters	
Powell Rouyn	Giant Yellowknife
Wasa	Cons. Discovery
Beattie	Camlaren
Francoeur	

ratio parameters (coefficient value = 0.01), while a moderate skewness (coefficient range 0.2–0.5) is noted for the other three geometric parameters.

Lognormal models were successfully fitted to the observed distributions of dimensional parameters, based on the results of the χ^2 goodness of fit test at the 0.05 confidence level. The models fitted at the same level of confidence to the observed distributions of attitudinal parameters include a normal model for dips and a circular normal model for strike directions.

Statistics summarizing the fitted models are listed in the upper half of Table 11.2. Among them are the geometric means of the fitted lognormal models,

TABLE 11.2
Statistical Modeling Summary for Vein-Gold Deposits of the North American Shield (Sample Size: 50 Deposits).

Statistic of fitted model	Length of horiz.sect. in feet	Breadth of horiz. sect. in feet	Shape ratio R=B/L	Strike direct. from T.North in degrees	Dip in degrees
25%ile	1060	18	0.007	55	66
95%L.c.l.	1466	21	0.012	72	77
Mean	1753	18	0.016	81	85
95%U.c.l.	2097	38	0.022	89	93
75%ile	2800	60	0.040	95	107
Disp. coef.	0.08	0.32	0.251	0.22	0.32
Statistic of observed data					
Arith. mean	2207	51	0.029	81	84
Mode	2200	38	0.028	70	70
Stand.dev.	1588	59	0.031	34	30
Skewness	0.01	0.21	0.032	0.33	0.50

arithmetic mean of normal model, and circular mean of circular normal model, along with the limits of their 95% confidence intervals. Other listed statistics include the first and second quartiles of the fitted distributions and the dispersion coefficients, which give an indication of the spread of the data about the mean of the fitted models. A small coefficient (0.1) indicating a low variability is shown for the length parameter, while a moderate variability affects the breadth and shape ratio parameters (coefficient ranging from 0.1 to 0.5).

11.4. CONSTRUCTION AND ORGANIZATION OF DETECTION PROBABILITY TABLES

11.4.1. Introduction

For the purpose of calculating detection probabilities, the vein-gold deposits of the North American Shield are considered as dipping, orientated tabular bodies with horizontal cross sections of elliptical shape. Therefore, target dimensions for detection purposes are not the same as that of the deposits; they have to be recalculated for each deposit by introducing a dip component which is computed as described in Sections 2.2.3. of Chapter 2, and 5.4.2 of Chapter 5. As a result of the recalculation, the shape coefficients are substantially inflated, thus increasing the probabilities of detection. For example, the mean shape ratio of the horizontal sections of vein gold deposits themselves (0.016) becomes equal to 0.09 for the targets in the context of vertical detection and is increased to 0.19 when considering 55-degree angled detection.

Since strike orientation data were available for most of the 50 deposits of the sample, the orientation factor is introduced in the construction of the tables to improve the probability of success of the various field programs. Based on the results of a dynamic programming study, it was found that the probabilities of vertical detection by geophysical surveys or drilling are maximized if the grid is laid at angle of 18 degrees ± 10 with the expected strike direction. If we consider the 55-degree angled detection, the optimal grid orientation is at 30 degrees ± 10 with the expected strike direction.

The latter may be estimated from the results of previous surveys or from photogeological studies. If no prior information is available, the mean of the fitted circular normal model is a statistically valid estimate to be considered for grid design (See Table 11.2, upper half).

11.4.2. Description of Probability Tables

Two similar table designs are used to display the probabilities of single detection and confirmed detection of vein gold deposits by ground surveys and drilling programs. The grid spacings are varied within a range extending from 100 to 1500 feet in steps of 50–100 feet, depending on the nature of the field programs. For each grid spacing, the probability of single or confirmed detection is calculated within a 95% confidence interval, which is based on the 95% confidence interval of the mean target length. (See Table 11.2, upper half.)

A second set of tables is constructed for exploration planning purposes. The tables display the grid spacings and corresponding coverage costs per unit of area which are required to provide prespecified detection probability levels. The range of probability levels extends from 0.05 to 0.95 using steps of 0.05.

11.4.3. Organization of Detection Probability Tables

No probability tables were constructed to cover the airborne geophysical detection of vein gold deposits, because no known methods are available at the present time. The probability tables are organized into two blocks. One block, comprising Tables 11.4–11.6 covers vertical detection by ground geophysical surveys or vertical drilling to a depth of 300 feet. The second block covers the 55-degree angled detection by drilling (Tables 11.8–11.10).

11.5. DESIGN OF THREE STRATEGIES FOR THE DETECTION OF ARCHEAN VEIN-GOLD DEPOSITS OF THE NORTH AMERICAN SHIELD

11.5.1. Detection Strategies

Three types of detection strategies are offered to explorationists in order to meet the requirements of three kinds of exploration situations which could arise in the search for vein-gold deposits of the North American Shield. The first one, the liminal option, should prove attractive when dealing with the coverage of large areas of little known potential, because it provides a detection probability level of at least 0.500, while minimizing the coverage cost. The maximal strategy is a high-cost approach (three to five times that of the liminal option) providing a high detection probability level (0.980), which should be restricted to the coverage of small areas (less than one square mile) of high potential in well-prospected regions. Finally, the optimal strategy is offered to explorationists as a more suitable compromise between the two extreme options mentioned above, when faced with the problem of covering moderately large areas of good potential within strict budget limits. The detection parameters pertaining to three strategies are assembled in Table 11.3 for the reader's convenience.

11.5.2. Liminal Strategy

The detection parameters for ground geophysical surveys and vertical and angled drilling in the context of the liminal option are listed in the leftmost portion of Table 11.3. The parameters include the grid spacings and corresponding coverage costs associated with the detection probability level of 0.500.

The cost multiplying factor required to ensure confirmation of single detection is 2.1 for ground geophysical surveys and 2.6 for both vertical and angled drilling to a vertical depth of 300 feet. A comparison of the detection performances of vertical vs. 55-degree angled drilling is of interest to explorationists for planning purposes. The most convenient yardstick is the coverage cost per unit of area since the detection probability is held at the constant level of 0.500. Angled drilling should definitely be favored over vertical drilling since it leads to a 37% coverage cost reduction. But, if no prior knowledge of the likely direction of the dip is available from previous surveys or photogeological studies, vertical drilling may be a safer option in order to avoid the risk of drilling down dip and possibly failing to recognize the target.

DETECTION OF VEIN-GOLD DEPOSITS 315

TABLE 11.3
Summary of Detection Characteristics of Vein-Gold Deposits of the North American Shield for Three Types of Detection Strategies.

Type of survey and grid geometry	Liminal detection				Optimal detection			Maximal detection			
	Coverage cost: US$/mile sq. Grid size: feet				Coverage cost US$/mile sq. Grid size: ft	Detection proba- bility		Coverage cost: US$/mile sq. Grid size: feet			
	Detection		Confirmed detection					Detection		Confirmed detection	
Ground geophysical square grid	$1840 580 ft		$3920 325 ft		$1946 550 ft	0.66		$4308 300 ft		$10,220 175 ft	
Vertical drilling square grid	$605,200 580 ft		$1,760,000 325 ft		$675,000 550 ft	0.66		$2,077,000 300 ft		$6,220,000 175 ft	
55-degree angled drilling square grid	$424,200 800 ft		$1,090,000 500 ft		$536,000 700 ft	0.90		$1,160,000 440 ft		$3,490,000 300 ft	
Optimal 45-deg angled drilling square grid	—		—		$491,000 800 ft	0.88		—		—	

11.5.3. Maximal Strategy

The detection characteristics pertaining to the maximal strategy are listed in the rightmost section of Table 11.3, in the same manner as for the liminal option. If we require confirmation of single detection, the cost multiplier is 2.6 for ground geophysical surveys, and it escalates to 3.3 for both types of drilling program. The maximal cost multipliers are substantially higher than their liminal counterparts, contrary to what has been noted for all types of ore targets previously covered in this book. This difference plainly underscores the considerable difficulties to be faced when searching for targets as small as the vein-gold deposits are. As was the case for all other types of dipping deposits previously covered, the choice of angled over vertical drilling to the same depth is advantageous since it leads to a 52% coverage cost reduction, providing some indication of the likely direction of dips is available.

11.5.4. Optimal Strategy

As described in Sections 4.4.2 of Chapter 4 and 5.5.4 of Chapter 5, a three-stage dynamic programming procedure was conducted to determine the optimal parameters of the control grids for the ground field programs required for the detection of vein-gold deposits. The parameters include (a) the optimal orientation of the grids with respect to the expected strike direction of targets, (b) optimal grid spacing and corresponding coverage cost per unit of area, as well as associated probability of detection, and finally (c) the optimal drilling angle when the likely direction of the dip is known. The optimal parameters are listed in the following tables: Tables 11.7 for ground geophysical surveys, and 11.11 for vertical and angled drilling programs.

In order to compare the detection performances of the two types of drilling program, we should use the "expected loss" criterion, as described in Section 6.5.4 of Chapter 6, rather than the coverage cost criterion previously used, because of differences in optimal probability levels in the two cases. We find that the appropriate choice is that of the optimally angled drilling (45 degrees), because it leads to reduction of expected loss of 74% as compared to the vertical drilling option.

11.6. DETECTION PROBABILITY AND OPTIMIZATION TABLES

The following section includes eight tables which cover the probabilities of detection and optimal designs of ground programs in the search for vein-gold deposits of the Shield (pages 317–323).

TABLE 11.4

Probabilities of Detection of Vein-Gold Deposits of the North American Shield by Ground Geophysical Surveys or Vertical Drilling to a Depth of 300 Feet.

```
Survey design: square grid with spacings S by S feet.
The expected shape of the ore deposits is elliptical
with major axis = L in the 95% confidence interval:
L.c.l. ( geometric mean ( U.c.l., and minor axis = B feet,
R is defined as the ratio B/L with geometric mean = 0.09
grid orientation 18 degrees + or - 10 from the expected
strike line.
```

	Probability of detection		
Spacing S in feet	L.c.l. L = 1466	G. mean 1753	U.c.l. 2097ft
250	1.000	1.000	1.000
300	0.986	1.000	1.000
350	0.919	0.990	1.000
400	0.803	0.947	0.995
450	0.685	0.860	0.974
500	0.579	0.759	0.920
600	0.420	0.576	0.756
700	0.310	0.439	0.600
800	0.237	0.339	0.478
900	0.188	0.268	0.383
1000	0.152	0.217	0.311
1100	0.126	0.180	0.257
1200	0.105	0.151	0.216
1300	0.090	0.129	0.184
1400	0.078	0.111	0.159
1500	0.068	0.097	0.138

TABLE 11.5
Determination of Grid Size for Specified Levels of Probability in the Detection of Vein-Gold Deposits of the North American Shield by Vertical Drilling to a Depth of 300 Feet.

Survey Design: Square Grid With Spacings S by S Feet
Unit Cost = $6000 Per Diamond Drill-Hole

Probability Level Of Detection	Grid Size In Feet	Drilling Cost Per Mile Square In U.S.$. Thousands
0.05	1950	82.93
0.15	1200	175.61
0.25	950	258.82
0.35	800	347.47
0.45	700	438.91
0.55	650	500.48
0.65	550	675.43
0.75	500	803.19
0.85	450	974.35
0.95	400	1211.54

TABLE 11.6
Probabilities of Confirmed Detection of Vein-Gold Deposits of the North American Shield by Ground Geophysical Surveys or Vertical Drilling to a Depth of 300 Feet.

Survey design: square grid with spacings S by S feet. The expected shape of the ore deposits is elliptical with major axis = L in the 95% confidence interval: L.c.l. (geometric mean (U.c.l., and minor axis = B feet, R is defined as the ratio B/L with geometric mean = 0.09 grid orientation 0 degrees + or - 10 from the expected strike line.

Spacing S in feet	Probability of confirmed detection		
	L.c.l. L = 1466	G. mean 1753	U.c.l. 2097ft
100	1.000	1.000	1.000
150	0.987	1.000	1.000
200	0.928	0.969	0.996
250	0.853	0.916	0.960
300	0.740	0.852	0.915
350	0.367	0.759	0.862
400	0.230	0.531	0.798
450	0.155	0.284	0.689
500	0.111	0.198	0.368
550	0.081	0.145	0.264
600	0.060	0.110	0.196
700	0.033	0.066	0.119
800	0.018	0.040	0.077

TABLE 11.7
Optimal Design of Ground Geophysical Surveys for the Detection of Vein-Gold Deposits of the North American Shield.

Survey Design: Square Grid With Spacing S by S Feet.
Confidence Interval: All results are reported as 95% confidence intervals respectively.
Expected Target Length in Feet: 1466   1753   2097
Expected Shape Ratio R = 0.09
Unit Cost: $6/Station

Optimal Grid Spacing in Feet:	500	550	700
Cost in $ Per Mile Square:	2189	1946	1463
Probability of Detection:	0.57	0.66	0.60
Optimal Grid Orientation:	20 degrees + or -10 to expected strike direction of target.		

TABLE 11.8
Probabilities of Detection of Vein-Gold Deposits of the North American Shield by 55-Degree Angled Drilling to a Vertical Depth of 300 Feet.

Survey design: square grid with spacings S by S feet.
The expected shape of the ore deposits is elliptical
with major axis = L in the 95% confidence interval:
L.c.l. (geometric mean (U.c.l., and minor axis = B feet,
R is defined as the ratio B/L with geometric mean = 0.19
grid orientation 30 degrees + or - 10 from the expected
strike line.

	Probability of detection		
Spacing S in feet	L.c.l. L = 1466	G. mean 1753	U.c.l. 2097ft
400	1.000	1.000	1.000
450	0.995	1.000	1.000
500	0.967	1.000	1.000
550	0.906	0.991	1.000
600	0.820	0.966	1.000
650	0.727	0.916	0.993
700	0.638	0.847	0.975
800	0.498	0.692	0.891
900	0.396	0.559	0.766
1000	0.321	0.457	0.640
1100	0.265	0.379	0.537
1200	0.223	0.318	0.454
1300	0.190	0.271	0.388
1400	0.164	0.234	0.335
1500	0.143	0.204	0.292

TABLE 11.9
Determination of Grid Size for Specified Levels of Probability in the Detection of Vein-Gold Deposits of the North American Shield by 55-Degree Angled Drilling to a Vertical Depth of 300 Feet.

Survey Design: Square Grid With Spacings S by S Feet
Unit Cost = $7325 Per Diamond Drill-Hole

Probability Level Of Detection	Grid Size In Feet	Drilling Cost Per Mile Square In U.S.$. Thousands
0.15	1700	124.09
0.25	1400	167.46
0.35	1200	214.39
0.45	1000	289.81
0.55	900	346.39
0.65	850	382.03
0.75	800	424.21
0.85	700	535.83
0.95	650	611.00

TABLE 11.10
Probabilities of Confirmed Detection of Vein-Gold Deposits of the North American Shield by 55-Degree Angled Drilling to a Vertical Depth of 300 Feet.

Survey design: square grid with spacings S by S feet.
The expected shape of the ore deposits is elliptical
with major axis = L in the 95% confidence interval:
L.c.l. (geometric mean (U.c.l., and minor axis = $\bar{B}$ feet,
R is defined as the ratio B/L with geometric mean 0.19
grid orientation 0 degrees + or - 10 from the expected
strike line.

	Probability of confirmed detection		
Spacing S in feet	L.c.l. L = 1466	G. mean 1753	U.c.l. 2097ft
250	1.000	1.000	1.000
300	0.977	1.000	1.000
350	0.909	0.985	1.000
400	0.818	0.933	0.996
450	0.672	0.865	0.960
500	0.455	0.774	0.910
550	0.350	0.591	0.849
600	0.268	0.451	0.771
700	0.152	0.290	0.482
800	0.081	0.182	0.333
900	0.043	0.109	0.225
1000	0.022	0.064	0.152
1100	0.011	0.038	0.098

TABLE 11.11
Optimal Designs for the Detection of Vein-Gold Deposits of the North American Shield by Drilling to a Vertical Depth of 300 Feet.

```
Survey Design:  square grid with spacing S by S feet
Confidence interval:  all results are reported as 95% confidence intervals
respectively
Expected target length in feet:      1466      1753      2097
Unit Cost = $20 per linear ft.
```

(a) Vertical drilling:			
Optimal grid spacing in feet:	500	550	700
Cost in $thou. per mile square:	803	675	439
Probability of detection:	0.57	0.66	0.60
Optimal grid orientation:	18 degrees to expected strike of target		

(b) Angled drilling at 55 degrees:			
Optimal grid spacing in feet:	700	700	900
Cost in $thou. per mile square:	536	536	346
Probability of detection:	0.76	0.90	0.77
Optimal grid orientation:	30 degrees to expected strike of target		

(c) Angled drilling at optimal angle = 45 degrees:			
Optimal grid spacing in feet:	700	800	1000
Cost in $thou. per mile square:	621	491	336
Probability of detection:	0.90	0.88	0.76
Optimal grid orientation:	30 degrees to expected strike of target		

REFERENCES AND SELECTED READINGS

1. AMIRYAN, S. O., 1960, Mineralogy of gold deposits, *Akad. Nauk. Arm. S.S.R.* **31**(1), 43–48.
2. BOYLE, R. W., 1961, The geology, geochemistry and origin of the gold deposits of the Yellowknife District, N.W.T., *Geol. Surv. Can. Mem.* No. 310.
3. COOKE, H. C., and JOHNSTONE, W. A., 1932, Gold occurrences of Canada, a summary account, *Geol. Surv. Can. Econ. Geol. Ser.* No. 10.
4. DOUGHERTY, E. Y., 1939, Some geological features of the Kolar, Porcupine, and Kirkland Lake gold fields, *Econ. Geol.*, **34**, 622–653.
5. ELEVATORSKI, E. A., 1981, *Gold Mines of the World*, MINOBRAS, Santa Monica, California.
6. EMMONS, W. H., 1937, *Gold Deposits of the World*, McGraw-Hill, New York.
7. HUTSON, R. J., 1980, Exploration models for gold in the Archean, *Can. Min. J.* **101**(4), 66–68.

8. KELLY, S. F., 1957, Resistivity and magnetic surveys in 1936 on the Broulan–Porcupine gold prospect, South Porcupine, Ontario, Canada, in *Methods and Case Histories in Mining Geophysics,* Sixth Commonwealth Mining and Metallurgy Congress Volume, Montreal, pp. 283–291.
9. MIDDLETON, R. S., and CAMPBELL, E. E., 1979, Geophysical and geochemical methods for mapping gold-bearing structures in Nicaragua, *Geol. Surv. Can. Econ. Geol. Rep.* No. 31, pp. 779–798.
10. MOORE, E. S., 1940, Genetic relations of gold deposits and igneous rocks in the Canadian Shield, *Econ. Geol.* **35,** 127–139.
11. SEIGEL, H. O., JOHNSON, I., and HENNESSEY, J., 1984, Geophysical aids to gold exploration, *Northern Miner* **March 1st,** B17–B18.
12. SIXTH COMMONWEALTH MINING AND METALLURGY CONGRESS VOLUME, 1957, Structural geology of Canadian ore deposits, Canadian Institute of Mining and Metallurgy, Montreal, Canada.
13. WHITE, W. H., 1943, Mechanism and environment of gold deposition in veins, *Econ. Geol.* **38,** 512–532.
14. WRIGHT, L. B., 1937, Gold deposits in the Black Hills of South Dakota and Wyoming, *A.I.M.E. Trans.* **126,** 390–425.

CHAPTER TWELVE

OPTIMAL SELECTION OF EXPLORATION TARGETS FOR DRILL TESTING

12.1. GENERAL STATEMENT

12.1.1. Introduction

As outlined in Section 1.2.3 of Chapter 1, the mineral exploration sequence starts with preliminary planning followed by field planning of surveys and data processing, and ending with the assessment of results for further decision making. The initial planning requires many decisions of critical importance, such as the choice of type(s) of targeted ore deposits, selection of regions for the search and prospecting areas within the selected regions, and finally the choice of methodology of coverage of prospecting areas, including technology, and particularly, survey design, whose optimization is the main topic of the book.

The following stage, field planning, deals largely with the scheduling and logistics required to organize and implement in an optimal manner the field programs that were planned in the previous stage. The purpose of the operation is the acquisition of new data, which are assembled and processed in order to delineate exploration targets for detailed investigation. Most field programs generate a large number of exploration targets of diverse merits from which only a limited number are to be selected for drill testing, because of time and budget constraints. The optimization of the screening of exploration targets is the topic of the present chapter.

12.1.2. Data Acquisition

There are three types of geodata, which have to be collected and processed separately because of their difference in structure and nature. The first type, geological data, is mostly qualitative in nature; the second type, geochemical data, is mainly quantitative and stochastic in nature, and the third one, geophysical data, is chiefly quantitative and deterministic.

Geological and geochemical processes are rather poorly understood and, for the most part, are not subject to reproduction by experimentation within present laboratory capabilities. As a result, there is a lack of measurable causative relationships between observed features and underlying processes, which hinders the construction of quantitative models. While geochemical data are essentially numerical, geological data are mainly descriptive and have to be transferred from the nominal scale of measurement to the "metric" scale. This process of "quantification" is necessarily fraught with difficulties such as lack of objectivity, risk of error, and loss of information. Certain types of multivariate statistical techniques such as factor analysis and cluster analysis are favored when dealing with rather loosely structured data of the geological or geochemical type.

Geophysical data are much more tightly structured than the two previous types. Mathematical models of most geophysical phenomena have been constructed and tested either in the laboratory or by computer simulation. Geophysical data, always of a numerical nature, comprise (1) discrete measurements taken at grid intersections on the ground, and (2) continuous readings along profiles by airborne surveys. The latter are "digitized," i.e., sampled at regular intervals, as required by statistical processing.

12.1.3. Delineation of Exploration Targets

The delineation of exploration targets is essentially sequential in nature. The procedure calls for the investigation of areas of decreasing size to be covered by means of technologies of increasing resolution, in order to delineate exploration targets of increasing merit. At each stage of the sequence, the first step of the procedure is the recording of the complexly interrelated geological, geochemical, and geophysical signatures which are diagnostic of the environment of deposits of specific genetic types.

The second step is the quantification of geo-signatures as "anomalies" which are defined with respect to "thresholds." The latter are statistically determined as a reflection of the variability of the measured property throughout the area of interest. Thresholds are often expressed in terms of multiples of the standard deviation of the measurements collected over the whole surveyed area.[14] A more flexible definition of thresholds is provided by areal trend analysis, which may be applied in three ways: (1) least-squares fitting (Chapter 12, Ref. 1), (2) moving average (Chapter 2, Ref. 11), and (3) block averaging (Chapter 12, Ref. 8).

The third step is the actual delineation of exploration targets. The most commonly used approach is a "univariate" one, requiring the physical superimposition of a number of anomaly maps, one for each type of survey.[14] The spatial coincidence of the maximum number of anomalies of different types leads to the delineation of exploration targets for selection and testing. The multivariate approach, which is gaining increasing attention, was originally advocated by

Agterberg[1] and Culbert [7] almost a decade ago. Instead of considering each type of measurement independently from the others, the multivariate approach deals with "statistical patterns" based on all types of measurements at each sampling point. The spatial characteristics of the patterns are analyzed statistically and the targets are delineated in an objective manner.

12.1.4. Selection of Exploration Targets for Testing

During the past two decades, explorationists have increasingly turned to large-scale integrated programs based on various combinations of sophisticated airborne and ground survey technologies in order to pinpoint exploration targets for test drilling. This approach obviously leads to a problem of optimal choice under uncertainty, when the exploration staff wish to select a limited number of drilling locations of greatest merit from a large number of potential targets.

The magnitude and difficulty of the task is well documented in the available literature.[18] From a total of 700 ground checked geophysical anomalies within a 30,000 square mile area of India (project Hardrock), 25 targets were selected for drill testing, from which six emerged as potentially commercial deposits. Similarly, only 20 drill testing targets, four of which yielded sulfide mineralization of commercial potential, were selected from a total of 950 ground confirmed anomalies during a large-scale search for volcanogenic massive sulfide deposits in the Shield area of Northwest Quebec during the period 1968–1970.

For a long time, the problem of target selection has been dealt with in a subjective manner by experts "eyeballing" the various univariate maps and relying on semiquantitative schemes.[14, 35] However, as the ore search becomes more complex and sophisticated, explorationists are increasingly turning to multivariate statistical decision techniques to guide objectively the selection of targets for testing.[1, 7, 8, 15, 16, 19]

Two multivariate approaches to the optimal selection of exploration target for testing are open to explorationists. The first one relies on "control locations," while the second one does not. In the former approach, multivariate patterns associated with the targets to be screened are statistically compared to the patterns associated with "control locations" known to be ore-bearing, and may be "recognized" as either potentially ore-bearing or barren. The "pattern recognition" approach[11] is gaining increasing attention among explorationists, as witnessed by the number of papers devoted to the topic which are listed at the end of the chapter (Ref. 20–39). The second approach, developed only very recently by the second writer, tackles the more difficult problem of optimally selecting exploration targets for drill testing in little-known regions, where no adequate control locations are available.

12.2. OPTIMAL SELECTION OF EXPLORATION TARGETS BASED ON CONTROL LOCATIONS

12.2.1. Summary of Procedure

The multivariate statistical approach described below proceeds sequentially in four stages, as illustrated by Figure 12.1, as follows.

Stage 1: Data Acquisition and Processing. The initial stage is carried out in four steps: (1) acquisition of field data pertinent to the targeted ore deposit type within the selected prospecting area, (2) acquisition of the same kind of data from known occurrences of the selected deposit type in neighboring regions of geological make-up similar to that of the prospecting area, (3) statistical processing of both sets of data, and (4) delineation of targets to be screened.

Stage 2: Selection of Variates Required for the Construction of a Geostatistical Model. First, the most appropriate economic parameter (gross value, metal tonnage, grade) is selected as dependent variate. As a second step, a multivariate screening of the most statistically significant independent variates is conducted. Finally, measurements of the selected variates are obtained for all control locations and all targets under study.

Stage 3: Construction of a Geostatistical Model. The first step is the construction of a classifying model based on the selected geovariates, which is "calibrated" on the ore and nonore control locations in order to provide the probabilities that targets under study belong to either class. Secondly, a regression model also based on the selected geovariates is constructed and calibrated on all control locations in order to provide estimates of the economic parameters represented by the dependent variate.

Stage 4: Construction of a Payoff Criterion for the Optimal Selection of Exploration Targets. A payoff criterion involving probabilities of success and failure, regression estimates of economic worth and an appropriate cost function is constructed to rank the targets under study in order of decreasing merit, leading to the selection of a number of top-ranking targets, subject to budget constraints.

12.2.2. Screening of Geovariates

12.2.2.1. Introduction. Dependent and independent geovariates are both required for the construction of a geostatistical model to be used for the screening of exploration targets. The nature of the dependent variate has to reflect quan-

SELECTION OF EXPLORATION TARGETS

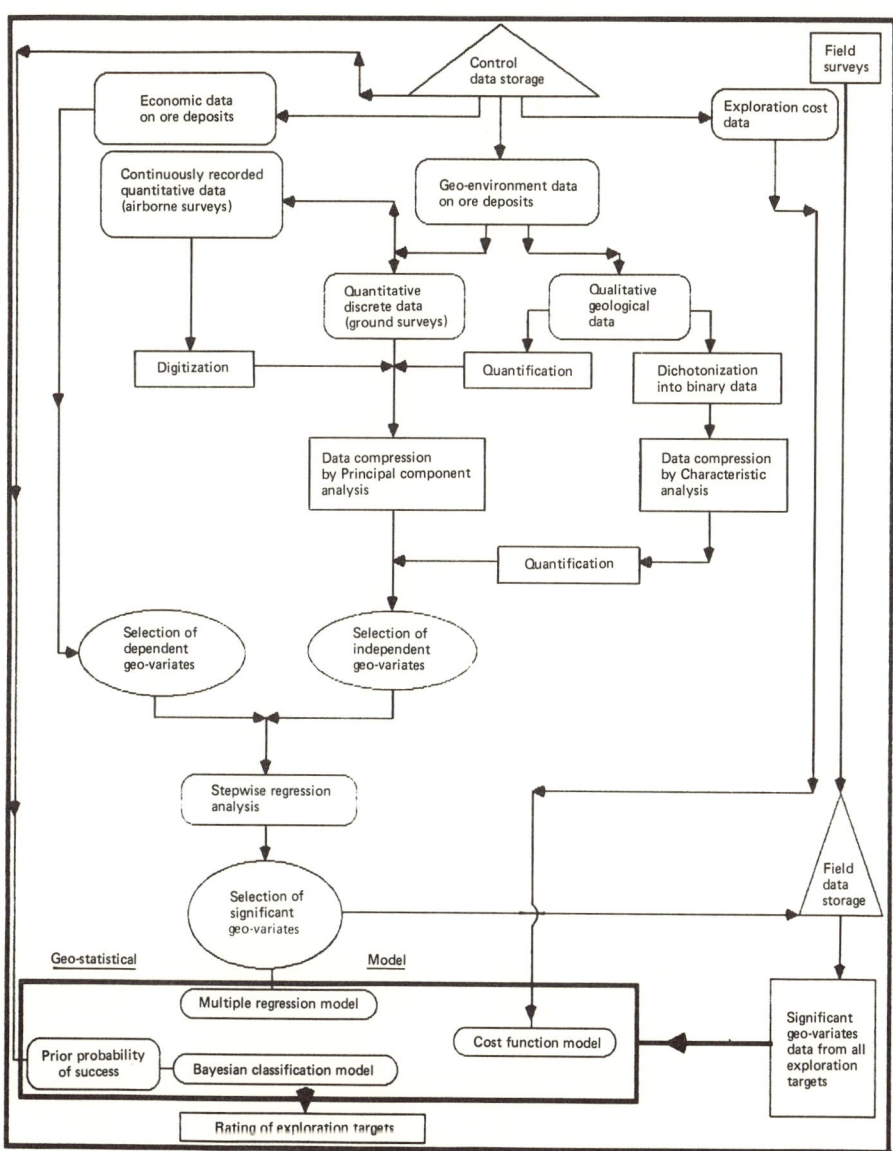

FIGURE 12.1
Flow Diagram of the Optimal Selection of Exploration Targets Based on Control Locations.

titatively the concept of "economic worth" in a manner which is dictated by the purposes of the study and the availability of data. Economic worth may be conveniently measured either by the expected gross value, or the discounted expected gross profit, or the total metal tonnage, or the average grade.

Independent variates quantitatively describe the geoenvironment of the various control locations and targets under investigation. Generally, one starts with a very large array of geovariates which are subjectively thought to be related to the occurrence of targeted ore deposit types in the geological environment prevailing in the prospecting area. Objective statistical procedures are then required to screen the geovariates in two steps for relevance and statistical significance. The first step is referred to as "data compression" and the second one as "feature selection," according to the general "pattern recognition" terminology.

12.2.2.2 Data Compression. Data compression is an initial screening procedure required to reduce objectively the large size of the array of subjectively chosen independent geovariates. Principal component analysis (PCA) is a standard statistical technique applicable to geophysical, geochemical, and quantified geological and other "metric scale" data. PCA is equivalent to a change of coordinates in the n-dimensional space, which works best when dealing with well-structured data.[1,4] However, when dealing with more loosely structured data as are most quantified geological and geochemical measurements, Culbert[7] and Klovan[16] recommend the use of the factor analysis technique, which improves on PCA by adding a "rotation" of the system of reference.

Compression of descriptive (nominal scale) geological data is rather troublesome because a partly subjective quantification procedure has to be carried out before going to PCA or factor analysis. Botbol[5,6] devised the technique of Characteristic Analysis to obviate the difficulty. First the descriptive data are simply and objectively quantified into a binary system: presence coded as 1, and absence as 0. The procedure of characteristic analysis, rather simple to handle by computer, is illustrated by Figure 12.2. The writer successfully used the technique to compress an array of 170 geological variates qualitatively describing ore-bearing porphyry-type deposits of the American Cordilleras into a more compact set of 44 most typical geological characteristics.[20]

12.2.2.3. Feature Selection. Feature selection is a crucial stage of the procedure leading to the construction of a geostatistical model. Its principal purpose is to select objectively among the previously compressed array of independent geovariates only those which are statistically correlated with the dependent variate. This is done by stepwise regression analysis based on a linear multiple regression function which is designed to approximate objectively and

SELECTION OF EXPLORATION TARGETS 331

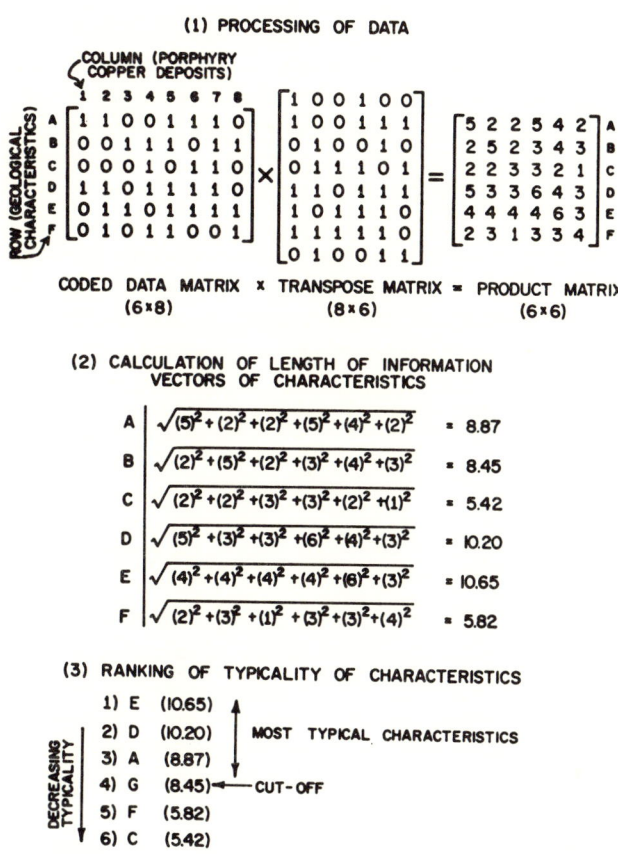

FIGURE 12.2
Flow Diagram of Characteristic Analysis.

quantitatively the complex relationship existing between the geoenvironment (independent variates) and economic worth (dependent variate).[1, 4]

The linear regression equation is written in its most general form as

$$y = a_1 x_1 + a_2 x_2 + \cdots + a_n x_n \qquad (1)$$

where y is the dependent variate, and a_i and x_i the coefficients and independent variates representing the geo-factors $i = 2 \ldots n$. The theory of linear regression requires that the variate y be normally distributed. This can be easily attained for y if we choose the logarithm of y as dependent variate because the frequency distributions of most economic parameters are generally lognormal in nature. A simple logarithmic transformation written as $\log(1 + x_i)$ is usually sufficient to "normalize" all independent variates with very skewed distributions. As a result the most general regression equation is written as

$$\log y = a_0 + a_1 \log(1 + x_1) + \cdots + a_n \log(1 + x_n) \qquad (2)$$

The measurements of the independent geovariates (x_i) obtained from the control locations are used to calculate least-squares estimates of the coefficients a_i and the standard error for each coefficient. A statistical test, known as the t-test, will show whether a coefficient is statistically significant. If not, this means that the corresponding geo-factor is not statistically correlated with economic worth, and thus may be dropped from the model. Figure 12.3 illustrates the step-by-step screening of a large array of 170 qualitative geological parameters of ore-bearing porphyry deposits of the American Cordilleras by characteristics and regression analyses into only 12 quantified, statistically significant geovariates which can be used to construct a geostatistical model.

12.2.3. Construction of a Classification Model

The outcome of the drill testing of any exploration target under study is either success in detecting mineralization of commercial importance (population 1) or failure to do so (population 2). The objective assignment of targets to either class can be done by using a statistical technique known as classification analysis. In 1966, Griffiths[13] proposed the use of discriminant functions which were generalized later by Harris[15] as multiple discriminant functions. More recently, Wignall[19] introduced a generalized Bayesian classification procedure for use in oil and mineral exploration.

The construction of the classification model is based on the significant independent geovariates which have been previously screened by compression and feature selection. The function is calibrated on two sets of control locations. One set comprises thoroughly investigated control locations which have proved to be ore-bearing and serve as control for population 1. The second set serving as control for population 2 is composed of locations which have proved devoid of commercial ore despite thorough investigations.

Anderson[3] describes the standards required for good classifications. There are, first, requirements to be met by control and study data alike, such as multinormality, and equality of the variance–covariance matrices of population

SELECTION OF EXPLORATION TARGETS

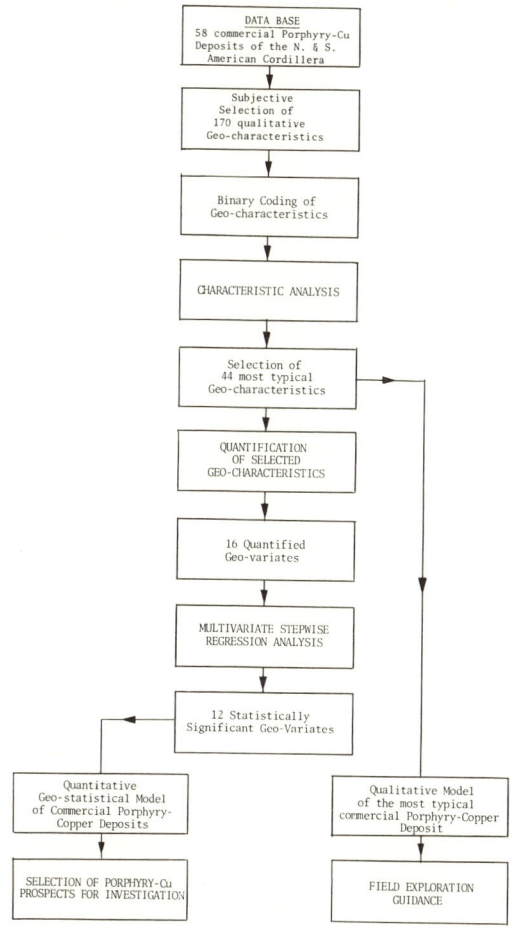

FIGURE 12.3
Flow Diagram of Statistical Screening of Independent Geovariates.

1 and population 2 (homoscedascity). Other requirements to be satisfied pertain to the classification methodology itself. A satisfactory classification procedure will minimize the probabilities of misclassification which are of two kinds. One kind arises when a location actually beongs to population 1 but is wrongly assigned to population 2; vice versa, a location may be wrongly assigned to 1, though actually belonging to 2. In each case, there is a cost attached to the misclassification, and the ratio of the two costs is what matters most to ensure a good classification.

Earlier workers used multiple discriminant analysis and discriminant scores for classification purposes. More recently, Bayesian classification analysis has attracted increasing attention. The scores are transformed into probabilities that a location belongs to a particular population, taking into account prior probabilities which are derived from the historical record. The success ratio as defined in Section 1.1.3 of Chapter 1 is a valid estimate of the prior probability of success. The Bayesian probability of success is formulated in terms of the probability density functions and prior probabilities for populations 1 and 2 in a manner that is illustrated by Figure 12.4.

Once the classification model is calibrated on the two kinds of control locations mentioned above, the measurements of all significant geo-variates are collected from each target under study and fed into the classification model to generate the probabilities that the target belongs either to population 1 or 2. All probabilities of success are ranked in order of decreasing magnitude, and all locations whose probability of success is greater than 0.500 are retained for further screening by means of the payoff criterion.

12.2.4. Design of a Criterion Optimizing the Selection of Exploration Targets for Further Testing

The maximization of the probability of success which is mentioned in Section 12.2.3 is one approach to the selection of targets for testing. A second approach would be the maximization of the regression estimates of the economic parameter represented by the dependent variate. These are obtained by feeding the measurements of significant geo-variates obtained from the locations under study into the multiple regression model described in Section 12.2.2. All estimates within their 95% confidence intervals are then listed in order of decreasing magnitude to enable the selection of the best targets.

However, the two previous approaches are not realistic because they do not take into account any cost consideration. Clearly, a criterion that would provide the maximization of economic reward in a probabilistic context (probability of

Probability density function

$$f_i(X) = [(2\pi)^p |S|]^{-1/2} \exp[-1/2 \, (X - \overline{X}_i) \, S^{-1} \, (X - \overline{X}_i)']$$

p = number of significant geo-variates

S = variance-covariance matrix

|S| = determinant of S

S^{-1} = inverse of variance-covariance matrix

$(X - \overline{X}_i)'$ = transpose

i = 1 for population of ore-bearing targets

i = 2 for population of barren targets

Bayesian probability of occurrence of commercial ore

$$P_0 = \frac{q_1 f_1(X)}{q_1 f_1(X) + q_2 f_2(X)}$$

q_1 = historical record of success for region

  = number of ore bodies/total number of drilled prospects

$q_2 = 1 - q_1$ = prior probability of failure

$f_1(X)$ probability density function for population 1

$f_2(X)$ probability density function for population 2

FIGURE 12.4
Model of a Bayesian Classification Function.

success) under cost constraints should be much more satisfactory. The payoff criterion which is described in Figure 12.5 fulfills these requirements. Payoff values obtained by feeding into the model the gross economic estimates, probabilities of success, and cost data for all targets under study are ranked in order of decreasing magnitude to allow the selection of a small number of the top-ranking targets, subject to budget constraints.

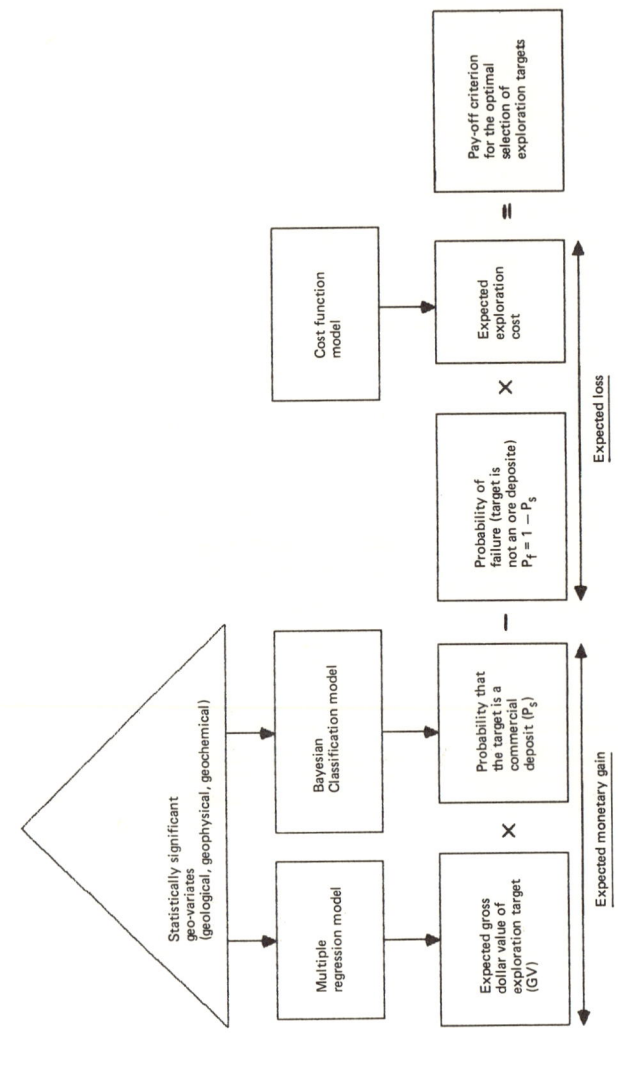

FIGURE 12.5
Design of a Payoff Model for the Optimal Selection of Exploration Targets for Drill Testing.

SELECTION OF EXPLORATION TARGETS 337

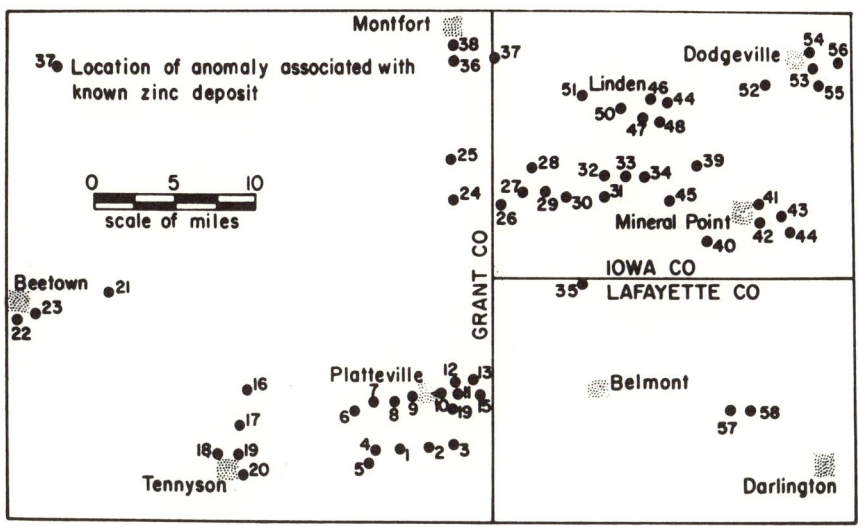

FIGURE 12.6
Selection of Exploration Targets in the Upper Mississippi Valley Zinc District: Control Locations.

12.2.5. Application of the Methodology to the Selection of Mississippi Valley-Type Pb–Zn Prospects in the Upper Mississippi Valley Pb–Zn District

The writers applied the method described above to the selection of Mississippi Valley-type Pb–Zn prospects in the Upper Mississippi Valley District of the North American Paleozoic Platform.[36] The dependent variate was the gross dollar value of the deposits; the independent variates were geological and geochemical in nature. The control area covering 56 control locations is shown in Figure 12.6 and the study area including 58 exploration targets to be rated for screening is shown in Figure 12.7.

The Bayesian classification scheme provided a clear cut separation of 22 locations showing Bayesian probabilities of success greater than 0.80 and moderate to large gross value estimates, from 33 locations with probabilities of success lower than 0.35 and low estimates. Only three of the 58 locations were found to lie between the 0.80 and 0.35 thresholds, and are likely to have been misclassified. A total of 14 among the 22 locations selected for testing are concentrated within the Darlington–Blanchardville area (See Figure 12.8), where

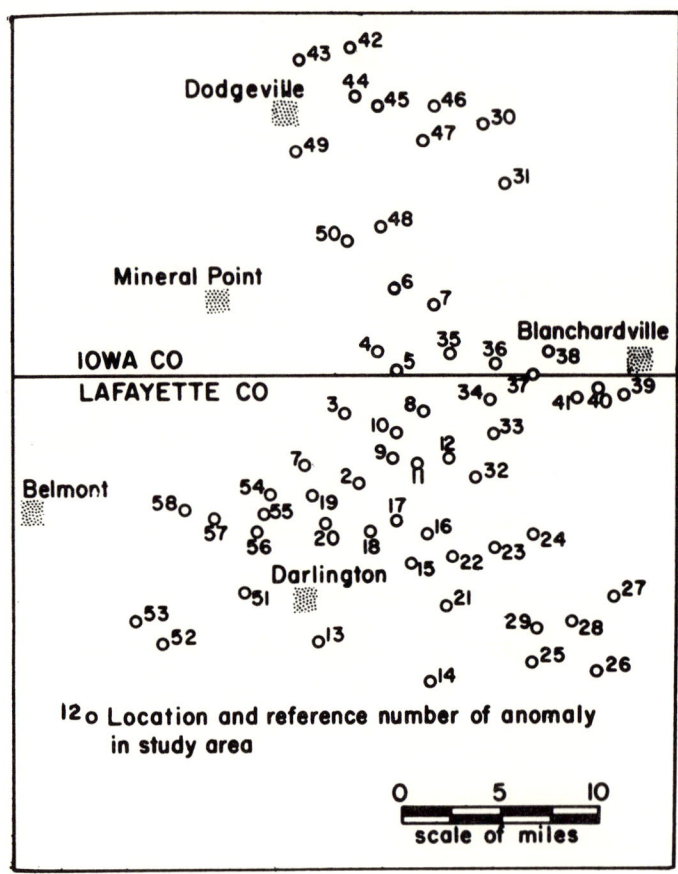

FIGURE 12.7
Selection of Exploration Targets in the Upper Mississippi Valley Zinc District: Location of Exploration Targets to be Rated.

shallow lead mineralization and deeper zinc mineralization were previously detected by trenching and drilling.

12.2.6. Critique of the Methodology

12.2.6.1. Merits of the Method. No claim is made that the sequential multivariate approach described above is an exploration panacea which should supplant the traditional role of exploration staff in the critical decision making

SELECTION OF EXPLORATION TARGETS

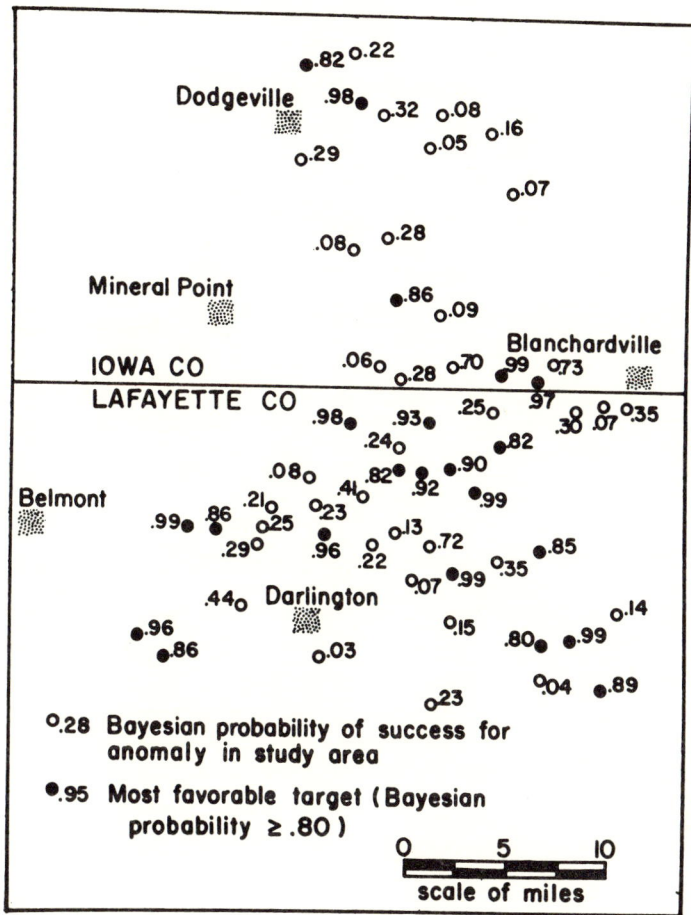

FIGURE 12.8
Selection of Exploration Targets in the Upper Mississippi Valley Zinc District: Location of Exploration Targets Selected on the Basis of Bayesian Probabilities of Success.

leading to the selection of targets for drill testing. The advantages of the statistical procedure are fourfold: The method provides (1) an objective rating of the selected targets, (2) the confidence intervals associated with the rating, (3) the probabilities involved, and (4) an accurate and speedy way of processing large amounts of data, and the possibility of updating the planning based on feedback information, as the field program progresses.

12.2.6.2. Methodological Difficulties. In the present mineral exploration climate, explorationists are increasingly turning their attention to remote and litte explored regions of the world which are unlikely to provide reliable control locations. The same problem arises in the search for new types of deposits within regions hitherto well explored for other types of ore deposits.

A second difficulty with serious statistical implications is caused by the bias in favor of population 1 (success) as against population 2 in the makeup of the control sample. Obviously, failure will always be less well documented than success.

12.2.6.3 Statistical Difficulties. The geostatistical model on which the methodology depends so much is affected by serious limitations; severe restrictions are imposed by the theory of the linear least-squares model on which both multiple regression and multiple discriminant analyses are predicated. The multinormality requirement can be satisfied by means of logarithmic or power transformations, which, in turn, introduce distortion, errors, and loss of information. Anderson and Bahadur[3] have devised a remedial approach when the variance-covariance matrices of populations 1 and 2 are not equal.

However, Agterberg (Chapter 1, Ref. 54) and Simard (Chapter 12, Ref. 24) indicate a way out of the difficulties mentioned above by the application of logistic regression and classification models instead of the restriction-ridden linear models used above.

12.3. OPTIMAL SELECTION OF EXPLORATION TARGETS WITHOUT CONTROL LOCATIONS

12.3.1. Rationale and Methodology

The present approach, which was only recently developed by the second writer, tackles the difficult problem faced by management when required to select optimally exploration targets for testing in little-known regions where adequate control is not available. The flow-chart shown in Figure 12.9 summarizes the method.

The first stage consists of the compilation of all available geographic, topographic, geologic, geophysical, and geochemical data relevant to the area to be investigated and ore deposit type to be searched for, followed by experimental design to quantify geologic and other qualitative relevant information. For example, supposing there are two rock types of interest and several others of no interest; then we could code them 1,0 for the first type, 0,1 for the second type, and $-1, -1$ for the others. Each location in the prospecting area will thus have

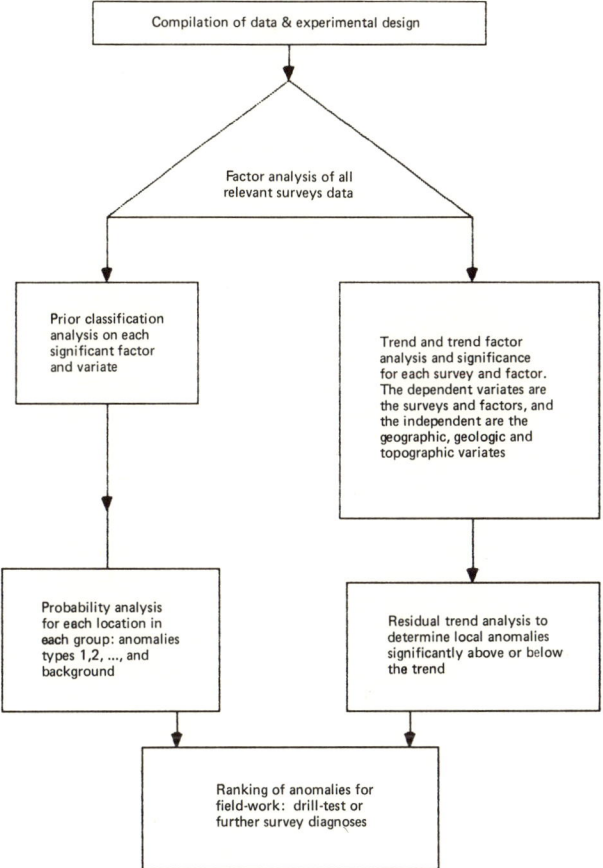

FIGURE 12.9
Flow Diagram of the Optimal Selection of Exploration Targets Without Control Locations.

a multivariate vector of survey information attached to it, and is given an identification number for greater convenience. The whole set of information is then stored on disk file or on file in the computer memory and given an appropriate file name. The file is then accessed and subjected to a series of optimal screening procedures to detect the best targets. All computer programs used in the optimal screening procedures, including TREND and POPMIX, are written in FORTRAN IV and are prepared for interactive use on computers and free format in reading from the data files.

12.3.2. Preclassification Analysis

This analysis is devised to detect anomalies on each survey variate, whether there are control locations or not. Fisher[43] formulated the method of moments to dichotomize sets of data into two populations. The POPMIX program devised by the second writer is based on Fisher's theoretical considerations. Our underlying assumption is that there is an anomalous set (population 2) and a background set (population 1) of locations on each quantitative survey. POPMIX provides the following output:

i. the proportion p of locations belonging to population 1 and population 2;
ii. the mean m_1 of population 1;
iii. the mean m_2 of population 2.

Thus, for each survey variate, we can select the anomalous set, since we know the proportion, p, and the total number, N, of locations in the study area; multiplication will yield the number of anomalous locations, and a biased division of $m_2 - m_1$ by m_1 will yield the value of the threshold separating the two populations.

12.3.3. Factor Analysis

Quite commonly there is a meaningful correlation, either direct or inverse, among the different survey variates. Factor analysis makes full use of these interactive relationships. Factor analysis is in fact the most powerful tool to yield linear sums and contrasts of the survey variates. Finally, factor analysis is more powerful than principal component analysis, because it incorporates PCA as a first step, prior to the maximization of the variability for which it accounts by "rotation."

Our analysis is carried out using the SPSS system, which has proved to be reliable and accurate. Among other output, the SPSS factor analysis yields the following:

i. Factor 1, which gives the linear combination and contrast of variates accounting for the greatest amount of variability among locations.
ii. Factor 2 giving the linear sum and difference, which account for the second largest amount of variability. Any other significant factors are also given.

iii. The proportion of variability accounted for by each factor.
iv. The correlation matrix giving the interactive relationship coefficients among the survey variate.

It is possible to determine which type of survey the various factors are highlighting from the examination of the coefficients of the survey variates. In the case of negatively correlated variates, one may determine which factor makes use of the contrast. This exercise will help at a later stage to decide why particular locations are to be considered as anomalous.

12.3.4. Reclassification of Locations Using Factors

Where there are two or more variates, the preclassification program, POPMIX, is used again, this time by redefining the variates to yield factor 1, factor 2, and any other significant factors. POPMIX now yields the anomalous set of locations on each factor. In most field projects, at least three sets of locations become apparent at this stage, namely, factor 1 anomalies, factor 2 anomalies, and background.

The POPMIX output yields at this stage a vector of factor scores for each location, the factor means, and the proportion of anomalies to background on each factor. The factor scores enable us to separate the locations into two anomalous sets and a background set. These three sets are stored in the memory of a new file along with their original information vectors.

12.3.5. Probability Analysis

The new file with three sets of locations including (1) anomalies on factor 1, (2) anomalies on factor (2), and background is now used as control set for the classification analysis. The whole set of locations are now probabilistically assigned to the three populations, namely, anomalies type 1, anomalies type 2, and background. The probabilities can be used as objective rating measurements; thus, the best targets can be selected and are identified by their geographic coordinates.

12.3.6. Trend and Trend-Factor Analyses

One of the aims of trend and trend-factor analyses is to screen out local anomalies from regional trends, because of the potential economic value of the former. This stage of the analysis makes use of the TREND program designed by the second writer and based on the multivariate general linear model (MGLM). We find the best fit for the vectors of survey variates on the vector of geographic coordinates, topographic and coded geological variates. The coefficients of the model are found by solving the matrix equation

$$Y (N \times p) = XB(N \times m) \cdot (m \times p)$$

where the dimensions of the matrices are given in brackets, and where B is the matrix of the best coefficients.

Once all coefficients are found, the data matrices Y and X are called again and read to calculate the trend estimates at each location for each of the survey variates. The analysis is repeated to find trend values for each factor at each location. The main output from TREND is

  i. The coefficient for each trend equation for each variate, and for each factor;
  ii. The trend and trend factor estimates for each location.

12.3.7. Residual Trends and Trend Factors

We now have the actual survey readings and factor scores on all variates and factors, as well as the trend and trend factor estimates at each location. Residuals are obtained at each location by subtracting the trend from the actual survey reading. The residual trends on the factors are particularly useful for the determination of local anomalies significantly above or below the trend. To determine the significance, the standard error of the differences is calculated on each factor. The residuals on each factor at each location are now screened to find the locations with significant residuals. These locations are local anomalies on the particular trend given in the output of the TREND program.

12.3.8. Summary–Conclusion

Since we do not have the advantage of having prior control locations to rely on for the selection of locations of greatest merit in the prospecting area, we proceed to construct our own set of control locations within the prospecting area itself by means of the preclassification analysis and factor analysis. As was the case in the first approach, we use the control set to obtain a probabilistic classification of the locations under investigation. Their screening is further refined in a spatial context by means of trend analysis, through which local anomalies are separated from the global ones. The local anomalies are ranked in order of merit, based on the magnitude of the associated residuals. It should be noted that the second approach described here involves a sequential maximization, first of the probabilities of belonging to classes 1 and 2, secondly of

SELECTION OF EXPLORATION TARGETS 345

the magnitude of the residuals, without constraints, as opposed to the optimization under cost constraints featured by the first approach.

12.3.9. Field Example: Mount Bundey, Northern Territory, Australia

Our field study is an exploration project in the Mount Bundey and Mount Guyder environs, in which the second writer was involved as field manager. We were seeking contact metasomatic U–Ag–Au or base metal prospects.

The Mount Bundey–Annabaroo area of the Northern Territory is situated some 70 miles ESE of Darwin. It includes most of the 1-mile geological sheet, D52 448. The rocks comprise Lower Proterozoic sedimentary formations that have been folded and weakly cleaved before being intruded by granitoids. The regional metamorphism is lowermost greenschist, and distinct contact aureoles surround the granitoids. In the northern contact aureole an iron mine has been abandoned, reportedly as uneconomic.

Previous work by the Bureau of Mineral Resources and Geopeko, Ltd. has highlighted the prospective nature of the region. Their aerial and ground geophysical and geological surveys gave the following information:

i. radiometric anomalies on the granitoids and immediate environs;
ii. magnetic anomalies over the northern Proterozoic outcrops, including the iron mine;
iii. correlation of the Golden Dyke Formation and Craig Creek member with the host rock mineralization in the Alligator Rivers U–Au field;

It should be noted that the Rum Jungle Uranium mine is 30 miles to the west of Mount Bundey, and the Alligator Rivers U–Au mines some 50 miles east.

Our party carried out a ground radiometric survey over the Mount Bundey and Mount Guyder areas, utilizing four-wheel-drive vehicles and GIS 3 spectrometers. In all, 247 ground spectrometer readings were taken, and total counts and uranium counts recorded.

The following survey data sets were then collated for each of the 247 locations: Geographical (latitude and longitude), ground and aerial radiometric (very highly correlated), magnetomic readings, topography, and geology. Experimental designs were applied to quantify the geological data.

Our computer software package described in Sections 12.3.2–12.3.7 was then employed to screen out 37 anomalous prospects: 31 being chosen by the probability classification analysis and a further 6 by the residual trend as locations significantly above the trend. The most northerly group of anomalies are on the iron mine, and four of the southerly ones are covered by Geopeko mining leases.

The inner group of anomalies in the northwestern aureole of Mount Bundey have since been drilled, and the result was a new silver mine grading up to 40 oz. to the ton being discovered. The authors have not had any reports on the other anomalous prospects as to whether any drilling program has been carried out. The silver discovery was classical: the anomaly in the Mount Bundey aureole is a radiometric high coinciding with a magnetomic low.

REFERENCES AND SELECTED READINGS

Optimal Selection of Exploration Targets Based on Control Locations: General Topics

1. AGTERBERG, F. P., 1974, *Geomathematics,* Chap. 14, Elsevier, Amsterdam.
2. AGTERBERG, F. P., 1978, Quantification and statistical analysis of geological variables for mineral resource evaluation, *Mem. BRGM,* No. 91, pp. 399–406.
3. ANDERSON, T.W., and BAHADUR, R. R., 1962, Classification into two multivariate normal distributions with different co-variance matrices, *Ann. Math. Stat.* **33,** 420–432.
4. BOLCH, B. W., and HUANG, C. J., 1974, *Multivariate Statistical Methods for Business and Economics,* Prentice Hall, Englewood Cliffs, New Jersey.
5. BOTBOL, J. M., 1971, An application of characteristic analysis to mineral exploration, *Can. Inst. Min. Metall. Spec. Vol.* **12,** 92–99.
6. BOTBOL, J. M., SINDING LARSEN, R., McCAMMON, R. B., and GOTT, G. B., 1978, A regionalized multivariate approach to target selection in geochemical exploration, *Econ. Geol.* **73,** 534–546.
7. CULBERT, R. R., 1976, A multivariate approach to mineral exploration, *Can. Inst. Min. Metall. Bull.,* **69,** 39–52.
8. DE GEOFFROY, J., and WIGNALL, T. K., 1970, Application of statistical decision techniques to the selection of prospecting areas and drilling targets in regional exploration, *Can. Inst. Min. Metall. Bull.* **63**(699), 893–899.
9. DE ST. JORRE, M. G. F., and WHITMAN, W. W., 1972, A probabilistic method of ranking underground exploration proposals, *Econ. Geol.* **67,** 789–793.
10. DUBOV, R. I., 1972, A statistical approach to the classification of geochemical anomalies, Proceedings of the London Symposium of the Institute of Mining and Metallurgy, pp. 275–285.
11. FUKANAGA, K., 1972, *Introduction to Statistical Pattern Recognition,* Academic, New York.
12. GREEN, P. E., 1978, *Analyzing Multivariate Data,* Dryden Press, Hinsdale, Illinois.
13. GRIFFITHS, J. C., 1966, Application of the discriminant functions as a classification tool in the earth sciences, *Kansas Geol. Surv. Comput. Contrib.* **7,** 48–51.
14. HARRIS, A., 1976, Definition and priority rating of exploration targets in the Cold Spring Pond area, Central Newfoundland, *J. Geochem. Expl.* **6,** 91–101.
15. HARRIS, D. P., 1965, Multivariate statistical analysis: A decision tool for mineral exploration, Symposium on Computer Application in Mining and Exploration Transactions, Vol. 1, pp. C1–C35, University of Arizona, Tucson.
16. KLOVAN, J. E., 1968, Selection of target areas by factor analysis, *Western Miner* **41**(2), 44–54.
17. MARIOTT, F. H. C., 1974, *The Interpretation of Multiple Observations,* Academic, London.

18. PETERS, W. C., 1978, *Exploration and Mining Geology*, Wiley, New York.
19. WIGNALL, T. K., 1969, Generalized Bayesian classification functions (K classes); Application to a regional geochemical survey in southeastern Pennsylvania, *Econ. Geol.* **64**, 570–576.

Optimal Selection of Porphyry–Cu–Mo Prospects

20. DE GEOFFROY, J., and WIGNALL, T. K., 1972, A statistical study of geological characteristics of porphyry–Cu–Mo deposits in the Cordilleran Belt of North and South America: Application to the rating of porphyry prospects, *Econ. Geol.* **67**, 656–668.
21. SINCLAIR, A. J., and GODWIN, C. I., 1979, Application of multiple regression analysis to drill target selection, Casino Porphyry–Cu–Mo deposit, Yukon Territory, Canada, *Trans. Inst. Min. Metall. London* **88**, B93–B106.

Optimal Selection of Ni–Cu–Ultramafic Prospects

22. BULL, A. J., and MAZUCHELLI, R. H., 1975, Application of discriminant analysis to the geochemical evaluation of gossans, in *Geochemical Exploration 1974*, pp. 219–226, Elsevier, Amsterdam.
23. CAMERON, E. M., SIDDELEY, G., and DURHAM, C. C., 1971, Distribution of ore elements in rocks for evaluating their ore potential: Nickel, copper, cobalt, and sulfur in ultramafic rocks of the Canadian Shield, *Can. Inst. Min. Metall, Spec. Vol.* **11**, 298–314.
24. SIMARD, R. G., 1980, Logistic model applied to geophysical data for sulfide nickel exploration, *Can. Inst. Min. Metall. Bull.* **73**(824), 96–100.

Optimal Selection of Volcanogenic Massive Sulfide Prospects

25. AGTERBERG, F. P., CHUNG, C. F., FABBRI, A. G., KELLY, A. M., and SPRINGER, J. S., 1972, Geomathematical evaluation of copper and zinc potential of the Abitibi area, Ontario and Quebec, *Geol. Surv. Can., Pap.* No. 71–41.
26. DIVI, S. R., THORPE, R. I., and FRANKLIN, J. M., 1979, Application of discriminant analysis to evaluate compositional controls of massive sulfide deposits in Canada, *J. Math. Geol.* **11**(4), 391–406.
27. FAVINI, G., and ASSAD, R., 1974, Statistical aeromagnetic and gravimetric criteria for massive sulfide districts in Greenstone areas of Quebec and Ontario, *Can. Inst. Min. Metall. Bull.* **67**(752), 58–63.
28. FAVINI, G., and ASSAD, R., 1979, An optimized decision model for area selection in massive sulfide exploration, *Can. Inst. M. Metall. Bull.* **72**(804), 118–126.
29. FOX, J. S., 1978, Interpretation of drill hole geochemical data from volcanic rocks of the Hanson Lake Mine, Saskt., *Can. Inst. Min. Metall* **71**(793), 111–116.
30. FOX, J. S., 1979, Host-rock geochemistry and massive volcanogenic sulfide ores, *Can. Inst. Min. Metall. Bull.* **72**(804), 127–134.
31. MARCOTTE, D., and DAVID, M., 1981, Target definition of Kuroko-type deposits in Abitibi by discriminant analysis of geochemical data, *Can. Inst. Min. Metall. Bull.* **74**(829), 102–107.
32. PIRIE, J. D., and NICHOL, I., 1981, Geochemical dispersion in wall-rocks associated with the Norbec deposit, Noranda, Quebec, *J. Geochem. Expl.* **15**, 159–180.
33. WHITEHEAD, R. E. S., and GOVETT, G. J. S., 1974, Exploration rock geochemistry: Detection of trace element halos at the Heath–Steele Mines (New Brunswick, Canada) by discriminant analysis, *J. Geochem. Expl.* **3**(4), 371–386.

Optimal Selection of Mississippi Valley-Type Pb–Zn Prospects

34. CAMERON, E. M., 1969, Regional geochemical study of Slave Point carbonates, Western Canada, *Can. J. Earth Sci.* **6,** 247–268.
35. DE GEOFFROY, J., and WU, S. M., 1968, Selection of drilling targets from geochemical data in the Southwest Wisconsin zinc area, *Econ. Geol.* **63**(8), 787–795.
36. DE GEOFFROY, J., and WIGNALL, T. K., 1970, Statistical decision in regional exploration: Application of regression and Bayesian classification analysis in the Southwest Wisconsin zinc area, *Econ. Geol.* **65,** 769–777.

Optimal Selection of Uranium Prospects

37. PATTERSON, D. A., PIRKLE, F. L., JOHNSON, M. E., BEMENT, T. R., STABLEIN, N. K., and JACKSON, C. K., 1981, Discriminant analysis applied to aerial radiometric data and its application to uranium favourability in South Texas, *J. Math. Geol.* **13**(6).
38. PIRKLE, R., CAMPBELL, K., and WECKSUNG, G. W., 1980, Principal component analysis as a tool for interpreting NURE aerial radiometric survey data, *J. Geol.* **88,** 57–67.

Optimal Selection of Oil and Gas Prospects

39. WIGNALL, T. K., 1968, Oil and gas exploration: Statistical decision criteria, S.P.E. Paper No. 223, Annual A.I.M.E./S.P.E. meeting. Houston, Texas.

Optimal Selection of Exploration Targets Without Control Locations

40. CONNOR, J. J., and MIESCH, A. T., 1964, Application of trend analysis to geochemical prospecting data from Beaver Country, Utah, *Computers in the Mineral Industry,* Part 1, *Geological Science,* Vol. 9, Stanford University.
41. DE GEOFFROY, J., and WIGNALL, T. K., 1973, Design of a statistical data processing system to assist regional exploration planning, *Can. Min. J.* Part 1, **94**(11), pp. 30–35.
42. DEMNATI, A., and NAUDY, H., 1975, Gamma-ray spectrometry in Central Morocco, *Geophysics* **40**(2), 331–343.
43. FISHER, R. A., 1936, The use of multiple measurements in taxonomic problems, *Ann. Eugenics,* **7,** 179–188.
44. GRAYSON, C. J., Jr., 1960, Decision under uncertainty: Drilling decisions by oil and gas operators, Harvard Business School Division of Research, Boston.
45. GRIFFITHS, J. C., 1962, Use of computers and statistics in the exploration and development of mineral resources, Symposium Proceedings, College of Mines, University of Arizona, pp. E.14–25.
46. KAUFMAN, G. M., 1962, *Statistical Decision and Related Techniques in Oil and Gas Exploration,* The Ford Foundation, Prentice Hall, Englewood Cliffs, New Jersey.
47. KRUMBEIN, W. C., 1959, Trend analysis of contour-type maps with irregular control point spacing, *J. Geophys. Res.* **64**(7).
48. MORRISON, D. F., 1967, *Multivariate Statistical Methods,* McGraw-Hill, New York.

49. NEEDHAM, R. S., WILKES, P. G., SMART, P. G., and WATCHMAN, A. L., 1973, Alligator Rivers region of the Northern Territory: Results of airborne geophysical surveys conducted by the Bureau of Mineral Resources of Australia; B.M.R. record 1973-208.
50. RAO, C. R., 1952, *Advanced Statistical Methods in Biometric Research,* Wiley, New York.
51. WIGNALL, T. K., 1969, Generalized classification functions: A Bayes-theoretic approach, Annual meeting of the Northwestern Science Association, Cheney, East Washington State College.

CONCLUSION

After two decades of intense activity and expansion, mineral exploration is at a crossroad. Slichter's timely call of 1960 for a new philosophy of mineral exploration has to be renewed now. The spectacular successes of the 1960s and the 1970s resulting from the massive injection of new technology were often accompanied by much waste of talent and resources and unnecessary duplication which cannot be repeated in the present circumstances of economic difficulties and financial stringency.

It is hoped that the call for a new philosophy of mineral exploration based on the systematic optimization of the sophisticated technology at hand which is advocated in the Preface and first chapter of this book will be heard by the mineral exploration fraternity, and put into practice as actively as has been done in the recent past by their mining engineering brethren. The methodology, mathematical and statistical apparatus, and the computational facilities are available and ready to be harnessed by explorationists for the "good cause" of optimizing mineral exploration.

By restricting the topic of the book to only two main items among some of the most important aspects of the optimization of mineral exploration—namely, the optimization of field program designs and selection of exploration targets for testing—we were able to delve into the two subject matters in greater detail and depth and, hopefully, demonstrate to explorationists the practical value of optimization. We hope that other workers will take on the task where we left it and extend the application of optimization to the many fields of mineral exploration which were briefly mentioned in Section 1.2.4 of Chapter 1.

It appears that the setting up of "think tanks" of experts fully devoted to the systematic optimization of mineral exploration, both within existing government organizations and large multinational exploration corporations, would be timely and highly beneficial. As a result of their efforts, the mineral exploration business would be urged to new peaks of efficiency and success in the 1980s and 1990s and beyond.

Looking beyond the next two decades through the proverbial crystal ball is a notoriously risky exercise. One could expect that an extensive restructuring of the mineral exploration methodology and technology will take place toward the end of the century, as most ore deposits detectable within the first 300 feet of the Earth's crust will have been located by means of the existing technology improved by optimization. The detection of deeper subjacent ore deposits will require some completely new technology, possibly not even heard of at the present, which will undoubtedly be very expensive, hence a greater than ever need for optimization. Therefore, optimization is here to stay, first as an improvement for the next two decades, and as a necessity beyond the end of the century.

APPENDIX

Measurement Conversion Table

Type of Measurement	Imperial to Metric	Metric to Imperial
Distance	1 mile = 1.61 kilometers 1 yard = 0.914 meter 1 foot = 0.305 meter	1 kilometer = 0.621 mile 1 meter = 1.09 yards 1 meter = 3.28 feet
Area	1 sq. mile = 2.60 sq. kilometers 1 acre = 0.405 hectare 1 sq. foot = 0.093 sq. meter	1 sq. kilometer = 0.386 sq. mile 1 hectare = 2.47 acres 1 sq. meter = 10.8 sq. feet
Volume	1 cubic yard = 0.765 cubic meter 1 cubic foot = 0.028 cubic meter 1 gallon = 4.55 liters	1 cubic meter = 1.31 cubic yards 1 cubic meter = 35.3 cubic feet 1 liter = 0.220 gallon
Mass	1 short ton = 1.02 metric tonnes 1 pound = 0.454 kilogram 1 ounce = 28.3 grams 1 ounce/sh. ton = 27.8 gr/metr. tonne	1 metric tonne = 0.980 short ton 1 kilogram = 2.20 pounds 1 gram = 0.035 ounces 1 gr/metr. tonne = 0.036 oz/sh. ton

353

SUBJECT INDEX

Alaska, 59, 110, 145, 148, 149
Algorithm, 85, 92
  recursive, 86
  SIMPLEX, 86, 87
Angled detection, 27, 68, 92, 100, 107
Anomalous set, 343
Anomaly, 326, 342–344
  geochemical, 17, 19
  geophysical, 17, 19, 30
  global, 344
  local 343, 344
Arctic Islands, 283–288

Background set, 326, 342, 343
Bayesian classification, 329, 332, 334, 337
Bayesian decision theory, 10
Bayesian probability, 4, 5, 334, 335, 337, 339
Black Hills, 306, 307
British Columbia, 144, 147, 148, 162, 176, 189

Calculus, 84–86, 90, 91, 106
California, 145, 148, 190, 191, 261, 262, 263, 305
Carbonate formation, 142, 148, 162, 163, 176, 284, 285, 287
Cavity filling, 307, 309, 310
Centrality, 98, 99
Characteristic Analysis, 330, 331
Chi-squared test, 25, 99, 100, 117, 150, 217, 244, 311
Circular normal model, 25, 100, 217, 218, 244, 310
Classifying function, 20, 332, 334, 335, 337, 342
Coefficient of
  dispersion, 99, 116, 150, 217, 244, 245, 264, 289, 310
  skewness, 99, 115, 150, 217, 244, 264, 289, 310

Color anomaly, 112, 237
Computer, 43, 81, 96, 97
  program, 43–45, 86, 91, 341–345
Conditional probability, 4, 5, 50, 51
Confidence interval, 32, 40, 48, 102, 103, 107, 249, 267, 339
Confirmed detection, 21, 35–37, 46–48, 101, 117, 151, 219, 245, 291, 313
Constraint, 80–84, 86, 104, 105
Contact metasomatic deposit, 96, 97, 142,
  detection, 146, 147
  geological setting, 141, 143–145
Control location, 10, 327, 332, 334, 337, 340
Cordillera Belt, 25, 96, 97, 109–111, 114, 145, 149, 163, 176, 177, 188–190, 236, 261, 262
Cost, 55–57
  analogy, 56
  component, 59, 60
  estimate, 56
  function, 60, 61, 64, 68, 79, 88–90
  indexation, 57
  multiplying factor, 121, 122, 198, 201, 220, 222, 246, 248, 265, 293, 314, 316
  schedule, 58
Course of action, 77, 80, 87
Criterion, 77–81, 88, 89, 106, 328, 334
Cutoff grade, 1, 24

Data
  acquisition, 6–8, 10, 17–19, 325, 328, 340
  base, 95, 96
  bit, 97
  compression, 328–331, 333
  processing, 10, 97, 98, 325, 328, 340
Decision tree, 10
Degree of belief, 3
Deposit
  mineral, 1, 17, 19, 21, 22, 24
  ore, 1, 9, 11, 22, 24, 96, 97, 325, 332

355

SUBJECT INDEX

Derivative, 84, 85, 90, 91
Detection, 17–21, 26–28, 32, 33, 43, 48
  range, 21, 27–29
  sensitivity, 48, 49
Deterministic, 22, 31, 84
Dichotomization, 329, 330, 340, 342
Digitization, 326, 329, 340
Dip
  angle, 22–25, 27–29, 31, 100, 149, 216, 217, 242, 289, 310
  component, 27–29, 100, 101, 117, 151, 218, 245, 290, 312
Direct approach, 90–92
Direct ascent, 85
Direct detection, 17, 112, 146, 213, 214, 239, 286, 308, 309
Discriminant analysis, 332, 334
Discrimination, 21, 22
Dolomitization, 284, 285
Drilling, 17, 18, 66–68, 72, 73, 91
  angle, 91, 92, 107, 201–203, 220–222, 246–249, 265–267, 314, 316
  churn, 66, 286
  diamond, 51, 66
  optimization, 7, 11, 91, 92, 106, 107, 123, 124, 200, 202, 203, 222, 248, 249, 267, 293, 316
  percussion, 66, 138, 302
  testing, 7, 10, 11, 325, 327, 339
Dynamic programming, 9–11, 86, 87, 89, 91, 105–107, 218, 222, 248, 267, 313, 316

Economic worth, 1, 5, 9, 11, 24, 96, 329–331, 334, 340
Effectiveness, 78, 88
Efficiency, 88
  criterion, 78, 79, 89, 90, 105, 122, 201, 222, 248, 293, 316
Electromagnetic survey, 20, 21, 28, 68, 113, 214, 215, 240, 287, 309
ELLIPGRID program, 32, 33, 43, 53
Exhalite, 236, 237, 239
Expected loss, 79
  criterion, 123, 203, 222, 248, 293, 316
Expected monetary value, 79, 334, 336
Expenditure, 56

Exploration target, 10, 11, 325–327, 334, 335, 340, 344
Extremal root, 85, 91
Extremum, 81, 82

Factor analysis, 326, 330, 342, 343
Feature selection, 330, 332, 333
Fiducial interval (*see also* confidence interval), 99, 102, 104, 105, 107, 249, 267, 279
Field
  artificial, 20, 28, 29
  potential, 28, 29
Filtering, 20
Flight elevation, 28, 34, 35
Frequency distribution, 98–100, 115, 150, 163, 217, 243, 289, 310, 311
Frequentist concept, 3

Geochemical
  detection, 18, 33, 113, 114, 147, 215, 287, 288, 309, 310
  signature, 18, 19, 325
Geographic areas
  Central Asia, 147
  Greenland, 284, 285
  Manitoba, 211–213, 216, 237, 238
  Mexico, 110, 146, 177
  Mississippi Valley, 283, 284, 337, 338
  Missouri, 283, 284
  Ontario, 30, 34, 35, 61, 211–213, 237–239, 306, 309
  Quebec, 4, 30, 212, 238, 239, 241
  Selwyn Mountains, 162, 189, 283–285, 288, 306
  Southwest U.S.A., 110, 111, 145, 148, 162, 176, 189, 190, 238
  Tri-State, 283, 284
  Ungava, 212, 213, 215
Geological age, 110, 145, 211, 235, 283, 284, 305
Geological detection, 17, 18, 112, 146, 213, 214, 238, 239, 286, 308, 309
Geological occurrence, 1, 4, 11, 13, 14
Geometric descriptor, 22, 23
Geometric factor, 24
Geometric probability, 31–33
Geophysical contrast, 29, 30, 61–65

SUBJECT INDEX

Geophysical survey cost, 58, 62, 63, 65
Geophysical detection, 17–21, 28, 112, 113, 147, 214, 215, 240, 287, 309, 326
  airborne survey, 34–36, 50, 61, 90, 103, 112, 113, 117
  ground survey, 42, 43, 45, 65, 91, 103, 113, 147, 215, 240, 287, 309
  signature, 19, 326, 327
Goal, 77, 78
Goodness of fit, 99, 100
Gossan, 112, 237
Gravimetric survey, 28, 113, 147, 215, 240, 287
Great Slave Lake, 283–285
Greenstone belts, 211, 213, 215, 236–239, 241, 249, 306–308, 327
Greisen, 189, 190
Grid
  control, 30
  geometry, 30, 35, 36, 38, 39, 42, 101
Gross National Product, 57

Halo, 17, 18, 113, 123, 124, 240, 241, 287, 288
Heuristic model, 6, 75, 78, 79, 85, 86
Histogram, 25, 26, 98, 99
Homoscedascity, 234
Hornfels, 141, 176, 189

Independence, 4, 5, 124
Indirect detection, 17–21, 112–114, 147, 214, 215, 240, 241, 286, 287, 308–310
Induced polarization, 30, 43, 68, 113, 147, 215, 240, 286, 309
Information, 10
  feedback, 8
  prior, 8
  theory of, 19, 52
Iron formation (*see also* exhalite), 215, 236, 237, 239, 240, 241, 307

Lateral coverage, 21, 33, 35, 36, 38
Leadville district, 162, 163
Liminal strategy, 104, 121, 198, 199, 220, 221, 246, 247, 264, 265, 291–293, 314, 315
Linear inequality, 82
Linear model, 331, 340, 343, 344

Linear programming, 86
Lithogeochemistry (*see also* geochemical detection), 18–20, 113, 114, 147, 236, 241, 287, 288
Logarithmic transformation, 99, 332, 340
Logistic model, 340
Logistics, 8, 59, 60
Lognormal model, 25, 26, 99, 100, 116, 117, 149, 163, 177, 190, 217, 243, 263, 289, 312
Loss function, 79

Magnetic induced polarization, 113, 240
Magnetic survey, 28, 65, 68, 113, 147, 214, 215, 240, 308, 309
Manto deposit, 162, 163
Marble, 141–143, 162, 163, 176, 189, 285
Mathematical programming, 84, 86
Maximal strategy, 104, 105, 118–121, 200, 201, 220–222, 247, 248, 265, 266, 292, 293, 315, 316
Mean, 99, 115, 150, 163, 217, 243, 244, 263, 264, 289, 290, 310, 311, 342
Median, 99
Message, 19, 20
Metasomatic deposits, 142, 144, 145
Metasomatism (*see also* contact), 141–143
Mineral exploration
  context, 2
  methodology, 1, 9, 10, 21, 22
  optimization, 9–11, 75–77
  purpose, 1
  sequentiality, 6–8
  success, 1, 4, 5
Minimax strategy, 81
Misclassification, 334, 337
Mississippi Valley-type deposit
  detection, 286–288
  geological setting, 283, 284
  statistical modelling, 289, 290
Mobile Belt, 211, 212
Mode, 99, 110
Model
  concept, 4, 5, 109, 143, 235, 238, 239
  fitting, 24, 25, 99, 100
Multiple Spectral Sensing, 17, 112, 214
Multipurpose survey, 249, 267

SUBJECT INDEX

Nickel deposit
  detection, 214, 215
  geological setting, 211–213
  komatiite series, 213
  spinifex texture, 214
  statistical modelling, 217, 218
  tholeiitic series, 213
Noise, 19, 20, 22
Normal model, 25, 99, 100, 217, 243, 244, 289, 310, 311
Northwest Territories, 215, 238, 239, 243, 284, 285, 306, 310, 311

Objective function, 81–83, 88–91, 105
Observation platform, 17, 28, 34
Open pit mining, 111, 189, 284
Operations Research, 10, 81, 87, 88
Ore body, 1, 11, 22–24
Ore treatment, 2
OPTGRID program, 43, 91, 92
Optimal choice, 9–11, 75, 77, 78
Optimal grid orientation, 10, 31, 40–44, 48, 49, 101, 151, 218, 245, 313
Optimal strategy, 101, 122, 123, 201–203, 246–249, 266, 267, 291, 293, 315, 316
Optimization, v, vi, 9, 75–77, 95, 104
  drilling angle, 106, 107, 203, 248, 249, 316
  survey grid, 9, 10, 77, 122, 202, 203, 222, 248, 249, 293, 316

Pattern
  drilling, 32, 33, 42, 43, 65, 66, 123, 124, 286, 309
  recognition, 21, 327, 330
Payoff criterion, 79, 330, 334
Permissible area, 82
Pine Creek District, 190
Plunge angle, 23, 24
POPMIX program, 342, 343
Porphyry-Cu-Mo deposit
  economic features, 110, 111
  detection, 112–114
  geographic distribution, 110
  geological features, 109
  statistical modelling, 114–117
Porphyry dykes, 307
Precision, 78, 79

Price, 2, 3, 55
Principal Component analysis, 330, 342
Probability, 3, 4, 6, 22, 84
  detection, 4, 11, 34, 35, 40, 43, 45, 48, 50, 51, 75, 76, 88, 89, 91, 92, 101, 102, 117, 118, 151, 165, 178, 191, 218, 219, 245, 246, 264, 290, 291, 312, 313
  economic worth, 11, 14, 15
  occurrence, 11, 13, 14
  success, 3–6, 79, 80
Probe, 20, 21, 27, 28
Prospecting
  area, 9, 10, 325
  detection, 17, 112, 146, 214, 238, 239, 286, 308, 309
  success ratio, 78, 334, 335
Pyritic halo, 19, 50, 51, 110, 112–116, 123, 124, 237, 240, 284, 309

Quantification, 3, 77, 78, 326, 340
Quartile, 98, 99

Range, 98, 99, 114, 115, 149, 163, 177, 217, 241, 289, 310
Ratio costing, 56, 59, 160
Reduction to Pole, 29
Refractory ore, 307, 308
Regression analysis, 330–332
Remote sensing, 17, 112, 214
Replacement, 307, 310
Residual, 344
Resolution, 21
Reward, 78, 88, 89, 106
Risk, 2, 5, 6, 10

Sampling
  continuous approach, 17, 18, 61, 90, 325
  discrete approach, 17, 18, 65, 91, 92, 288
  size, 98, 288
Scaling factor, 89
Scanning, 8
Scheduling 1–8, 10, 325
Sensitivity analysis, 48–50, 87
Sensor, 18, 19, 21, 22
Sequential approach, 6–8, 20, 50, 51, 123, 124, 326, 328, 341
Shape ratio, 23, 27, 28, 31, 211–213

SUBJECT INDEX 359

Shield region, 61, 96, 97, 211, 214, 215, 235, 238, 306, 307
Signal, 19, 20
Signature, 19, 21, 326
Simpson's rule, 44, 45
Skarn, 141–143, 148, 149, 176, 189, 190, 305
Skimming, 8
Spatial coincidence, 30, 326
Standard deviation, 99, 115–117, 150, 165, 217, 243, 289, 311, 326
Statistical decision theory, 80, 81
Statistical occurrence, 3–5, 79
Statistical pattern, 326, 327
Statistical sampling, 98, 288
Statistical shape, 99, 115, 150, 151, 165, 217, 243, 289, 310, 311
Statistical summarization, 98, 99
Stochastic variate, 84, 325
Stratabound deposit, 235–237, 283
Strategy, 80, 81, 104, 105, 118, 121–123, 198, 201, 202, 219, 220, 222, 246, 248, 265, 267, 291, 293, 314, 316
Stratiform deposit, 235, 283
Strike direction, 23, 90–92, 105, 106, 148, 217, 243, 244, 286, 311
Structural hinge, 285
Structural setting, 109, 110, 144, 145, 211, 235, 238, 285, 307
Student-t test, 99, 344
Success, 1, 3–5, 88
 ratio, 4, 329, 333, 335
Supergene enrichment, 111, 176, 237

Tactite, 189, 190
Target
 definition, 326
 geometry, 27, 28, 31–35, 42, 43, 100, 101
 selection, 2, 325, 327, 328, 340, 341
Telluride, 307
Tintic District, 162–164

Trend analysis, 326, 343, 344
Threshold, 104, 326, 334, 337, 342
Trial repetition, 5, 6

Uncertainty, 2, 3, 5, 8, 10, 100
Unit costing, 56–58, 60, 65, 66

Variability, 99, 116, 150, 151, 165, 217, 218, 289, 290, 312, 342, 343
Variance, 90, 332, 335
Variate, 81, 82, 84, 85, 328–331, 333, 342, 343
Vein gold deposit
 detection, 308–310
 economic geology, 305, 306
 geological setting, 307, 308
 stastical modelling, 310–312
Vertical continuation, 29, 30
Vertical detection, 27–29
Volcanogenic massive sulfide deposit, 235–238
 alteration pipe, 236, 237
 calc-alkaline series, 236
 dalmatianite, 237
 detection, 238–240
 exhalite, 236, 239
 geological setting, 236, 237, 261, 262
 massive ore, 237
 metallogeny, 237, 238
 occurrence mode, 238
 ophiolitic type, 235
 pyroclastic series, 236, 262
 stringer ore, 236, 238
 tholeiitic series, 236

Wholesale Price Index, 57
Word Processor, 97

Yellow Pine District, 190
Yukon Territories, 110, 145, 161, 162, 176, 189

Zoning, 109, 110, 141–145, 148, 189, 190

AUTHOR INDEX

Abrams, M. J., 112
Agterberg, F. P., 25, 77, 326, 327, 330
Agocs, W. B., 32, 35, 77
Albers, J. P., 262
Anderson, T. W., 332, 340

Ball, C. W., 189
Barragar, W. R. A., 85
Bellman, R. E., 85
Bolch, B. W., 330
Boldy, J., 239, 240
Botbol, J. M., 330
Brant, A. A., 8, 80
Butler, B. S., 162

Callahan, W. H., 284, 286
Cameron, E. M., 212, 285
Campbell, N., 285
Cathro, R. J., 189
Celasun, M., 10, 34, 42
Chung, C. F., 33, 34, 35, 38, 40
Coad, P. R., 214
Cobb, H., 77
Coyle, R. G., 87
Cranstone, D. A., 57
Cribb, J. L., 29
Culbert, R. R., 327

Davis, J. C., 25
De Geoffroy, J. G., 25, 77, 80, 288, 326, 327, 337
Dowsett, J. S., 215
Drew, L. J., 33, 42, 49, 67, 77
Duckworth, E., 80

Einaudi, M. T., 142, 175, 176

Elevatorsky, E. A., 305
Elliot, I. L., 19

Fisher, R. A., 43
Fox, J. S., 236, 239, 240
Fukanaga, K., 327

Gilmour, P., 236
Griffiths, J. C., 332
Guilbert, J. M., 109

Harbaugh, S. W., 82, 83, 84, 86
Harris, A., 327
Harris, D. P., 327
Henderson, R. G., 29, 35
Hewlett, R. F., 11, 79
Hollister, V. F., 98
Holmes, R., 249
Hutchinson, R. W., 235
Hutson, R. J., 308

Jackson, S. A., 285
Johnson, N. I., 25

Kelley, J. C., 77, 79
Kelly, S. F., 309
Kendall, M. G., 32, 77
Kerr, P. F., 141, 145, 190
Kinkel, A. R., 262
Klichinov, V. A., 147
Klovan, J. E., 327
Knopf, A., 190
Koch, G. S., 25
Koulomzine, T., 4
Krumbein, W. C., 25
Kuzwart, M., 19

Lee, Y. W., 21
Levinson, A. A., 19, 326

Lovering, T. S., 146
Lowell, J. D., 109

Macqueen, R. W., 285
Marcotte, D., 237
Marriot, F. H. C., 40
McCammon, R. B., 33, 35, 39
Metz, P. A., 56, 59
Mickey, M. R., 42, 48
Miller, C. P., 215
Morris, H. T., 147
Murphy, J. D., 285, 288

Naldrett, A. J., 211

Obial, R., 215

Parasnis, D. S., 19, 43
Paterson, N. R., 22, 28, 34, 38, 240
Peters, C. W., 4, 19, 28, 34, 327

Raisbeck, G., 19, 21
Reedman, J. H., 19
Rennie, C. C., 162, 189
Riddler, R. H., 236, 239
Ridge, J. D., 146, 147, 162, 190
Roscoe, 77
Rosenberg, 21
Ross, J. R., 211
Roubens, M., 11, 79

Sangster, D. F., 148, 235, 237, 238
Savinskii, I. D., 32, 33, 43
Seigel, H. O., 113, 240, 309

Shurygin, A. M., 33, 42, 77
Sillitoe, R. H., 110, 114
Sinclair, A. J., 33
Singer, D. A., 32, 33, 42, 43
Slichter, L. B., v, 32, 34, 77, 78, 80, 141

Smirnov, D. F., 141, 142, 175
Smith, C. L., 285, 288
Solomon, H., 32
Spector, A., 29

White, L. G., 189
Wignall, T. K., 327

Wilde, D. J., 84
Wilson, H. D. B., 212
Wright, L. B., 306

Young, G. A., 148

Zurbrigg, H. F., 212
Zurflueh, E. G., 21

LISTED JOURNALS

Annals of Eugenics, 348
Annals of Mathematical Statistics, 346
Biometrics, 53
Biometrika, 13
Bulletin of the Australian Society of Exploration Geophysicists, 51
Bulletin of the Geological Society of America, 233
Bulletin of Mineral Industry Experimental Station of the University of Pennsylvania, 13
Canadian Institute of Mining and Metallurgy Bulletin, 12, 14, 15, 52, 74, 93, 139, 140, 210, 233, 279, 280, 281, 346, 347
Canadian Journal of Earth Sciences, 279, 303, 348
Canadian Journal of Sciences, 233
Canadian Mining Journal, 12, 14, 74, 139, 303, 323, 348
Canadian Petroleum Geologists Bulletin, 303
Computers & Geosciences, 14, 53
Computer Journal, 93
Cost Engineering, 74
Earth and Planetary Sciences Letters, 139
Economic Geology, 14, 53, 74, 93, 140, 209, 210, 233, 280, 281, 303, 323, 324, 346–348
Geoexploration, 74
Geologisch Rundschau, 210
Geophysics, 52, 53, 93, 139, 140, 281, 303, 348
Geosciences of Canada, 303
Journal of the American Statistical Association, 12
Journal of Geochemical Exploration, 93, 281, 303, 346, 347
Journal of Geology, 281, 348
Journal of the Geological Society of the Philippines, 233
Journal of Geophysical Research, 348
Journal of Mathematical Geology, 13, 14, 53, 93, 94, 347, 348
Journal of Research of the United States Geological Survey, 53
Management Sciences, 14
Marine Geology, 94
Mining Congress Journal, 74, 93
Mineralium Deposita, 13, 74, 140
Mining Engineering, 13, 93, 139
Mining Geology, 280
Mining Magazine, 74
Mining World, 93
Operations Research, 92
Operations Research Quarterly, 93

Photogrammetric Engineering, 52
Quarterly of Colorado School of Mines, 14, 53
Science, 13
Transactions of Institute of Mining and Metallurgy, London, 13, 93, 347
Western Miner, 209, 346